Principles of
Stratigraphic
Analysis

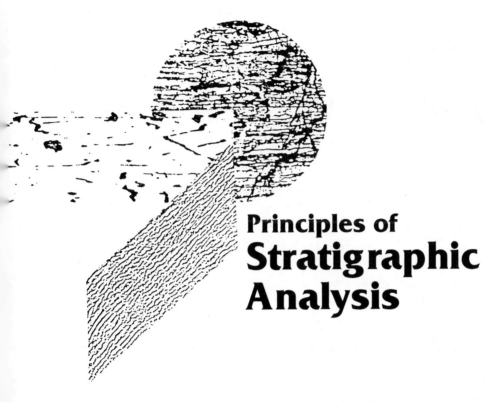

Principles of
Stratigraphic
Analysis

Harvey Blatt .

School of Geology
and Geophysics

The University of
Oklahoma, Norman

William B.N. Berry .

Department of Geology
and Geophysics

University of
California-Berkeley

Scott Brande

Department of Geology

University of
Alabama at Birmingham

Boston

Blackwell Scientific Publications

Oxford / London / Edinburgh / Melbourne / Paris / Berlin / Vienna

Blackwell Scientific Publications

Editorial offices:

Three Cambridge Center, Cambridge, Massachusetts 02142, USA
Osney Mead, Oxford OX2 0EL, England
25 John Street, London, WC1N 2BL, England
23 Ainslie Place, Edinburgh, EH3 6AJ, Scotland
54 University Street, Carlton, Victoria 3053, Australia

Distributors:

USA
Blackwell Scientific Publications, Inc.
Three Cambridge Center
Cambridge, MA 02142
(Orders: Tel. 800-759-6102   617-225-0401)

Canada
Oxford University Press
70 Wynford Drive
Don Mills, Ontario M3C 1J9
(Orders: Tel. 416-441-2941)

Australia
Blackwell Scientific Publications (Australia) Pty Ltd
54 University Street
Carlton, Victoria 3053
(Orders: Telephone: 03-347-0300)

Outside North America and Australia
Blackwell Scientific Publications, Ltd.
Osney Mead
Oxford OX2 0EL
England
(Orders: Telephone: 011-44-865-240201)

Typeset by G & S Typesetters, Inc., Austin, Texas
Printed and bound by The Maple Vail Press

Blackwell Scientific Publications, Inc.
© 1991 by Blackwell Scientific Publications, Inc.
Printed in the United States of America
90 91 92 93 94 5 4 3 2 1

Library of Congress Cataloging in Publication Data
Blatt, Harvey.
    Principles of stratigraphic analysis / Harvey Blatt, William B. N. Berry, Scott Brande.
        p.   cm.
    Includes bibliographical references and index.
    ISBN 0-86542-069-6
    1. Geology, Stratigraphic. 2. Paleontology, Stratigraphic. I. Berry, William B. N. II. Brande, Scott.
III. Title.
QE651.B68   1991
551.7—dc20
                                                          90-43784
                                                               CIP

Dedicated to Midge and Suzanne
Harvey Blatt and William B. N. Berry

Dedicated to Harold and to the memory of Selma
Scott Brande

— ● This book will probably change my life.
At least I think it will.
Maybe it won't.
Charles M. Schulz

If they don't want to read it,
nobody's going to stop them.
Yogi Berra

• Contents

6 • Evolution and Biostratigraphy   145

7 • Depositional Environments and Facies I: Continental Environments   195

11 • Physical Framework for Stratigraphic Analysis 357

12 • Economic Stratigraphy I: Primary Deposits 399

13 • Economic Stratigraphy II: Deposits Formed by Secondary Mineralization 443

• Preface

The *Glossary of Geology* (1987) defines stratigraphy as the science concerned with all attributes of sedimentary rocks, including their form, distribution, lithology, fossil content, depositional environment, vertical sequence, and age. In brief, a stratigrapher must truly be a person "for all seasons." A stratigrapher's abilities must include intensive grounding in geologic fundamentals, including field work at the Earth's surface and methods of interpretation of remotely sensed data obtained from satellites, airplanes, or wireline logs dropped into boreholes extending more than 10 km below the surface. For this reason, stratigraphy is one of the most demanding and fascinating of geologic disciplines.

How do practicing stratigraphers ply their trade? What questions do they ask of the rocks they study? How do they go about obtaining adequate answers? Most stratigraphers, indeed most geologists, are employed by the petroleum industry in the search for oil and gas. These resources are, of course, located below ground level, so a central part of the search involves the gathering and interpretation of data obtained from the subsurface through seismology or by the use of probes of one type or another that are lowered into boreholes on wires—wireline logs or "electric logs." Although small photographic cameras have been used, essentially all data obtained from wireline logs is indirect. The geologist or geophysicist determines the electrical or nuclear properties of the subsurface units and interprets these properties in terms of rock features such as lithology, directional structures, and porosity. These data are then combined with information from surface outcrops near the well site to generate a three-dimensional picture of the local or regional stratigraphy.

The discipline called stratigraphy is thus a synthetic subject concerned with the origin and temporal and spatial distribution of layers of sedimentary rock. It is closely related to its sister discipline, sedimentology, but the two specialties can usually be distinguished in the following ways: (1) Stratigraphers always deal with ancient rocks rather than modern sediments. For this reason, most of the discussion and examples in this book are concerned with ancient rocks. Modern sediment studies are considered only as needed to explain features seen in ancient rocks. (2) Stratigraphers usually deal with packages of rocks, or at least with the relationship of a particular unit to adjacent units. Sedimentologists may deal with sequences, but it is not a requirement. Much sedimentologic research deals with either geochemical or geophysical (hydraulic) problems of sediment deposition and lithification.

The rock sequences studied by stratigraphers are most commonly bounded by lithologic, biologic, or erosional surfaces, but in some circumstances it may be useful to define rock packages on the basis of chemical variations or on changes in response to induced electrical or physical energy (seismic stratigraphy). It is apparent from the differing character of the bounding surfaces that not all of them are visible to the unaided eye; some can be recognized only by laboratory analyses. Because of this, stratigraphers must possess a wide variety of skills involving both field work and an understanding of the capabilities of several types of sophisticated analytical equipment.

Because stratigraphy is synthetic, constructed from data contributed by other subdisciplines of geology, chemistry, physics, and biology, and because the trend in all sciences is toward increased specialization ("more and more about less and less until you know everything about nothing," as some say), stratigraphy might be compared to chariot racing—a field that has had its day and is now on the way out. This view is reflected in the fact that only four stratigraphy textbooks were published in the United States during the past 25 years: *Dynamic Stratigraphy* by Matthews (1974, 1984), *Sedimentation and Stratigraphy* by Boggs (1987), *Integrative Stratigraphy* by Brenner and McHargue (1988), and *Basics of Physical Stratigraphy and Sedimentology* by Fritz and Moore (1988). Yet stratigraphy is a popular course with students, is taught at almost all schools that offer an undergraduate degree in geology, and is commonly a required course.

Why does the subject matter of stratigraphy often have a poor image in the academic community? Why does it seem so old-fashioned to many geologists? Perhaps the problem lies in the way in which it is presented. It is this thought that prompted us to write this book. In recent years, many new ideas have affected stratigraphy, ideas such as the relationship between plate tectonics and sediment accumulations, magnetic stratigraphy, seismic stratigraphy, selective preservation of taxa, organic matter biomarkers, correlation by trace element geochemistry, cathodoluminescence petrography, and stable isotopes, to cite a few. It is our hope that this book will contribute to a renaissance of stratigraphic thought among the current generation of students. The whole (stratigraphy) may indeed be greater than the sum of the parts (paleontology, sedimentology, ecology, etc.).

Each of the parts, however, has a voluminous literature, and we were early faced with the question of book length. Geology texts have become enormous in recent years, as if the reading and retention abilities of the current generation of undergraduates were markedly greater than those of its predecessors. For example, one of us determined that the average number of pages of the mineralogy, petrology, paleontology, and structural geology texts he used as an undergraduate was 493. The current equivalent texts average 852 pages (when adjusted for the larger page size of two of them). Can modern undergraduates absorb 73% more material than yesterday's, material that is, in addition, more abstract? It seems doubtful to us, and so we determined to keep our stratigraphy text to less than 500 pages. Achieving this goal was not without its difficulties, and some items that we wished to consider had to be omitted when hard decision time arrived. There were no perfect solutions, but we did our best. We encourage our readers, both students and faculty, to let us know of omissions that they consider intolerable in a textbook designed to introduce undergraduate geology majors to the subject of stratigraphy. All suggestions will be considered when we prepare a second edition of this book. We ask only that each suggestion for an addition be accompanied by a suggestion for a deletion.

———— • **Acknowledgments**

Special thanks are due to the following scientists, who have read and improved some of the chapters: J. L. Ahern, K. D. Crowley, R. D. Elmore, C. W. Harper, Jr., J. E. Hazel, J. Jell, D. London, P. K. Sutherland, and P. D. Ward. Errors of fact or interpretation that may remain should be attributed to the authors.

1
Accumulating Stratigraphic Data

The process of accumulating stratigraphic data has undergone enormous changes in recent years. The older techniques, described by Compton (1985), consist essentially of measuring stratigraphic sections and describing them, collecting megafossils and identifying them to determine geologic age, and, finally, constructing a geologic map. Almost all data were obtained from outcrops viewed at close range. However, this method, while still essential in areas of adequate outcrop, has been increasingly supplemented in recent years by remotely sensed data from borehole logging, exploration seismology, and satellite imagery. Borehole logging and seismic reflection, in particular, have had an explosive increase in usefulness to stratigraphers. Satellite imagery appears to have a bright future for stratigraphic applications, but in most published images the resolution is more suitable for large-scale uses in tectonics, structural geology, and environmental geology than for detailed stratigraphic work. The present capabilities of aerial photography and more sophisticated remote sensing techniques are classified for reasons of national security, but stories have circulated about being able to see such things as a package of cigarettes on a parking lot from tens of kilometers above the surface. Might geologists be able to determine degrees of grain sorting in a sandstone from a suitably equipped airplane?

As described by Miall (1984), five fundamental changes have occurred in sedimentary geology since 1960 that have revolutionized the field of stratigraphic analysis.

1. The development of sedimentology from a field dominated by grain size, shape analysis, and mineral identification in thin section to a discipline capable of

1

explaining the origin of sedimentary packages through facies studies and facies models. It is now possible to interpret and predict the composition, geometry, and orientation of most stratigraphic units by using the facies approach, based on a wide range of studies in modern environments and ancient rock units.

2. An extension of the facies approach has been the recognition of a depositional system, a complete package of depositional environments and their sedimentary products. Each system is bounded by unconformities or by lateral facies transitions into a genetically unrelated system, such as the change from a deltaic system to a carbonate system parallel to the shoreline or to a pelagic system offshore in deep water.

3. The evolution of modern seismic stratigraphic techniques has greatly expanded our understanding of three-dimensional packages of sediment. Seismic reflections derived from sedimentary rocks are reasonably close to "time lines," particularly when geologically rapid eustatic sea level changes are the major control of transgressions and regressions. Modern computer processing techniques enable the delineation of major depositional systems and their relationship to each other. By being able to visualize the big picture the stratigrapher can avoid becoming bogged down in detail of little significance. The major features of stratigraphic packages commonly can be more easily seen in seismic data than in the partially exposed, incomplete sections that are available to the surface geologist.

4. Plate tectonic theory has revolutionized our understanding of the development of depositional basins through geologic time. Tectonics is the underlying cause of the basins in which stratigraphers ply their trade, and different tectonic styles generate different types of stratigraphic accumulations. For example, flysch and turbidites are much more common along convergent plate margins than along divergent ones.

5. In recent years, time correlation has been greatly refined by increased precision in radiometric dating techniques and by the development of magnetic reversal stratigraphy. For some Mesozoic and Cenozoic strata, numerical ages of individual stages are now reliable to within 10^5 or even 10^4 years.

Surface Observations

Measured Sections

The ability to describe an outcrop well is a learned skill, and, as with all learned abilities, some people are better at it than others. If you doubt this, recall your own impressions the first time you were taken to an outcrop on a field trip for your beginning geology course. What were you able to observe? If you were an average student, as were the authors of this book, you probably noticed only a general coloration, perhaps some layering, and possibly grain size variations if a conglomeratic unit was present. On closer approach you may have found some megafossils such as clams, brachiopods, or trilobites. But it is less likely that you noticed any cross-bedding, graded bedding, grain size sorting within layers, or pink feldspar grains. And if you noticed things such as grain rounding, differential cementation, or differences in weathering character between limestone and dolomite, you were indeed most unusual.

In measuring a stratigraphic section in outcrop, all of these things need to be noted, as well as strike and dip, faulting, unconformities, the presence of intrusive or extrusive rocks in the sequence, rhythms in the lithologic sequence (for example, sandstone beds are always overlain by red shales, but carbonate beds are overlain by evaporites), and many other features. The object is to enable someone reading your field notes to visualize the outcrop without having to visit it. The ability to write such descriptions requires that you have either a mental or a written list of features to look for, some understanding of what each feature might mean in terms of the history of the unit, and lots of practice. There is a great difference between understanding theory and seeing its results in the field. Faults rarely appear in outcrop as those razor-sharp lines featured in elementary textbooks. Graded bedding is commonly subtle rather than appearing as pebbles at the base of a bed and gradually changing to fine sand or mud at the top. Indeed, even mudrocks can be size-graded. Description of outcrops is certainly one of those activities in which practice makes better.

Fossil Collections

Fossils are collected for three basic reasons:

1. To determine geologic age and sequence of rocks (Are the rocks overturned?)
2. To correlate the rocks with other fossiliferous units (Is there an unseen fault between two outcrops?)
3. To help determine the environment of deposition of the unit (nonmarine or marine, shallow water or deep water)

In this regard it is useful to be aware of the geologic ranges and environmental tolerances of at least the major groups of organisms. Are all ammonites Mesozoic? Are belemnites present in Tertiary rocks? Are all clams marine? Can corals tolerate brackish water? Obviously, the more information of this type that you know, the more accurate can be your field work. Specialists are able to identify many organisms to the level of genus and species and thus achieve great precision in the field in both age determination and depositional environment, but for most geologists, such determinations based on fossils are necessarily postponed until reference books are available. Much information about depositional environment can also be obtained from sedimentary structures seen in outcrop (Chapters 7–9), and a mental list of these structures and their environmental meaning is perhaps easier to retain than the appearances and ranges of a few hundred fossils.

Fossils can occur in all types of rocks, from evaporites and conglomerates to limestones and shales. Limestones are essentially biologic in origin and so generally contain more fossils per unit volume than other rock types, but wide variations occur. Oolitic limestones may be unfossiliferous, although interbedded shales are rich in fossils. Conglomerates can be veritable graveyards of land vertebrates. Cherts of Mesozoic and Tertiary age may contain abundant, identifiable microfossil remains.

Also to be noted during fossil collecting are their orientations (Are turritellids aligned on the bedding plane? Are most clam shells oriented concave downward?), state of articulation and fragmentation (Are bivalves articulated? Are the bryozoans abraded?), and association with other fossil groups (Are trilobites, extinct since Permian time, found in the same bed as ammonite fragments with intricate suture patterns, which are Mesozoic?). Such observations indicate whether the organisms were indigenous to the depositional site, were transported an appreciable distance, were deposited by strong and unidirectional currents, or were reworked from geologically older formations. Compton (1985) provides a useful discussion of the significance of these variations, as does Shrock (1948).

Mapping

Commonly, the result of surface stratigraphic analysis is construction of a geologic map. Two types of maps are normally made based on collection and analysis of standard field data: outcrop maps and bedrock maps. Outcrop maps show exactly what is visible in the field. They are factual and not interpretive. With such a map, one has only the data and can decide for oneself the locations of pinchouts, facies changes, or whether concealed faults or folds are present. Unfortunately, outcrop maps are rarely published and exist only in field notes or preliminary drafts of published maps. Perhaps the best examples of outcrop maps are those published by the Canadian provinces that are extensively covered by glacial debris. Others have been published by the Geological Survey of Great Britain and a few by the U.S. Geological Survey (Kupfer, 1966).

The geologic maps with which we are all familiar are interpretive maps of rock ages, types, and distributions as visualized by the map's author. Many a geologist has visited an area with geologic map in hand, expecting to see a particular stratigraphic relationship or facies change, only to end up in a plowed wheat field or 100 m from shore in a rowboat. The multi-colored and attractive geologic maps that we appreciate are interpretive, and this must not be forgotten in our search for answers using published geologic maps. We might favor an alternative interpretation of the stratigraphy if we were aware of the limited area of outcrop on which the map is based.

Fundamental Units.   The level of rock unit most referred to by stratigraphers is the **formation**, which is defined simply as a mappable lithologic unit. Obviously, to be mappable, a formation cannot be defined on the basis of microfossils, thin-section petrology, or permeability. Only features clearly seen in the field setting are permissible, such as a single lithology or an assemblage of lithologies that form a genetic unit distinct from overlying and underlying lithologies. Well-recognized examples are the Tapeats Sandstone (Cambrian) of the Grand Canyon region, underlain by Precambrian crystalline rocks and overlain by the Bright Angel Shale; or the black Utica Shale (Ordovician) of northern New York, underlain by the Trenton Limestone and overlain by the Loraine Sandstone. In these two cases, each formation is of uniform lithology and so is designated "shale" or "sandstone." Units termed "formation" are usually composed of several lithologies, such as the Duchesne River Formation (Eocene) of northern Utah, a genetically related mixture of conglomerates, sandstones, and mudrocks. Unfortunately, formational nomenclature is not perfectly consistent, and some lithologically homogeneous units are also termed "formations" rather than limestones or sandstones.

The term **group** is used for two or more contiguous formations that seem to be genetically related. An example is the Delaware Mountain Group (Permian) in west Texas, composed of Brushy Canyon, Cherry Canyon, and Bell Canyon sandstones, each a recognized formation in its own right. Lithostratigraphic nomenclature for units smaller in scale than a formation is rather arbitrary, and the term **member** is used at the discretion of the field geologist. One geologist's member may be another's formation. A third geologist may feel that the unit or bed under consideration is not deserving of special mention at all. Members are commonly named locally and unofficially (not registered with the American Stratigraphic Commission). Formations may have several named members interspersed with unnamed parts of the unit.

Tongues are wedge-shaped lithologic units that extend laterally outward from the next larger stratigraphic unit, a formation. A tongue may be a member, the distinction being either arbitrary or based on the observation that the unit termed a tongue is continuous with its parent formation in one lateral direction but abuts a different formation in the other direction. A **bed** is the smallest lithologic subdivision, usually ranging in thickness from a centimeter to a few meters. Customarily, only distinctive **marker beds**, useful for correlation, are given proper names. Examples include bentonite beds and coal beds. A typical example of the usage of lithostratigraphic terms is shown in Figure 1-1, illustrating the complex and sometimes inconsistent use of the terms group, formation, member, and tongue.

Subsurface Usage. Lithostratigraphic terminology used by petroleum geologists doing subsurface studies based on well logs is somewhat looser than the terminology of stratigraphers studying surface exposures, and the relationship of informally named subsurface lithologic units to surface outcrops of the same age may not be known. Also, in contrast with surface procedures, subsurface units are often given genetic names such as "granite wash" (feldspathic coarse clastics resting on uplifted granitic basement) or "reef rock" (bioherm and possibly its surrounding talus apron).

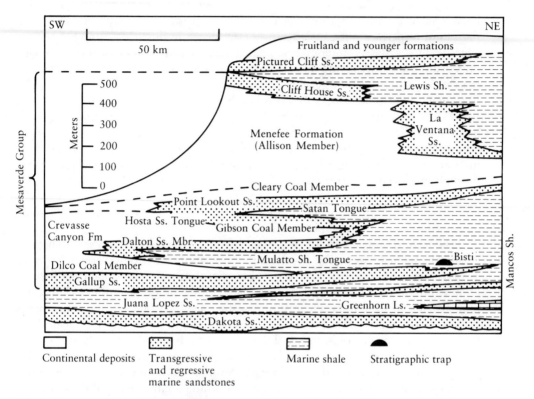

FIG. 1-1 ● Upper Cretaceous stratigraphic cross-section of the San Juan Basin, New Mexico, illustrating usage of lithostratigraphic terminology. (Reprinted by permission of the American Association of Petroleum Geologists from Beaumont, Dane, and Sears, *American Association of Petroleum Geologists Bulletin*, v. 40, 1956.)

Subsurface Observations

Approximately two-thirds of the Earth's land surface is underlain by sedimentary rocks, with thicknesses of 20,000 m in some deep basinal areas such as the Gulf Coastal region of the United States. Reliable estimates of total sediment volume in the crust exceed 10^8 km³, enough material to keep stratigraphers and sedimentologists occupied for some time to come. Unfortunately, about 99% of this volume is hidden from our view beneath the surface. Despite this fact, few academic stratigraphers regularly include subsurface data as an integral part of their analyses. It is largely in the petroleum industry that subsurface data are continuously used by sedimentary geologists, in part because their companies regularly and routinely obtain such data and in part because oil and gas deposits are not exposed at all at the

Earth's surface (except tar sands and an occasional oil seep). Needless to say, all sources of relevant data should be exploited in geologic investigations, and the existence of many hundreds of thousands of deep wells drilled in the continuing search for oil and gas reserves has provided abundant subsurface data of several types.

Cuttings

In the process of drilling a well, drilling mud is pumped down the inside of the drill pipe as a lubricant and to act as a seal against the pressure of fluids seeking to enter the sides of the hole from the rocks being penetrated. This mud circulates downward, through the pipe and drilling bit, and outward into the space between the drill pipe and the rock wall and returns upward to the surface, carrying with it rock fragments produced by the drill bit. The fragments are normally 1–5 mm in diameter, are recovered at the drilling site by the geologist assigned to the well, and are stored by the oil company or the state for subsequent study by stratigraphers and sedimentologists (Schmoker et al., 1984). North American practice is to collect samples every 3 m (10 ft) of drilling depth. There are several difficulties associated with stratigraphic interpretation of cuttings. One problem is that the stratigraphic resolution they provide is commonly even less accurate than the 3-m sample interval. During drilling in soft lithologies, rock may cave into the mud stream from the side of the hole many meters or more above the drill bit, contaminating the sample received at the surface. A second problem is that cuttings of different density may rise in the mud stream at different rates, further blurring lithologic contacts. As an added difficulty, drilling rates can change markedly in units of different resistance (for example, quartz-cemented sandstones as contrasted to shale or carbonate rock), so that a correction must be made to properly relate the cuttings received to the depth from which they came. Hence the stratigraphic data obtainable from well cuttings are far from ideal; they can be very valuable, nevertheless, because cuttings may be the only pieces of rock available from the subsurface.

Cores

Two types of rock cores can be obtained from subsurface units during the search for oil and gas: sidewall cores and bottom-hole cores. Sidewall cores are taken of selected horizons in perhaps 5–10% of wells, normally in horizons that have at least a trace of hydrocarbon content based on analysis of well cuttings or where shows of oil are anticipated. Sidewall cores are quite small, about 3 cm in width and 5 cm in length, and tend to have disturbed textures because of the explosive method used to drive the coring barrel into the wall of the hole. Also, because the cores are so short they are unlikely to include bedding contacts. Nevertheless, these cores provide a three-dimensional piece of rock of significant size for stratigraphic or sedimentologic studies, such as petrographic and microfossil analysis. As with cuttings, the cores may be available for study by geologists not affiliated with the company that obtained the core. Sometimes, however, the cores are considered proprietary and are inaccessible to those outside the company.

Bottom-hole cores are of greater value to the stratigrapher than sidewall cores because of their greater width, about 10 cm, and much greater length, commonly a few tens of meters. The greater width of these cores permits recovery of macrofossils, trace fossils, and small to medium-size sedimentary structures, although the width of the cores is not adequate for evaluating the scale significance of the structures. For example, an erosion surface may be either a local scour or a major regional disconformity. Further, nearly all cores retrieved during drilling are unoriented with respect to compass direction. But the taking of oriented cores is possible and is occasionally done (Nelson et al., 1987). Unfortunately, the procedure is very expensive. The advantage of oriented cores is their usefulness in the determination of microfabric directions (bedding, ripples, flow textures, stylolites, etc.). Accumulated error in orientation resulting from sampling and measurement inaccuracies is $\pm 11°$. The bottom-hole core compared to a sidewall core has the obvious advantages of much greater length and of being taken generally normal to bedding (assuming no structural deformation) rather than parallel to bedding as in a sidewall core. Unfortunately for the stratigrapher, bottom-hole cores are by far the more expensive to obtain and are much less common.

Electric Logs

In addition to rock samples in the form of cuttings or cores, the drilling of holes by explorationists generates a wide variety of indirect data in the form of geophysical logs ("wireline logs"). The various types of geophysical tools in common use for well logging do not determine lithology directly, but determine rock characteristics related to lithology, from which lithology can be inferred (Table 1-1), particularly when several types of logs are used in combination (crossplots). Log analysis and cutting or core analysis are complementary and should be used together for stratigraphic interpretations. Some type of geophysical log is run in all holes drilled by the petroleum industry, and in most cases, several different types are obtained.

Spontaneous potential (SP) logs identify differences in electrical potential between a movable electrode in the borehole and the fixed potential of a surface electrode. Voltage differences result mainly from the movement of ions from the formation waters into the drilling mud but partly from invasion of the drilling mud into the formation. The potential when the electrode in the well is opposite a bed of shale is called the **shale base line** (Fig. 1-2) and is taken as zero. Variations from the base line are usually negative (current flow from the formation into the drilling mud, more saline to less saline) and are plotted to the left. Opposite permeable units, deviations from the shale base line frequently tend to reach an essentially constant deflection, defining a **sand line**. For the stratigrapher the SP curve is useful to detect permeable beds, to locate their boundaries and permit their correlation, and to give a qualitative indication of bed shaliness (clay content).

TABLE 1-1 • Log types and their uses

Log	Property Measured	Units	Geologic Uses
Spontaneous potential	Natural electric potential (compared to drilling mud)	Millivolts	Lithology (in some cases), correlation, curve shape analysis, identify porous zones
Resistivity	Resistance to electric current flow	Ohm-meters	Identification of coals, bentonites, fluid evaluation
Gamma-ray	Natural radioactivity related to K, Th, U	API units	Lithology (shaliness), correlation, curve shape analysis
Neutron	Concentrations of hydrogen (water and hydrocarbons in pores)	Percent porosity	Identification of porous zones, crossplots with sonic, density logs for empirical separation of lithologies
Sonic	Velocity of compressional sound wave	Microseconds/ meter	Identification of porous zones, coal, tightly cemented zones
Caliper	Size of hole	Centimeters	Evaluate hole conditions and reliability of other logs
Density	Bulk density (electron density) includes pore fluid in measurement	Kilograms per cubic meter $(= g/cm^3)$	Identification of some lithologies such as anhydrite, halite, nonporous carbonates
Dipmeter	Orientation of dipping surfaces by resistivity changes		Structural analysis, stratigraphic analysis

(Reproduced with permission of the Geological Association of Canada from Cant, *Geoscience Canada,* v. 10, 1983.)

Environmental determination can also be made from SP log responses. For example, a distributary channel fill is characterized by an abrupt lower contact of the basal sand with the underlying lithology and a gradual upward decrease in grain size (increasing shaliness), resulting in an upward increase in the SP value from the sand line toward the shale base line.

Resistivity logs depict the resistance of a rock unit to an applied electrical current. When dry, most rock types do not transmit electrical currents and are highly resistive; examples include dense carbonates, evaporites, coal, and rocks saturated with hydrocarbons. Low resistivity is generated by the presence in the rock of saline waters (electrolytes), present either in shales (which average 60% hydrous clay minerals) or in permeable, water-saturated sandstones and carbonate rocks.

In a resistivity survey, electrodes of known spacing are lowered into the hole, and the voltage difference is measured between the rock adjacent to the borehole invaded by mud-filtrate and the rock more distant from the borehole. The depth of mud-filtrate penetration ranges from less than 1 mm in impermeable units to more than 10 cm in permeable ones. The wider the spacing of the electrodes, the deeper

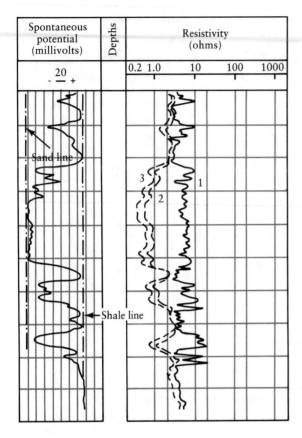

FIG. 1-2 • Relationship between spontaneous potential and resistivity logs in a sand-shale sequence. The three traces on the resistivity log show the traces at increasing distances from the borehole, from 1 to 3. The sandstones have low resistivities because they are filled with salt water. (Reprinted by permission of the publisher from *Log Interpretation, Volume I—Principles*, Schlumberger Technology Corporation, 1972.)

the penetration of current into the bed (Fig. 1-2). Comparison is made between the resistivity of mud-invaded and uninvaded zones. The spacing of the electrodes determines not only the depth of penetration of the electric current but also the stratigraphic resolution of the tool. The closer the spacing, the thinner the bed that can be detected. For best resolution a specialized resistivity tool called the **laterolog** is used. In this tool the electrodes are closely spaced so that the electric current is forced to flow horizontally outward from the borehole as a thin sheet so that beds less than 1 m in thickness can be recognized.

Gamma-ray logs measure the natural radioactivity of the lithologic unit and are used commonly for correlation and facies studies. In sedimentary rocks, radioactive elements (mostly ^{40}K and uranium) tend to concentrate in clay minerals and organic matter, so that the gamma-ray log provides a measurement of the muddiness of the unit. Pure quartz sandstones and clean carbonate rocks give a low log response, whereas mudrocks, volcanic ash, and richly feldspathic sandstones give a high log response (Fig. 1-3). The stratigraphic resolution of the gamma-ray log is somewhat better than that of the resistivity laterolog. SP and gamma-ray traces are commonly interpreted as continuous tracings of grain size variation in detrital units (Fig. 1-4), although in fact only clay content is being recorded. An increase in clay does not necessarily correlate with decreasing mean size of the nonclay fraction, for various depositional or diagenetic reasons (Rider, 1986).

Russell (1985) provides an excellent field example of correlation of a black shale unit of Devonian age in the northeastern United States and Canada. A stratigraphic section about 70 m thick was clearly divisible into six subunits based on its gamma-ray response over a distance of several hundred kilometers.

Ettensohn et al. (1979) used a scintillometer to construct radioactivity profiles of surface outcrops of Devonian-Mississippian black shales in eastern Kentucky and were able to correlate their results with gamma-ray logs of the same units in the subsurface.

Gamma-ray spectral logs may be used for elemental analysis and geochemical logging.

Neutron logs measure the gamma radiation emitted when a rock unit is bombarded by neutrons from a radioactive source. The neutrons collide with elemental nuclei in the minerals and liquids of the rock and lose energy, the greatest loss occurring during collision with hydrogen atoms. Hence the intensity of gamma radiation recorded is a measure of hydrogen concentration. Hydrogen in rock units is located almost entirely in the fluids (hydrocarbons, water), so neutron logs are used as a measure of porosity. Care must be taken when shaley or gypsiferous units occur in the stratigraphic section because the hydrogen atoms in the water that occurs as an integral part of clay minerals and gypsum also cause loss of energy of the impacting neutrons.

Sonic logs consist of a set of transmitters and receivers for emitting and recording sound pulses. The fastest path for sound waves to travel is through the rock rather than through the borehole mud, so the time of first arrival of sound pulses is a measure of the acoustic properties of the unit, which depend on lithology and porosity. The sonic log is normally run with a gamma-ray log (Fig. 1-5). Sonic logs are particularly useful in identifying horizons with marked velocity contrast that have acted as reflectors in seismic reflection profiles.

Caliper logs record the diameter of the borehole, which varies downhole as a function of differences in lithology of the beds traversed. For example, fractured shales will cave into the hole, enlarging the borehole diameter at that stratigraphic level. The caliper log is usually plotted on the same track as the gamma-ray log with hole size increasing from left to right.

Density logs are usually run with a gamma-ray log and measure the electron density of rock units by bombarding them with gamma rays and analyzing the resultant scattered radiation. The electron density is proportional to the bulk density of the rock, which is affected by the specific gravity of the minerals and the amount and type of fluids in the rock pores.

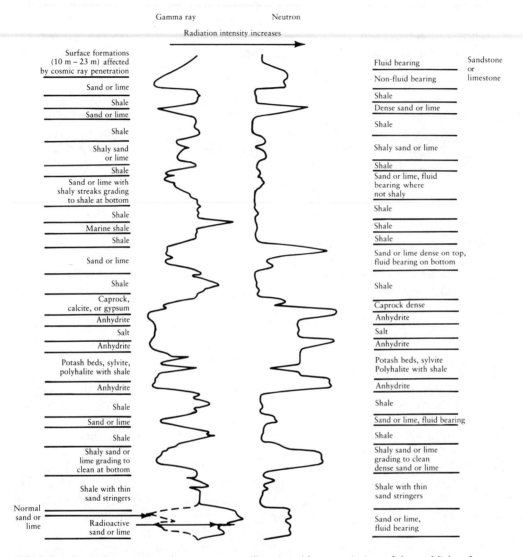

FIG. 1-3 • Typical gamma-ray log responses. (Reprinted by permission of the publisher from *Well Logging and Interpretation*, Part 8, Dresser Atlas, 1982.)

Dipmeter logs are microresistivity devices designed to determine the dip of discontinuities (such as stratification, cross-bedding, and faults) in subsurface rocks. Three or four electrodes are oriented radially in a plane normal to the borehole. Where a dipping bed is encountered, the response to the lithologic change (actually the change in a detectable rock characteristic) occurs at a slightly different elevation for each electrode, and the strike and dip of the discontinuity surface can be determined (the "three-point problem" in structural geology). Dips of less than 2° are

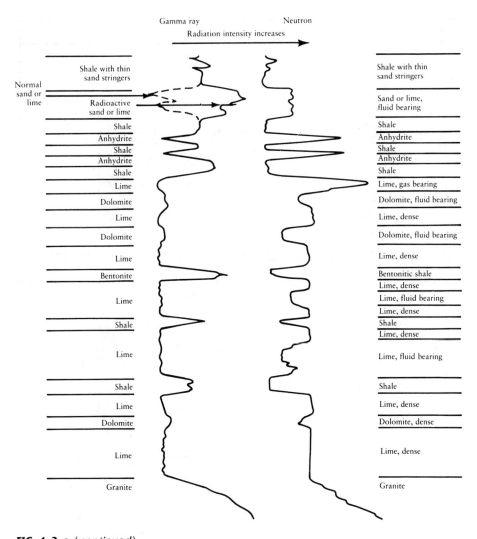

FIG. 1-3 • *(continued)*

recognizable. Some boreholes are drilled at an angle to the horizontal, which must be taken into account in determining the dip of the stratigraphic discontinuity. The azimuth and dip of the hole deviation usually are also shown on the plot.

Dipmeter data are normally presented as a "tadpole plot" (Fig. 1-6) in which the head of the tadpole indicates the dip angle and the tail is oriented to indicate azimuth, north always being toward the top of the plot. The dipmeter responds to all dipping surfaces in rocks and therefore has considerable value in stratigraphic

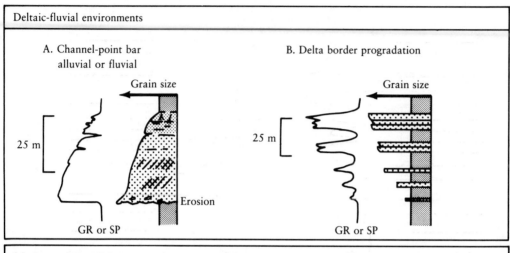

Deltaic-fluvial environments

A. Channel-point bar
   alluvial or fluvial

B. Delta border progradation

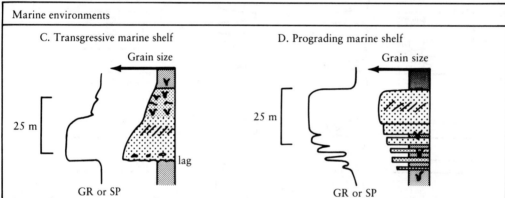

Marine environments

C. Transgressive marine shelf

D. Prograding marine shelf

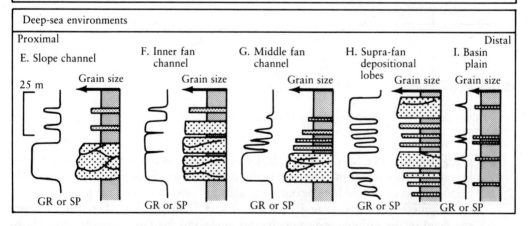

Deep-sea environments

Proximal   Distal

E. Slope channel

F. Inner fan channel

G. Middle fan channel

H. Supra-fan depositional lobes

I. Basin plain

FIG. 1-4 • Facies indications from gamma-ray (or SP) log shapes. These are idealized examples both of log shape and sedimentologic facies. (Reprinted by permission of John Wiley & Sons, Inc., from Rider, *The Geological Interpretation of Well Logs*, copyright © 1986.)

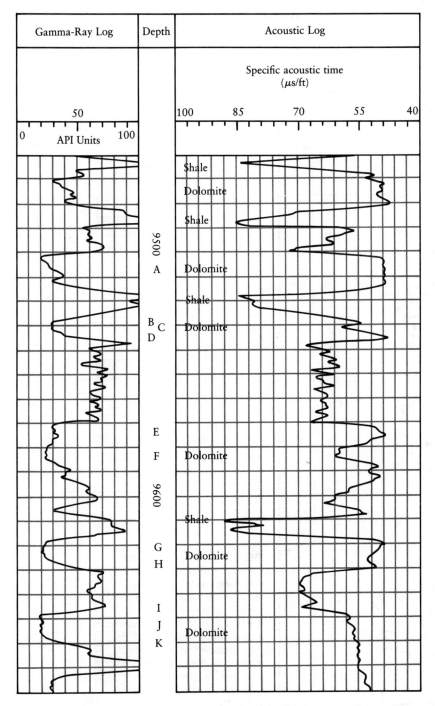

FIG. 1.5 • An acoustic log in a dolomite-shale sequence. (Reprinted by permission of the publisher from *Well Logging and Interpretation*, Part 10, Dresser Atlas, 1982.)

Dip angle and direction

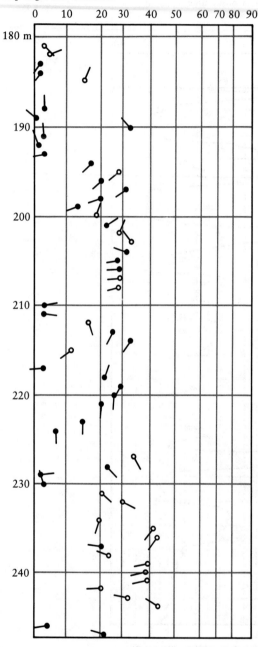

FIG. 1.6 • Example of a dip-meter log indicating direction and magnitude of dip. The section from 205 to 209 m shows uniform dip, and the section 219 to 225 m shows a downward lessening of dip. (Reproduced with the permission of the Geological Association of Canada from Cant, *Geoscience Canada*, v. 10, 1983.)

TABLE 1-2 • Generalized relationship between rock lithology and electric logs

Lithology	Resistivity	Spontaneous Potential
Clay, shale	Low	Low
Sand, salt water	Low	Very high
Sand, fresh water	High	Medium
Sand, oil or gas	Very high	Very high
Limestone, compact	High	Low
Limestone, porous	Low	Very high
Limestone with oil	Very high	Very high

(Reproduced by permission of the publisher from Hobson and Tiratsoo, *Introduction to Petroleum Geology*, Scientific Press, 1975.)

work involving structural, facies, and paleocurrent analyses. Very few dipmeter patterns have a unique interpretation, however, so they cannot be correctly interpreted without lithologic data. Schlumberger Well Services (1981) illustrates many distinctive dipmeter patterns, which can be interpreted in terms of faults, folds, unconformities, channels, depositional dips, or drape over massive sedimentary bodies such as reefs or buried topography. Some of the more common examples are discussed by Gilreath (1977).

Application and Interpretation of Log Patterns   Table 1-2 and Figure 1-7 illustrate the idealized responses of the various major types of well logs to lithologies and subsurface pore fluids and how the logs can be interpreted in terms of depositional environments and possibly also tectonics. When combined with dipmeter data, these geophysical techniques can provide fairly detailed subsurface stratigraphic information and lithostratigraphic correlations. The combined use of surface outcrops and subsurface data generated in the search for hydrocarbons provides the best three-dimensional information about sedimentary basins. Many examples are discussed by Miall (1990).

Log patterns may be used at three levels of interpretation (Galloway and Hobday, 1983): (1) determination of vertical sequence and bedding architecture, (2) recognition and mapping of log facies, and (3) interpretation of depositional environment (Krueger, 1968). Examples of stratigraphic-sedimentologic features that are routinely identified from log traces include proportions of different rock types, thicknesses of lithologic units, character of bedding contacts, grain size variations in sandstones, fining-upward and coarsening-upward sequences, progradational and regres-

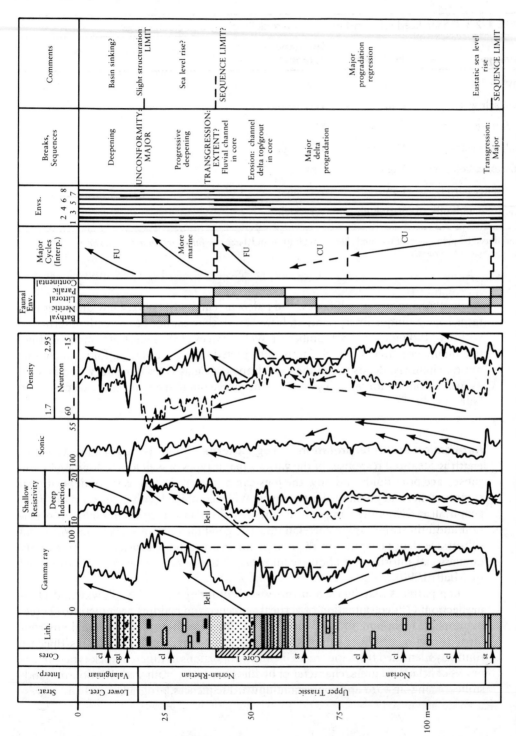

FIG. 1-7 ● A completed sequential analysis. Lithology, electrofacies, and sample data are brought together to give a vertical sequence of depositional environments and an indication of the depositional sequences and major breaks. Environments: 1 = deeper marine; 2 = shallower marine; 3 = prodelta; 4 = delta front; 5 = delta top; 6 = channel; 7 = beach-littoral; 8 = continental; *CU* = coarsening upward, *FU* = fining upward. (Reprinted by permission of John Wiley & Sons, Inc., from Rider, *The Geological Interpretation of Well Logs*, copyright © 1986.)

sive sequences, and the organic content of mudrocks. These attributes reflect the processes that are operative during sediment deposition and permit recognition of the depositional environments and facies represented in the well. Correlation of such data among adjacent wells leads to regional patterns, and, when the data are combined with paleontologic and petrographic information from cuttings and possibly cores, rather complete paleogeographic and paleogeologic descriptions are possible.

Seismic Stratigraphy

All stratigraphers, and perhaps other sedimentary geologists as well, should be familiar with the use and interpretation of seismic data (Cross and Lessenger, 1988). Such data are produced in enormous quantities by the petroleum industry in the search for oil and gas and are available for the rocks beneath the oceans as well as for those on the continents. The most familiar seismic cross-sections are those oriented normal to the Earth's surface, but, thanks in large part to advances in computer processing of seismic signals, it now is possible to produce horizontal seismic sections as well (Brown, *in* Berg and Woolverton, 1985). When modern seismic information is supplemented with wellbore data from cores, cuttings, and electric logs, it becomes possible to make reliable and detailed three-dimensional interpretations of features such as folds, faults, lithologic changes, depositional environments, and eustatic sea level changes—in short, most of the things in which stratigraphers are interested.

Whether or not features can be seen in a seismic section depends on their magnitude compared to the wavelength of the seismic pulse.

$$\text{Wavelength} = \text{velocity} \times \text{period} = \text{velocity/frequency}$$

Velocities in sedimentary basins are usually in the range 2000–2500 m/sec, and the dominant frequency of reflections is commonly about 40 Hz, so the common wavelength is 50–60 m. Wavelength generally increases with depth because velocity increases and frequencies become lower; wavelengths of 250–300 m are typical. Because wavelength limits resolving power, deeply buried features must be much larger than shallow features to produce the same appearance on the seismic cross-section.

Resolution is defined as the minimum distance between two features necessary to permit the two features to be "seen." This distance is approximately 1/4 wavelength, so with a wavelength of 60 m a bed 15 m thick can be distinguished from an adjacent thick bed, provided sufficient acoustic contrast exists. Another characteristic of a seismogram is the **detection limit**, the minimum thickness for a layer to give a reflection. This thickness is approximately 1/30 wavelength. Thus at shallow depth it is possible to "see" a bed 15 m thick and obtain a reflection from a bed 2 m thick. An idealized seismic section and the terminology used to describe it are shown in Figure 1-8.

Seismic reflections are generated by discontinuities in the subsurface. Within a sedimentary section the major discontinuities are the sharp lithologic contrasts, such as those between a sandstone and an overlying shale or between a limestone and an

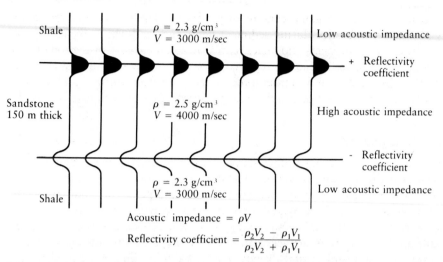

FIG. 1-8 • Idealized seismic section illustrating the pattern generated by the contact between a sandstone and enclosing shales. The seismic trace is vertical within each of the three beds because the reflectivity coefficient is zero. If the sandstone should become shaley, reflectivity coefficients would converge toward zero as the bedding contacts become less well defined lithologically. The numbering convention for subscripts is that the first bed encountered by the wave is 1.

underlying evaporite. Lateral lithologic changes within individual beds, such as a facies change from fine grained sandstone to silty shale, will cause a change in reflectivity coefficient and shape of the positive and negative bulges (Fig. 1-8) that different coefficients cause, but the dominant coefficient contrasts will be where different lithologies overlie each other with a sharp acoustic impedance contrast between them. As a result, seismic reflections follow chronostratigraphic boundaries rather than lithostratigraphic ones. Thus one can recognize on a seismic cross-section packages of sediment that are bounded by approximate time surfaces, and the resulting correlations from one seismic section to a neighboring one show relationships between packages of sediment deposited during the same time intervals. These packages are termed **seismic sequences** (Figs. 1-9, 1-10). They are composed of a relatively conformable succession of genetically related strata interpreted as a **depositional sequence**, for example, the paralic, near-shore marine, clinoform, and basinal facies of a deltaic depositional sequence. The upper and lower boundaries of depositional sequences are unconformities or their correlative conformities. The time interval represented by strata of a given sequence may differ from place to place, but the range is confined to synchronous limits marked by ages of the sequence boundaries where they become conformities. Depositional sequence boundaries are recognized in seismic data by identifying reflections caused by lateral terminations of strata. By correlating seismic sections worldwide, it is possible to recognize eus-

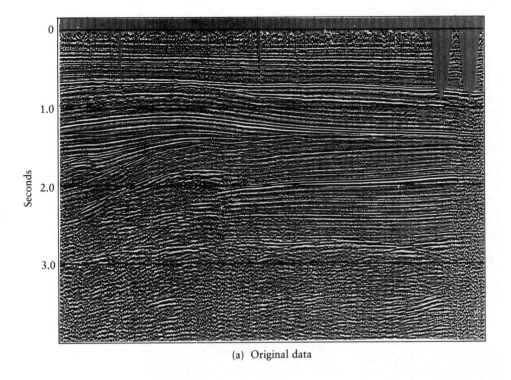

(a) Original data

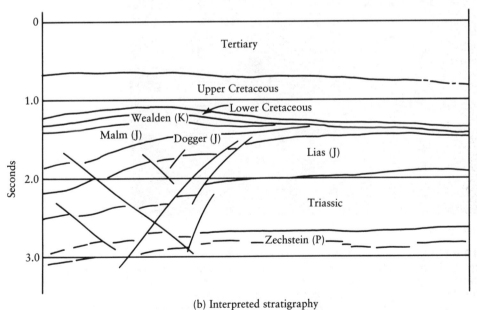

(b) Interpreted stratigraphy

FIG. 1-9 • Seismic record and interpretive geologic cross-section through the Hohne oil field, Gifhorn Basin, West Germany. Oil is produced from the Dogger beds immediately below the Cretaceous unconformity. (Reprinted by permission of the publisher from Dobrin, *Introduction to Geophysical Prospecting*, McGraw-Hill, 1976.)

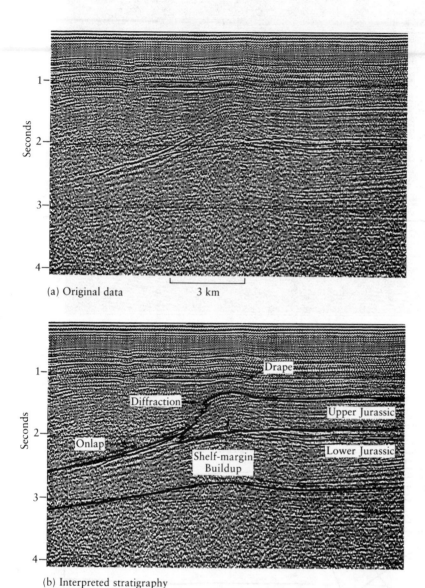

(a) Original data

3 km

(b) Interpreted stratigraphy

FIG. 1-10 • Shelf-margin carbonate buildup can be seen by (1) reflection from top and front of buildup, (2) onlap of cycles onto buildup, (3) change from continuous, parallel reflectors into discontinuous reflectors, (4) numerous diffractions, (5) drape over buildup, and (6) abrupt change in dip of reflectors. Wells encountered a series of Mesozoic shelf-margin buildups along eastern Atlantic continental margin off Africa. Buildup displayed on this line is interpreted as Late Jurassic. (Reprinted by permission of the American Association of Petroleum Geologists from Bubb and Hatledid, *in* Payton, *American Association of Petroleum Geologists Memoir*, v. 26, 1977.)

TABLE 1-3 • Geologic interpretation of seismic facies parameters

Seismic Facies Parameters	Geologic Interpretation
Reflection configuration	Bedding patterns Depositional processes Erosion and paleotopography Fluid contacts
Reflection continuity	Bedding continuity Depositional processes
Reflection amplitude	Velocity-density contrast Bed spacing Fluid content
Reflection frequency	Bed thickness Fluid content
Interval velocity	Estimation of lithology Estimation of porosity Fluid content
External form and areal association of seismic facies units	Gross depositional environment Sediment source Geologic setting

(Adapted from Mitchum et al., *in* Payton, 1977, p. 122.)

tatic changes in sea level, worldwide transgressions, and regressions. Local variations from the regional or worldwide pattern can then be attributed to relatively local tectonic events.

Within each seismic sequence are a variety of reflection geometries, continuities, amplitudes, frequencies, and interval velocities (Table 1-3); these are the subject of **seismic facies** analysis. Seismic facies units are groups of seismic reflections whose characteristics differ from adjacent groups. After seismic facies units are recognized, their limits defined, and areal associations mapped, they are interpreted in terms of environmental settings and lithologies. The various types of seismic facies patterns and their interpretation are extensively described and illustrated by Mitchum et al. (*in* Payton, 1977).

Figure 1-11 illustrates the main styles of seismic facies reflection patterns, most of which are best seen in sections parallel to depositional dip. Parallel or subparallel reflections indicate uniform rates of deposition. Divergent reflections result from differential subsidence rates such as across a shelf-margin hinge zone. Prograding reflections are particularly common on continental margins, where they represent deltaic or continental slope outgrowth. Variations in progradational patterns reflect different combinations of depositional energy, subsidence rates, water depth, and sea level position. The other seismic facies patterns can be similarly interpreted in terms of depositional stratigraphy and sedimentation. Nearly all of the petroleum indus-

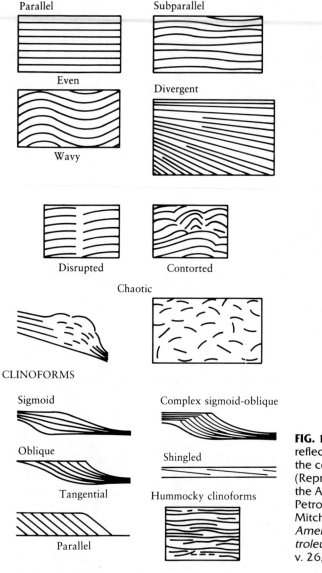

FIG. 1-11 • Typical seismic reflection patterns, illustrating the concept of seismic facies. (Reprinted by permission of the American Association of Petroleum Geologists from Mitchum et al., *in* Payton, *American Association of Petroleum Geologists Memoir*, v. 26, 1977.)

try's expenditures for surface geophysical work are devoted to seismic exploration (compared to gravimetric or magnetic surveys), so such data are abundant. To the extent that they are made available by the individual companies, they should be used regularly by all stratigraphers.

Most seismic stratigraphic or seismic lithologic interpretations have been applied to clastic depositional systems. Carbonate rocks, because they consist of only one or, at most, two minerals, are more difficult to study using seismic data. Only within the last few years has knowledge of the petrophysical characteristics of carbonate rocks, combined with advances in the technology of geophysical tools, made possible accurate interpretation of seismic images of subsurface carbonate sections (Fontaine et al., 1987). For example, chalk deposits display continuous high-amplitude reflections at top and base with an internal reflection-free zone. Inner shelf strata are characterized by parallel, low-frequency continuous reflections. Dolomitized zones exhibit a "marbled" zone, a practically reflection-free zone with a few discontinuous reflections. The various seismic facies in carbonate sections and their interpretation in terms of depositional environment are illustrated in Figure 1-12.

The interpretation of seismic data is often like much 20th century art, in that it consists of recognizing patterns and exercising imagination. In the hands of a skilled and insightful interpreter, the variety and accuracy of stratigraphic information obtainable from seismic cross-sections are truly extraordinary, as exemplified by American Association of Petroleum Geologists Memoirs 26 and 39 (Payton, 1977; Berg and Woolverton, 1985). The important factors for successful interpretation of seismic data are an understanding of fundamental principles (rock physics, computer processing), experience, and imagination.

Relating Borehole Logs to Seismic Traces

Seismic reflectivity depends on velocity and density of rocks, quantities measured in boreholes by sonic and density logs. Thus it should be possible to relate well logs to seismic traces. A seismic trace calculated from well log data is called a **synthetic seismogram**, and a well log calculated from a seismic trace is called a **seismic log**. The mathematical conversion from one of these data sources to the other is only approximate, however (Sheriff, *in* Payton, 1977) because

1. Logs are plotted in depth; seismic traces in two-way travel time. A commonly used conversion relationship is that a seismic sample interval of 2 milliseconds (msec) corresponds to a depth sample interval of about 3 m at 3000 m/sec velocity. Velocities in unlithified or poorly lithified sediments are typically only 1000–2000 m/sec.
2. Logs respond to the magnitudes of velocity and/or density; seismic reflectivities depend on differences in the product of velocity and density.
3. The frequencies used to record sonic logs are very high, whereas those recorded from seismic surveys are very low. Wavelengths seen on electric and sonic logs resolve bed boundaries to within fractions of meters; the resolution on seismic traces is tens or hundreds of meters (Fig. 1-13).
4. Because of the high frequencies used, electric logs in wellbores "see" (penetrate) only 1–100 cm laterally from the hole, whereas seismic response is from a reflecting surface tens of meters in extent.
5. The "noise" in an electric log results from different factors than the noise in a seismic trace. Noise effects are removed from both well logs and seismic traces by computer processing.

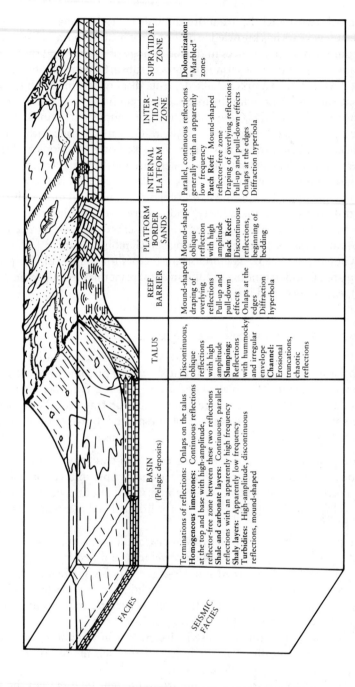

FIG. 1-12 ● Seismic facies of carbonate depositional environments. (Reprinted by permission of the American Association of Petroleum Geologists from Fontaine et al., *American Association of Petroleum Geologists Bulletin*, v. 71, 1987.)

SEISMIC FACIES	BASIN (Pelagic deposits)	TALUS	REEF BARRIER	PLATFORM BORDER SANDS	INTERNAL PLATFORM	INTER-TIDAL ZONE	SUPRATIDAL ZONE
	Terminations of reflections: Onlaps on the talus **Homogeneous limestones:** Continuous reflections at the top and base with high-amplitude, reflector-free zone between these two reflections **Shale and carbonate layers:** Continuous, parallel reflections with an apparently high frequency **Shaly layers:** Apparently low frequency **Turbidites:** High-amplitude, discontinuous reflections, mound-shaped	Discontinuous, oblique reflections with high amplitude **Slumping:** Reflections with hummocky and irregular envelope **Channel:** Erosional truncations, chaotic reflections	Mound-shaped draping of overlying reflections Pull-up and pull-down effects Onlaps at the edges Diffraction hyperbola	Mound-shaped oblique reflection with high amplitude **Back Reef:** Discontinuous reflections, beginning of bedding	Parallel, continuous reflections generally with an apparently low frequency **Patch Reef:** Mound-shaped reflector-free zone Draping of overlying reflections Pull-up and pull-down effects Onlaps at the edges Diffraction hyperbola		**Dolomitization:** "Marbled" zones

FIG. 1-13 • Scale of a typical seismic wave form as compared to an outcrop. (Reprinted by permission of the author and the publisher from Miall, *Principles of Sedimentary Basin Analysis*, Springer-Verlag New York, 1990.)

6. The seismic trace at the well location may relate to a different part of the Earth than that seen by the well unless **migration** has been performed. In this process, the positions of the reflections as received are relocated to the geographic position at which they were generated. To do this, it is necessary to estimate the velocity distribution along the raypath, a process involving approximations.

Remote Sensing

The most widely used data for stratigraphic studies are those obtained from examination of surface outcrops, borehole logging, and seismic reflection cross-sections. There are, however, other techniques that may be useful to stratigraphers. They are collectively termed **remote sensing** and include aerial photography, Landsat images, radar sensing, and various types of devices that are sensitive to various wavelengths of the electromagnetic spectrum. The subject of remote sensing is currently in an explosive phase of development because of its military usefulness, and there is no doubt that its capabilities by the end of the century will dwarf present ones, which already are substantial. The pace of development of remote sensing techniques is indicated by the flood of texts that have appeared in the past few years—at least six in 1986 and 1987 alone (Hord, 1986; Lo, 1986; Richards, 1986; Sabins, 1986; Drury, 1987; Lillesand and Kiefer, 1987).

Aerial Photographs

Geologic interpretation of aerial photographs is based on differences in photographic tone, color, texture, and pattern and the relation to the size and shape of associated features. Photos taken from an elevation of a few thousand meters and printed at a scale of $1:10^4$ to $1:10^5$ are used most commonly. The amount of stratigraphic information that may be obtained from aerial photographs depends primarily on the types of rocks present (lithostratigraphy), local climate, and stage of geomorphic development. Except for high mountainous areas where vegetative growth is restricted by altitude, arid and semiarid regions generally will have the largest proportions of rock outcrop, and tropical regions the least. Because features are more readily recognized where strong differences exist in the erosional resistance of adjacent rocks, sedimentary terrains yield the greatest amount of information from aerial photographs.

The presence of bedding in most sedimentary rocks is fundamental to their interpretation from aerial photos. Layering stands out because more resistant beds are brought into relief and less resistant beds are lowered as a result of weathering and transport of the debris downslope. In the absence of topographic expression, banding due to differences in vegetation or soil may delineate beds. Thus it is clear that grain size, composition, cementation, and permeability are important controls of the overall texture of an aerial photograph. From these photos there is the potential to determine stratigraphic features such as unconformities, bed thickness and lithology, facies changes, faults, and folds. Aerial photographs should not be neglected in regional stratigraphic studies.

Radar Geology

Determining surface geology from radar imagery is a widely used technique that produces pictures similar in appearance to standard aerial photographs but with lesser resolution. Original image scales are less than $1:10^5$. The advantage of radar imagery lies in the ability of radar waves to penetrate clouds, fog, and most precipitation so that images of "photographic" quality can be obtained independent of visibility or weather conditions. In preliminary stratigraphic work the smaller-scale images such as those produced by radar have the added advantage of convenience. A tenfold difference in scale translates to a need for only one photo rather than one hundred.

Other Remote Sensing Techniques

More recent developments in remote imagery include daytime photographic infrared imagery, which is used to delineate soils, rock formations, or structural features defined by vegetation. Best results are obtained in the spring or fall in temperate climates, when major vegetative changes occur. A variety of scales is available.

Thermal infrared scanning senses variations in emission of heat, and because thermal emissions from certain surfaces and substances are distinctive, the imagery may show the distribution of certain kinds of rocks, vegetation, and surficial deposits. It is particularly useful in hot spring and volcanically active areas. Scales range from very large to very small, depending on the reason the images were taken.

Landsat multispectral scanning is imagery transmitted from orbiting satellites. The scale varies from $1:10^6$ to $1:10^5$. In multispectral scanning, green, red, and two infrared wavelength bands are composed, and the images in either black and white or various colors are used for study of major topographic variations, geologic structures, and vegetation differences. In general, satellite images have too small a scale to be very useful in stratigraphic work except in the sense that some data are better than none at all.

Summary

The ability of earth scientists to unravel the history of our planet depends on their perspicuity and on the state of technology. Earlier generations of geoscientists were essentially restricted to a hammer, hand lens, polarizing microscope, and X-ray diffractometer. Today's researchers have access to a much wider variety of laboratory techniques as well as an ever-increasing mass of both direct and indirect data from the subsurface and from satellites. There is no doubt that the sophistication of tomorrow's electronics will dwarf that available today, and current technology will seem primitive within the lifetime of today's undergraduate students. Explorationists and professional academics alike will need to be very broadly trained in the more basic sciences that form the underpinnings of geologic research.

References

Anstey, N. A., 1982. Simple Seismics. Boston, International Human Resources Development Corporation, 168 pp.

Asquith, G. B. and Gibson, C. R., 1982. Basic Well Log Analysis for Geologists. Tulsa, American Association of Petroleum Geologists, 216 pp.

Bally, A. W., 1987. Atlas of Seismic Stratigraphy. Tulsa, American Association of Petroleum Geologists, Studies in Geology No. 27, 124 pp.

Beaumont, E. C., Dane, C. H., and Sears, J. D., 1956. Revised nomenclature of Mesaverde group in San Juan Basin, New Mexico. Amer. Assoc. Petroleum Geol. Bull., v. 40, pp. 2149–2162.

Berg, O. R. and Woolverton, D. G. (eds.), 1985. Seismic Stratigraphy, II. Tulsa, American Association of Petroleum Geologists, Mem. 39, 276 pp.

Burton, R., Kendall, C. G. St.C., 1987. Out of our depth: On the impossibility of fathoming eustasy from the stratigraphic record. Earth-Sci. Rev., v. 24, pp. 237–277.

Cant, D. J., 1983. Subsurface sedimentology. Geoscience Canada, v. 20, pp. 115–121.

Compton, R. R., 1985. Geology in the Field. New York, John Wiley & Sons, 398 pp.

Cross, T. A. and Lessenger, M. A., 1988. Seismic stratigraphy. Ann. Rev. Earth and Planet. Sci., v. 16, pp. 319–354.

Davis, T. L., 1984. Seismic-stratigraphic facies models. In R. G. Walker (ed.), Facies Models, 2nd ed., Toronto, Geological Association of Canada, pp. 311–317.

Dobrin, M. B., 1976. Introduction to Geophysical Prospecting, 3rd ed. New York, McGraw-Hill, 630 pp.

Dobrin, M. B. and Savit, C. H., 1988. Introduction to Geophysical Prospecting, 4th ed. New York, McGraw-Hill, 867 pp.

Dresser Atlas, 1982. Well Logging and Interpretation Techniques: The Course for Home Study, Parts I–XI. Houston, Dresser Industries, Inc.

Drury, S. A., 1987. Image Interpretation in Geology. Boston, Allen & Unwin, 243 pp.

Ettensohn, F. R., Fulton, L. P., and Kepferle, R. C., 1979. Use of scintillometer and gamma-ray logs for correlation and stratigraphy in homogeneous black shales. Geol. Soc. Amer. Bull., v. 90, pp. 421–423, pp. 828–849.

Fontaine, J. M., Cussey, R., Lacaze, J., Lanaud, R., and Yapaudjian, L., 1987. Seismic interpretation of carbonate depositional environments. Amer. Assoc. Petroleum Geol. Bull., v. 71, pp. 281–297.

Galloway, W. E. and Hobday, D. K., 1983. Terrigenous Clastic Depositional Systems. New York, Springer-Verlag, 423 pp.

Gilreath, J. A., 1977. Dipmeters. In L. W. LeRoy, D. P. LeRoy, and J. W. Raese (eds.), Subsurface Geology. Golden, Colorado School of Mines, pp. 389–396.

Haq, B. U., Hardenbol, J., and Vail, P. R., 1987. Chronology of fluctuating sea levels since the Triassic. Science, v. 235, pp. 1156–1167.

Hobson, G. D. and Tiratsoo, E. N., 1975. Introduction to Petroleum Geology. Beaconsfield, England, Scientific Press, 300 pp.

Hord, R. M., 1986. Remote Sensing Methods and Applications. New York, John Wiley & Sons, 362 pp.

Krueger, W. C., Jr., 1968. Depositional environments of sandstones as interpreted from electrical measurements—An introduction. Gulf Coast Assoc. Geol. Soc. Trans., v. 18, pp. 226–241.

Kupfer, D. H., 1966. Accuracy in geologic maps. Geotimes, v. 10, no. 7, pp. 11–14.

Lang, H. R., Adams, S. L., Conel, J. E., McGuffie, B. A., Paylor, E. D., and Walker, R. E., 1987. Multispectral remote sensing as stratigraphic and structural tool, Wind River Basin and Big Horn Basin areas, Wyoming. Amer. Assoc. Petroleum Geol. Bull., v. 71, pp. 389–402.

Lillesand, T. M. and Kiefer, R. W., 1987. Remote Sensing and Image Interpretation, 2nd ed. New York, John Wiley & Sons, 720 pp.

Lo, C. P., 1986. Applied Remote Sensing. New York, Longman, 393 pp.

Merkel, R. H., 1979. Well Log Formation Evaluation. Tulsa, American Association of Petroleum Geologists, Continuing Education Course Note Series No. 14, 82 pp.

Miall, A. D., 1986. Eustatic sea level changes interpreted from seismic stratigraphy: A critique of methodology with particular reference to the North Sea Jurassic record. Amer. Assoc. Petroleum Geol. Bull., v. 70, pp. 131–137.

Miall, A. D., 1990. Principles of Sedimentary Basin Analysis, 2nd ed. New York, Springer-Verlag, 664 pp.

Nelson, R. A., Lenox, L. C., and Ward, B. J., Jr., 1987. Oriented core: Its use, error, and uncertainty. Amer. Assoc. Petroleum Geol. Bull., v. 71, pp. 357–367.

Nurmi, R. D., 1988. Geologic interpretation of well log and seismic measurements in reservoirs associated with evaporites. *In* B. C. Schreiber (ed.), Evaporates and Hydrocarbons, New York, Columbia University Press, pp. 405–459.

Payton, C. E. (ed.), 1977. Seismic Stratigraphy—Applications to Hydrocarbon Exploration. Tulsa, American Association of Petroleum Geologists, Mem. 26, 516 pp.

Richards, J. A., 1986. Remote Sensing Digital Image Analysis: An Introduction. New York, Springer-Verlag, 281 pp.

Rider, M. H., 1986. The Geological Interpretation of Well Logs. New York, John Wiley & Sons, 175 pp.

Russell, D. J., 1985. Depositional analysis of a black shale by using gamma-ray stratigraphy: The Upper Devonian Kettle Point Formation of Ontario. Bull. Canadian Petroleum Geol., v. 33, pp. 236–253.

Sabins, F. F., Jr., 1986. Remote Sensing, 2nd ed. New York, W. H. Freeman & Co., 449 pp.

Schlumberger Well Services, 1972. Log Interpretation. Vol. I: Principles. New York, Schlumberger Technology Corporation, 113 pp.

Schlumberger Well Services, 1974. Log Interpretation. Vol. II: Applications. New York, Schlumberger Technology Corporation, 116 pp.

Schlumberger Well Services, 1981. Dipmeter Interpretations. Vol. 1: Fundamentals. New York, Schlumberger Technology Corporation, 61 pp.

Schmoker, J. W., Michalski, T. C., and Worl, P. B., 1984. Nonprofit Sample and Core Repositories Open to the Public in the United States. Washington, D.C., U.S. Geological Survey, Circ. 942, 102 pp.

Scholle, P. A. (ed.), 1977. Geological Studies on the COST No. B-2 Well, U.S. Mid-Atlantic Outer Continental Shelf Area. Washington, D.C., U.S. Geological Survey, Circ. 750, 71 pp.

Shrock, R. R., 1948. Sequence in Layered Rocks. New York, McGraw-Hill, 507 pp.

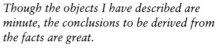

Though the objects I have described are minute, the conclusions to be derived from the facts are great.

Henry Sorby

2
Sedimentary Textures, Mineralogies, and Structures

Three classes of sedimentary rocks form about 95% of the stratigraphic record: conglomerates and sandstones, 20–25%; mudrocks, 60%; and carbonate rocks, 10–15%. The remaining 5%, although not abundant volumetrically, can supply valuable information about things such as the character of the depositional site (for example, evaporites and cherts), oceanic current patterns (phosphate rocks), or the abundance of gaseous oxygen in the environment (bedded iron deposits). So all types of sedimentary rocks need to be described in as much detail as seems reasonable, using both field observations and laboratory techniques. The sedimentary features that can be observed easily and that are meaningful to a stratigrapher will, of course, differ with the type of rock being described.

Conglomerates and Sandstones

Conglomerates and sandstones are the most easily described of the three major rock groups because of their relative coarseness and because of the contrast among adjacent grains in color and mineralogy. Important textural features include grain size and sorting, grain shape, quartz grain surface features, mineral composition, and clay content.

Grain Size and Sorting

The average grain size and spread of sizes in a coarse clastic rock can supply information about current strength at the site of deposition, the type of transporting

TABLE 2-1 • The standard grain-size scale for clastic sediments

	Name	Millimeters	Micrometers	ϕ
Gravel	———————	4096		−12
	Boulder			
	———————	256		−8
	Cobble			
	———————	64		−6
	Pebble			
	———————	4		−2
	Granule			
	———————	2	—— ——	−1
	Very coarse sand			
Sand	———————	1		0
	Coarse sand			
	———————	0.5	500	1
	Medium sand			
	———————	0.25	250	2
	Fine sand			
	———————	0.125	125	3
	Very fine sand			
	———————	0.062	—— 62 ——	4
	Coarse silt			
	———————	0.031	31	5
	Medium silt			
Mud	———————	0.016	16	6
	Fine silt			
	———————	0.008	8	7
	Very fine silt			
	———————	0.004	—— 4 ——	8
	Clay			
		↓	↓	↓

medium, and the environment of deposition. Sometimes the relationships are obvious. Desert sand dunes always contain a narrower range of grain sizes than a glacial till or an alluvial fan. Deposits of mountain streams are normally coarser grained than those of coastal deltas. But most commonly the grain-size differences among environments are more subtle, and it becomes necessary to use quantitative measures of size and sorting for comparative purposes.

The grain-size scale used by geologists was devised by C. K. Wentworth in 1922, and the **phi scale** was appended to Wentworth's scale by W. C. Krumbein in 1934 to facilitate statistical manipulation of grain-size data; $\phi = -\log_2(d_{mm}/1 \text{ mm})$ (Table 2-1). During field work, the sizes of pebbles, cobbles, and boulders can be determined by using a tape measure and the sand sizes by using a grain-size comparator. The mean size (average) may be estimated at the outcrop and determined more precisely if necessary by laboratory techniques such as sieving or settling tube analysis. The laboratory data are usually plotted as a cumulative curve, from which points are chosen to arrive at a mean grain size. The formula used by most sedimentologists is

$$\text{Mean size} = \frac{\phi_{16\%} + \phi_{50\%} + \phi_{84\%}}{3}$$

where $\phi_{16\%}$ is the phi value at 16% on the cumulative curve, and so on. The three phi values represent the lower, middle, and upper thirds of the cumulative curve.

The spread of grain sizes is given as the range in phi units that includes two-thirds of the grains. This range is twice the **standard deviation**. For example, if the mean size is 2.5ϕ and two-thirds of the grains have sizes between 2.0ϕ and 3.0ϕ, then the standard deviation is 0.5ϕ. Standard deviation is the accepted measure of the sorting of the sediment (Fig. 2-1). The formula for graphic standard deviation is

$$\text{Standard deviation} = \frac{\phi_{84\%} - \phi_{16\%}}{4} + \frac{\phi_{95\%} - \phi_{5\%}}{6.6}$$

An additional grain-size statistic that is commonly calculated and that is sometimes useful for environmental discrimination is **skewness**, a measure of the lop-

FIG. 2-1 • Hand-lens view of degrees of sorting. Values of standard deviation that divide each class of sorting are also shown. (Reprinted by permission of John Wiley & Sons, Inc., from Compton, *Manual of Field Geology*, copyright © 1985.)

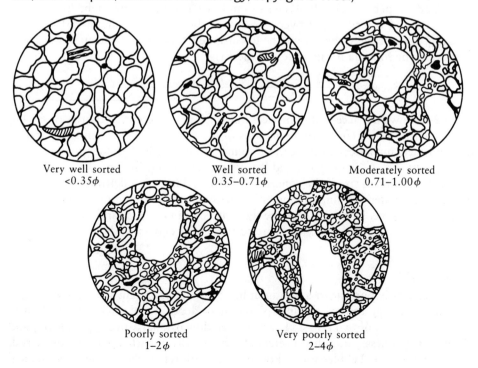

Very well sorted
$<0.35\phi$

Well sorted
$0.35–0.71\phi$

Moderately sorted
$0.71–1.00\phi$

Poorly sorted
$1–2\phi$

Very poorly sorted
$2–4\phi$

TABLE 2-2 • Classes of sorting and skewness

Classes of Sorting	
Less than 0.35ϕ	Very well sorted
0.35−0.50ϕ	Well sorted
0.50−0.71ϕ	Moderately well sorted
0.71−1.0ϕ	Moderately sorted
1.0−2.0ϕ	Poorly sorted
2.0−4.0ϕ	Very poorly sorted
Greater than 4.0ϕ	Extremely poorly sorted

Classes of Skewness	
−1.0 to −0.3	Strongly coarse skewed
−0.3 to −0.1	Coarse skewed
−0.1 to +0.1	Near-symmetrical
+0.1 to +0.3	Fine skewed
+0.3 to +1.0	Strongly fine skewed

sidedness or nonsymmetrical nature of the grain size distribution curve. If a sediment contains 80% pebbles and 20% coarse sand, it has a tail to the right (toward the finer sizes) and is termed positively skewed. If the sediment contains 80% coarse sand and 20% pebbles, the tail will be toward the left and the sediment has negative skewness. The formula for graphic skewness is

$$\text{Skewness} = \frac{(\phi_{95\%} + \phi_{5\%}) - 2(\phi_{50\%})}{2(\phi_{95\%} - \phi_{5\%})} + \frac{(\phi_{84\%} + \phi_{16\%}) - 2(\phi_{50\%})}{2(\phi_{84\%} - \phi_{16\%})}$$

Table 2-2 shows the accepted numerical limits for the different categories of sorting and skewness. More complete discussions of the significance of grain-size parameters and their numerical values can be found in most sedimentology texts, for example, Blatt (1982, Chapter 15) and Pettijohn et al. (1987, Chapter 3).

Related to grain size is the concept of matrix. **Matrix** in a detrital rock refers to the granular material filling the interstices between the larger grains in a conglomerate or sandstone. In a conglomerate the matrix commonly is sand; in a sandstone the matrix usually is clay minerals, which can be either detrital or diagenetic in origin.

Grain Shape

Grain shape is important to a stratigrapher because, like grain size and sorting, it can point the way toward the interpretation of depositional environment. Differences in shape are of two types: (1) gross three-dimensional aspect (grain form) and (2) grain roundness. Grain form refers to whether the particle approximates a rod, a disc, or a sphere. Pebbles formed of material with relatively isotropic mechanical properties, such as quartz, tend to be either rods (in rivers) or discs (on beaches), although shape overlap is common between the two environments (Dobkins and Folk, 1970). Strongly anisotropic pebbles, such as schist fragments, have a shape controlled largely by their anisotropy rather than by environment of deposition.

Gravel-size grains are nearly always well rounded because they are transported in traction along the stream bottom or beach face. Sand grains, however, remain angular if they spend much time in either saltation or suspension; silt grains travel in suspension and are invariably angular. The only environments that seem able to round quartz sand are desert dunes, beaches, and near-shore barrier bars. It must be kept in mind, though, that sand grains can be recycled from older sedimentary rocks upstream that may have been deposited in a different environment from that of the rock you are examining. For example, you may be examining Devonian river sand containing rounded quartz sand grains recycled from a Cambrian beach deposit exposed higher in the Devonian drainage basin.

Quartz Grain Surface Features

Although they are generally much too small to be visible by using a hand lens, a wide variety of types of grooves, gouges, etch pits, and miscellaneous markings occur on the surfaces of quartz sand grains. These surface irregularities have been intensely studied over the past 20 years with the use of scanning electron microscopy (Krinsley and Doornkamp, 1973) in an attempt to relate the various types of markings to depositional environment, with considerable success. Environments that produce distinctive patterns include glacial, eolian, and shallow marine. Unfortunately, the flow of water through sandstones after burial of the sediment tends to remove mechanically produced surface markings on quartz grains and to replace them with chemically produced etch pits, and in pre-Mesozoic rocks, study of surface texture is generally ineffective for environmental discrimination.

Clay Content

The presence of clay in a detrital rock may or may not be a good indicator of weak currents at the site of deposition. This uncertainty arises because in some rocks the clay can be clearly shown to be diagenetic. Unfortunately, the distinction between primary and secondary clay cannot be made at the outcrop and sometimes cannot be made even in the laboratory by using sophisticated analytical equipment. In general, conglomerates and sandstones deposited by turbidity currents in deep marine environments along convergent plate margins are more likely to contain abundant secondary clay than detrital rocks deposited by other mechanisms and in other environments, but small amounts of authigenic clay can occur in any rock.

Textural Maturity

Textural maturity is a concept developed in 1951 by R. L. Folk, before the abundance of secondary clay in many sandstones was widely recognized. **Textural ma-**

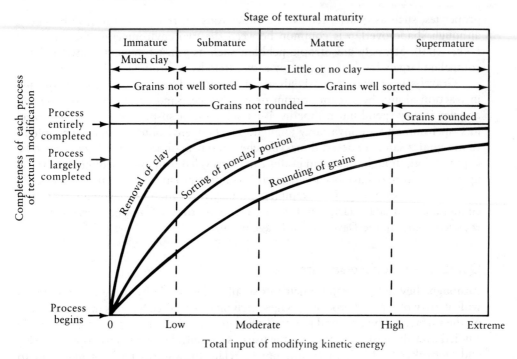

FIG. 2-2 • Textural maturity of sands as a function of input of kinetic energy. (Reprinted by permission of the publisher from Folk, *Journal of Sedimentary Petrology*, v. 21, Society of Economic Paleontologists and Mineralogists, 1951.)

turity is defined in terms of clay content, uniformity of sand or gravel grain sizes (sorting), and the perfection of rounding of quartz sand grains. Four stages of textural maturity are recognized (Fig. 2-2), and although none of them has a unique interpretation in terms of depositional environment, recognition of the stage of maturity does help to narrow down the possible choices (Fig. 2-3). If a sediment is texturally immature, and assuming that the clay is detrital rather than secondary, only weak currents were present at the depositional site, and the most likely environments are deep marine and fluvial overbank. If the quartz grains are well rounded, only environments dominated by high kinetic energies need be considered, such as beaches, offshore bars, or eolian dunes. But even in the case of texturally supermature sediments the environmental interpretation is not certain. Rounded quartz grains can be recycled into a river from an older beach deposit exposed upstream; that is, the rounding can be inherited.

Mineral Composition

The mineral composition of conglomerates and sandstones is complex and often extremely difficult to interpret accurately. Although the mineral composition of gravel-size particles can be made in the field as easily as one can do it in a hand-specimen petrology class laboratory, correct identification of sand size grains requires considerable experience. Even then, examination of the rock in thin section using a polarizing microscope is required to achieve adequate precision. But areal changes in mineral and rock fragment composition are very important in paleogeographic work, and as much as possible should be done in the field setting. The question of where to go next for a meaningful rock sample can commonly be answered by mineralogic trends pieced together from individual observations made at scattered outcrops.

The basic mineralogic distinctions that should be made at the outcrop are (1) rock fragments, (2) feldspars, and (3) quartz grains. Rock fragments (or lithic fragments) are pieces of igneous, metamorphic, or sedimentary rock, for example, a piece of schist, granite, or chert. The abundance of the different types of rock fragments depends in part on grain size. Granite fragments are most common in pebble and coarser sand sizes; the abundance of felsite grains is unaffected by grain size because of the microcrystalline nature of felsite. All rock fragments are more abundant closer to their source, normally a highland area. It sometimes is possible to determine the distance to the mountain front by plotting the decrease in maximum size of gravel-size grains versus distance (Pelletier, 1958).

Feldspars in a hand specimen are recognized (ideally) by their colors (mostly pink or white), cleavage, and opacity compared to quartz grains. But as everyone who has tried to make estimates of feldspar percentages from a sandstone hand specimen is aware, such estimates are notoriously imprecise and are usually underestimates.

Quartz grains are perhaps the easiest type of particle to identify in hand speci-

FIG. 2-3 ● Relationship among sedimentary volumes, environments of deposition, and textural maturity. The diagram is based on perceptions of sedimentologists from field studies; adequate numerical data do not exist. (Reprinted by permission of the publisher from Blatt, *Sedimentary Petrology*, W. H. Freeman & Co., 1982.)

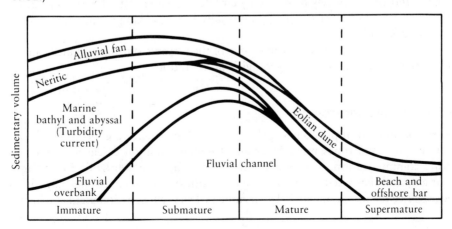

mens. The average sandstone contains about two-thirds quartz, but the range of variation among sandstones is from zero to 100%, so trends may be easily determined in the field if a gradient exists, either laterally or vertically in a stratigraphic section. Changes in the degree of rounding of the quartz grains in a sandstone may also be observed and used to interpret depositional environment, as has been demonstrated by Folk (1960) in the Tuscarora Formation (Silurian) in West Virginia.

Provenance Studies

The detrital mineral composition of a conglomerate or sandstone is important in determining the lithologic character of source areas. The object of a provenance investigation is to construct a map of upstream or upcurrent paleogeology that is as accurate as a geologic map of modern rock distributions, a goal never reached but worth striving toward. Sedimentary structures locate the upstream direction, but only detailed study of detrital minerals can reveal the types of rocks exposed and their distributions. An enormous literature exists that deals with sandstone petrology and provenance (Blatt et al., 1980, Chapter 8; Pettijohn et al., 1987, Part II); much of it is peripheral to the major concerns of stratigraphy, and most of it is concerned with thin-section petrography. The principles on which provenance determinations are based can be summarized as follows.

1. Determine the percentage of each detrital mineral and rock fragment type in the rock using an appropriate combination of standard thin-section work, accessory mineral concentration and study, and other techniques such as X-ray diffractometry or electron microprobe examination.
2. During the petrographic part of the study (at least), note alteration characteristics in all particle types and the relationship, if any, between particle type and grain size. Classify the roundness of grains, also as a function of grain size.
3. Note any diagenetic changes in the rock that may have changed the original mineral composition (Hutcheon, 1983), leading to a distorted picture of the source area. For example, feldspars and accessory minerals commonly are etched and/or completely dissolved by circulating waters after burial as revealed by the presence of remnants or by unusually loose packing of the grains in the thin-section slide.
4. Based on these data, existing knowledge of the tectonic setting, and men's intuition, make a paleogeographic/geologic interpretation.

Classification of Sandstones

As is true of all natural phenomena, the variations in mineral composition of conglomerates and sandstones must be grouped into meaningful categories to aid in scientific communication. The best groupings are those that provide insight into rock origin, and many tens of mineralogic classifications of sandstones have been proposed since the initial one by P. D. Krynine in 1948. Most are triangular schemes

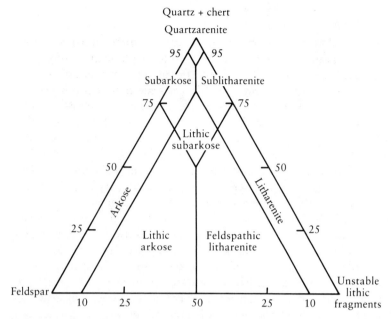

FIG. 2-4 • One of many mineralogic classifications of sandstones in common use. (Reprinted by permission of the publisher from McBride, *Journal of Sedimentary Petrology*, v. 33, Society of Economic Paleontologists and Mineralogists, 1963.)

with the corners of the triangle being quartz, feldspars, and rock fragments (Fig. 2-4). During field work it is not possible to be very precise in mineral identifications.

One of the more common ways to describe the texture and mineral composition of conglomerates and sandstones is by combining Folk's scheme of textural maturity with a mineralogic classification such as the one in Figure 2-4, perhaps appending a term for the type of cement or the presence of unusual constituents. For example, a measured section of the coarser clastics might be described as follows (youngest at top).

1. Immature, hematitic, coarse-grained granitic litharenite (possibly alluvial fan)
2. Submature, chert- and limestone-bearing, medium-grained arkose (possibly fluvial)
3. Submature, calcite-cemented limestone pebble conglomerate (possibly alluvial fan)
4. Supermature, medium-grained, quartz-cemented quartzarenite (possibly beach)

The sequence might represent an alluvial fan sequence produced by block-fault uplift of Precambrian granite that was overlain by Paleozoic shallow-water sandstone and limestone. Sequences similar to this are common in rocks of Pennsylvanian age in Colorado and Oklahoma.

Sedimentary Structures

The textural features of sedimentary rocks are small-scale geometrical irregularities, the smaller ones (such as grain surface features) requiring magnifications of several hundred times to be seen, the larger ones visible to the unaided eye (grain sorting). Sedimentary structures, on the other hand, are of larger scale than textures and are formed typically by juxtaposition of contrasting textures; for example, a fine-grained and poorly sorted sandstone overlying a well-sorted gravel produces a discontinuity termed **bedding**. The upper size limit of a sedimentary structure is similarly vaguely defined. Features of sedimentary rocks that are too large to be easily visible in outcrops are usually described under the general term **sediment geometry**; for example, the spatial relationship between lenticular sand bars and their surrounding shale beds is part of the geometry of the sedimentary sequence.

Because of their relatively large size, sedimentary structures are normally easily seen, either in a subsurface core or in an outcrop, but not always. Ripple cross-stratification and cross-bedding are visible in either a sidewall or a bottom-hole core, but flute casts require a larger surface to be detected. Some structures are poorly defined in a core but show up clearly on an outcrop surface, where the natural etching of the weathering process exaggerates the contrast between adjacent textures. In some rocks the beds appear structureless even on an outcrop surface but reveal hidden structures under X-radiography. It is worth going to great lengths to detect structures because they help to establish the hydraulic or biologic conditions prevailing at the time of deposition and hence the depositional environment. Also, structures are unlike textures in that they cannot be recycled. A rounded sand grain can be inherited from a previous environment and be unrepresentative of processes in the depositional environment of the sandstone being examined. But a sedimentary structure is destroyed when eroded; it cannot be recycled.

Because of the easy visibility and importance of sedimentary structures for the interpretation of depositional environments, many books have appeared that deal with the origin and environmental significance of such features in detrital rocks (Pettijohn and Potter, 1964; Reineck and Singh, 1980; Collinson and Thompson, 1988; Bouma, 1969; Conybeare and Crook, 1968; Allen, 1970). We will focus our attention on those sedimentary structures that are most common and/or useful to a stratigrapher (Table 2-3).

The most common sedimentary structure is **stratification**, which can occur on scales ranging from millimeters to tens of meters. Causes include grain size or compositional changes, grain shape or orientation changes, change in content of organic matter, change in amount of hematitic coloring matter, or change in fossil content. A useful scale and terminology for changes in the thickness of strata was suggested by Ingram in 1954 (Table 2-4). Sometimes a later current will blur or totally conceal the bedding surface between successive deposits (Fig. 2-5). When such minor diastems are detected, they are termed **amalgamation surfaces**. They represent events separated in time by perhaps only a few days or weeks, and it seems likely that most of them are visibly recorded in outcrop. The possibility of the deposits matching perfectly above and below the amalgamation surface seems remote.

TABLE 2-3 • Sedimentary structures in detrital rocks

Stratification

---

1. Relative abundance of conglomerate, sandstone, and mudrock.
2. Degree of intermixing: Is lower third mudrock, middle third conglomerate, and upper third sandstone? Or are rock types intimately intermixed?
3. If intermixed, is there a characteristic sequence, such as conglomerate units normally followed by sandstones?
4. Thickness and lateral extent of beds.
5. Nature of contacts with beds above and below: sharp or gradational.

Bedding Surfaces

---

1. Planar, wavy, or irregular; relation to lithology.
2. Ripple marks: symmetry, orientation, height, distance between crests.
3. Parting lineation: orientation.
4. Load casts.
5. Flute casts: scale and orientation.
6. Groove casts: description and orientation.
7. Organic tracks and trails: description.
8. Mud cracks, raindrop impressions, or other features.
9. Cut-and-fill structures.

Internal Character of Beds

---

1. Laminations: cause (e.g., organic matter, grain-size change), thickness, continuity; fissility.
2. Organic burrows: abundance, type, relation to lithology.
3. Cross-bedding: scale, orientation, angle of dip.
4. Graded bedding: thickness of graded unit, frequency, grain-size variation.
5. Convolute bedding, intraformational breccias.
6. Orientations of fossils (e.g., preferred orientation of elongate shells; brachiopod shells mainly concave up or down) or pebbles.
7. Imbrication: orientation.

(Reprinted by permission of the publisher from Blatt, *Sedimentary Petrology*, W. H. Freeman & Co., 1982.)

An example of an analysis and interpretation of bedding that produced results relevant to paleogeography is the study of Tertiary rocks in Spitsbergen by Atkinson (1962). The rock sequence consists of 2000 m of detrital conglomerates, sandstones, and shales, which occur with varying bed thicknesses and varying order of superposition. Statistical analysis of these changes revealed that conglomerates were preferentially overlain by sandstones, and sandstones by shales. An analysis of the ratios of thicknesses of beds in these cycles suggested short-term meteorologic variations as the cause of the observed preferences in lithologic association. Larger-scale litho-

TABLE 2-4 • Scale of stratification thickness

Very thickly bedded	Thicker than 1 m
Thickly bedded	30–100 cm
Medium bedded	10–30 cm
Thinly bedded	3–10 cm
Very thinly bedded	1–3 cm
Thickly laminated	0.3–1 cm
Thinly laminated	Thinner than 0.3 cm

(Reprinted by permission of the author from Ingram, *Geological Society of America Bulletin*, v. 65, 1954.)

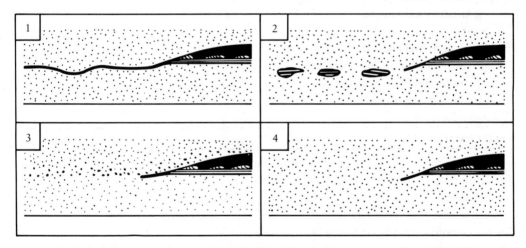

FIG. 2-5 • Amalgamation of the graded divisions of two successive currents. (1) The line of amalgamation is shown up by a horizontal joint, sometimes with small load casts. (2) The line of amalgamation is shown up by a row of angular or rounded mud flakes. (3) The line of amalgamation is shown up by a change of grain size. (4) The line of amalgamation is invisible due to perfect annealing between the two beds. (Reprinted by permission of the publisher from Walker, *Journal of Sedimentary Petrology*, v. 36, Society of Economic Paleontologists and Mineralogists, 1966.)

logic sequences also were found and were attributed to variation in the balance between the rate of upward movement of the source area and the rate at which erosion reduced the resulting topographic relief.

Some **bedding surfaces** display no visible markings other than randomly distributed swells and bumps that are, at present, uninterpretable. Other surfaces are so planar that a carpenter's level can be laid on them with the bubble centered in the glass tube. But many stratification surfaces exhibit features that are directional, vectorial, or patterned in a manner that reflects distinctive characteristics of the depo-

FIG. 2-6 • Parting lineation, Whirlpool Sandstone (Silurian), near Georgetown, Ontario, Canada. (Photo courtesy of R. G. Walker.)

sitional environment. Such features need to be described and interpreted in the field, where their spatial distribution and stratigraphic significance can be sensed most clearly.

Primary sedimentary structures are produced at and immediately adjacent to the sediment-water interface, penecontemporaneously with deposition of the sediment. But although they are produced at the upper surfaces of the sediment layer, some of them are expressed best (or exclusively) in ancient rocks on the base (sole) of the overlying bed. That is, the markings on the depositional surface are preserved as casts on the underside of the next sedimentary layer. This happens most commonly when the structure was created in a mud and the overlying layer was a sand.

Ripple marks form whenever noncohesive clastic sediment is in motion, and they occur in all depositional environments characterized by moving fluids. The presence of ripples does not define a specific environment, although they are more common or of greater size in some environments than in others. For example, ripple marks are more prominent in fluvial deposits than in deep-sea turbidites, although they occur in both environments. The explanation for the difference in frequency of occurrence lies in the fluid mechanics of sediment transport in the two environments, and is considered in great detail in sedimentology texts (see Blatt et al., 1980, Chapter 4; Reineck and Singh, 1980, Part 1). In stratigraphic work, probably the most important fact about ripple marks is that they are usually asymmetrical and that the steeper side of the ripple faces downcurrent.

Parting lineation is seen where fine grained sandstones with otherwise planar bedding surfaces separate along their shared bedding plane (Fig. 2-6). This parting surface displays a faint lineation related to alignment of the long axes of sand grains during deposition. The parting is slightly imperfect, so the surface has plasterlike remnants clinging to the bedding surface, and these patches are elongate in the direction of the lineation and sediment movement during deposition.

FIG. 2-7 • Flute casts origi-
nating from burrows (holes at
deep end of flutes), Upper
Devonian, near Ithaca, New
York. (Photo courtesy of R. G.
Walker.)

Flutes are asymmetrical structures preserved as casts on the base of sandstone beds, particularly in turbidity current deposits (Fig. 2-7). Because such deposits are found most commonly in deep-water marine sediments, flute casts are associated in many minds with deep-water deposits, but this is a dangerous attitude. Shallow-water lacustrine beds, for example, can also contain flutes. All that is required is for a turbid water-sediment mixture to flow rapidly over a muddy substrate on a slight incline. The deep part of the flute points upcurrent.

Grooves are long, straight features, also usually preserved as casts on the base of turbiditic sandstone beds (Fig. 2-8). They are essentially scratches into the mud produced by a tool of some sort that was carried in the turbid sediment load. Related markings include **chevron marks**, which look like an army sergeant's stripes with the Vs pointing downcurrent; and a variety of types of gouges termed **prod marks**, **bounce marks**, and **brush marks**. Most of these supply vectorial information about the currents that formed them.

The vagrant benthonic organisms that live in both fresh and saline environments produce marks on the sediment surface as they feed or burrow into the mud for protection. These **trace fossils** or **ichnofossils** tend to group by depth zones in the marine environment (Crimes, *in* Frey, 1975). In general, the deepest burrows are found in the shallowest water, and surface traces dominate in bathyl and abyssal depths.

Internal Bedding Structures

Vertical sections of sandstones commonly show parallel **laminae**, usually only a few grain diameters thick, sometimes barely exceeding the thickness of the coarsest grains. The laminae are defined by slight grain-size differences or by concentrations of mica. They are formed when the power of the moving fluid exceeds the stability field of ripples, and they occur in any environment where suitable flow conditions exist. Lamination may be prominent in stream deposits or turbidites.

Cross-bedding is perhaps the most common of all sedimentary structures, other than planar stratification, and a large experimental and descriptive literature exists describing it. Most cross-bedding is produced by avalanching of sand over the crest of a granular sediment pile, such as a ripple, dune, or bar, and in all cases the toe of the cross-bed points downstream. Many varieties of cross-beds have been described (see, for example, Fig. 2-9) and result from different combinations of grain size,

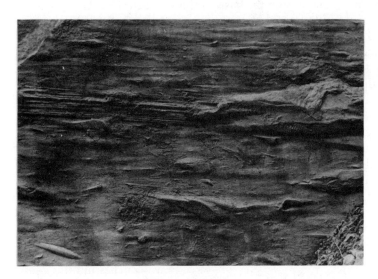

FIG. 2-8 • Long, thin, groove casts with numerous prod and bounce casts oriented as small angles to the grooves, Annot Sandstone (Eocene-Oligocene), southern France. (Photo courtesy of R. G. Walker.)

FIG. 2-9 • Herringbone cross-stratification caused by reversing tidal currents, Gates Formation (Lower Cretaceous), British Columbia, Canada. (Photo courtesy of Dale A. Leckie, McMaster University.) (Reprinted by permission of the American Association of Petroleum Geologists from *American Association of Petroleum Geologists Bulletin*, v. 66/2, 1982.)

amount of sediment being deposited, and the velocity of the moving fluid (Blatt et al., 1980, Chapter 5).

Flaser bedding (Fig. 2-10) is produced by rippled sand migrating across a muddy substrate. The flasers are incomplete mud laminae trapped in ripple troughs during periods of slack water (mud-draped ripples), and all transitions occur from predominantly sandy beds, with ripple cross-laminated sets separated by flasers, to predominantly shaly beds, with isolated ripple form sets (lenticular bedding). Flaser bedding is fairly diagnostic of tidal deposits, although it has been reported from other environments as well.

Graded bedding is defined by an upward decrease of grain size (Fig. 2-11) within a bed and can have a wide lateral extent in turbidite units; some graded beds have been correlated for many kilometers on the basis of this characteristic. Graded beds are produced almost exclusively by turbidity currents as the turbid sediment-water mixture slows toward the foot of the continental slope or lake center. Lateral as well as vertical grading occurs within a bed as the coarser sediment is deposited nearer to the shore (proximal) and only finer particles remain for deeper-water deposition (distal). Graded bedding is particularly valuable as a criterion for determining the "up direction" in structurally deformed sequences, which are so common along convergent plate margins.

In many turbidity current sandstones the graded bed is overlain sequentially by a sandstone with plane parallel laminae, a rippled unit, a second laminated sand or coarse silt, and, finally, a mudstone unit that may contain cryptic grading. Such a sequence of sedimentary structures is termed a **Bouma sequence** (Fig. 2-12) after the sedimentologist who first pointed out its generality. The graded, coarse-grained sandstone at the base represents the fastest current and deposition near the head of the turbidity current, and the overlying units of the sequence contain the sedimentary

FIG. 2-10 • Lenticular, flaser, and wavy bedding in Westphalian A (Carboniferous), Pembrokeshire, Wales, UK. The scale is 80 cm long. (George deV. Klein, *Clastic Tidal Facies*, © 1977. Reprinted by permission of Prentice-Hall, Inc., Englewood Cliffs, NJ.)

FIG. 2-11 • Graded bedding in a pebbly sandstone and associated flame structure in a midfan lobe channel sequence of a lower Mesozoic submarine fan, Mineral King caldera complex, Sierra Nevada, California. (Reprinted by permission of the publisher from Busby-Spera, *Journal of Sedimentary Petrology*, v. 55, Society of Economic Paleontologists and Mineralogists, 1985.)

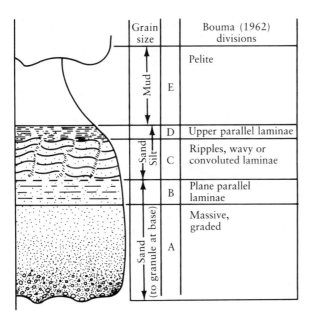

Grain size		Bouma (1962) divisions
Mud	E	Pelite
Sand—Silt	D	Upper parallel laminae
	C	Ripples, wavy or convoluted laminae
	B	Plane parallel laminae
Sand (to granule at base)	A	Massive, graded

FIG. 2-12 • Ideal sequence of structures in a turbidite bed. (Reprinted by permission of the publisher from Bouma, *Sedimentology of Some Flysch Deposits*, Elsevier Science Publishers B.V., 1962.)

FIG. 2-13 • An imbricated cobble sequence. The clasts are predominantly oriented with their *a*-axes parallel to flow and dipping upstream and their *b*-axes transverse to flow. This is from a fan delta unit, the Stadium Conglomerate, which here is eroded into paralic units of the Scripps Formation. This Middle to Late Eocene interval crops out northeast of La Jolla, California. (Photograph by Jeff May; reprinted by permission of the publisher from *Journal of Sedimentary Petrology*, v. 51, Society of Economic Paleontologists and Mineralogists, 1981, p. 1296.)

structures that have been shown in laboratory experiments to be formed as both the current strength and sediment size decrease. The uppermost muddy unit in the Bouma sequence results from either the tail of the turbidity current or pelagic sedimentation. Using a scanning electron microscope, O'Brien et al. (1980) have shown that the clay flakes in a turbiditic mud have a random orientation; those in a pelagic deposit have a more parallel orientation.

The **orientation** of concavo-convex fossils such as pelecypods can also be useful as an up-direction indicator. The stable position of such shells in a moving fluid is concave downward into the sediment, but care must be exercised in using this criterion. If the shell settles in relatively quiet waters, it may come to rest concave upward. The balance between shell size and current strength is the critical factor here.

Imbrication is a stacking of grains with their flat surfaces at an angle to the major bedding surface (Fig. 2-13) and is most clearly seen in conglomerates lacking interstitial sand. The flattened fragments of gravel dip upcurrent because this is the stable position in the moving fluid. Imbrication occurs in sandstones as well as conglomerates but cannot be detected without examination of thin sections cut normal to stratification.

Lineations of grains or fossils also are most easily seen when they are large enough to be easily visible to the unaided eye. In such cases the long dimension of the particle is parallel to paleocurrent direction, and if the particle has a larger or heavier end, that end will be upcurrent. Small linear fossils are often located in ripple troughs and are thus oriented normal to the current rather than parallel to it. The key element here is the spacing between the ripple crests relative to the length of the elongate fossil.

Mudrocks

Adequate description of mudrocks requires laboratory study to a greater degree than that of sandstones because of the much finer grain size of most mudrocks. A wide variety of techniques is commonly used (Blatt, 1982), including X-ray diffraction for clay mineral identification, sodium bisulfate fusion for segregation of quartz and feldspars from the matrix of clay minerals, electron microprobe for identification of feldspar types, polarizing and scanning electron microscopes for textural studies, and X-radiography of mudrock slabs for examination of subtle sedimentary structures such as cross-bedding, ripples, lamination, or burrowing by organisms (bioturbation).

Textures

Mudrocks are defined as detrital rocks that are finer grained than sandstones, and a convenient method of subdividing them during stratigraphic work is shown in Table 2-5. The principle on which the subdivision is based is that mudrocks are composed almost entirely of quartz (30%) and clay minerals (60%), and because quartz is harder than apatite (human teeth), the grit/slime ratio as detected by nibbling on a piece of mudrock can give a semi-quantitative estimate of the quartz/clay ratio. The importance of such determinations to a stratigrapher has been demonstrated by Blatt and Totten (1981), who were able to determine distance offshore in an epicontinental sea by the progressive decrease in quartz percentage and corresponding increase in clay minerals.

Because of the small sizes of the particles in the average mudrock, about 15 μm (6ϕ) for the quartz and less than 1 μm for the clay minerals, laboratory techniques are necessary for most textural work. But the quartz grains are always platy to elongate in shape and quite angular. This results from the fact that most quartz silt originates as flattened grains in fine-grained metamorphic rocks, and these tiny grains are transported in suspension by either water or wind. Hence they suffer few grain-to-grain impacts and cannot have their sharp corners chipped off to produce

TABLE 2-5 • Classification of mudrocks

Ideal Size Definition	Field Criteria	Fissile Mudrock	Nonfissile Mudrock
>⅔ silt	Abundant silt visible with hand lens	Silt-shale	Siltstone
⅓–⅔ silt	Feels gritty when chewed	Mud-shale	Mudstone
>⅔ clay	Feels smooth when chewed	Clay-shale	Claystone

(Blatt, Middleton, and Murray, *Origin of Sedimentary Rocks*, 2nd ed., © 1980, p. 382. Reprinted by permission of Prentice-Hall, Inc., Englewood Cliffs, NJ.)

rounding. Clay minerals are generally platy in shape because of the sheeted structure of clay minerals (phyllosilicates) and are always very small in size because of their defect-laden crystal structures.

Other than grain size, the only other textural feature of mudrocks that can be determined in the field is color. Nearly always, the color is reddish, greenish, or in the gray-to-black range, although yellows, shades of white, and other colors are possible in unusual circumstances. The red color reflects the presence of a few percent of ferric oxide (hematite) in the rock and usually reflects deposition in an oxidizing environment. The color may, however, be of early diagenetic origin, even in a seemingly impermeable mudrock. The green color reflects the absence of hematite and is caused by the naturally greenish hue of illite, the most abundant clay mineral. Gray and black colors result from the presence in the rock of a few percent of incompletely decomposed organic matter, although finely divided iron sulfides can also cause the rock to appear dark. The presence or absence of organic matter in a rock reflects the balance between the amount of organic matter available to be decomposed in the depositional environment and the amount of gaseous oxygen available to accomplish the decomposition (oxidation). In most depositional environments the circulation of water is adequate to continually renew the supply of dissolved oxygen, so organic matter is completely oxidized and leaves no residue. Poor water circulation can be caused by a variety of factors, such as the existence of a topographic/structural barrier at the entrance to a basin (the Black Sea, many fjords), the growth of an organic reef through time, a shift in water current patterns so that oxygen-rich waters no longer circulate, or a sudden increase in the amount of organic matter to be decomposed. In ancient rocks it commonly is difficult to determine which of the several possible causes is responsible for the onset of an oxygen-deficient environment.

Mineral Composition

As noted earlier, two groups of minerals dominate the average mudrock, quartz (including chert) and clay minerals. Most of the quartz and chert are detrital. Some of the finer silt-size silica particles, however, may be derived from diagenesis of clay minerals (see below), but the proportion of total quartz originating in this way is very small.

Clay minerals in mudrocks (and sandstones) are mostly of three types: illite, montmorillonite, and kaolinite. The major differences among the three can be expressed in terms of the major alkali and alkaline earth cations dominant in each: potassium in illite and sodium or calcium in montmorillonite; kaolinite is composed of only silica and alumina. Many clays when deposited are composed of interlayered illite and montmorillonite, but abundant field and laboratory evidence has shown that at about 3000 m burial depth (temperatures between 50°C and 100°C) the montmorillonite layers are converted to additional illite layers in a reaction interpreted to be

$$Montmorillonite + Al(OH)_4^{-1} + K^{+1} \rightarrow illite + H_4SiO_4$$
$$+ Ca^{+2} + Na^{+1} + Mg^{+2} + Fe^{+2,+3} + H_2O$$

and

$$H_4SiO_4 \rightarrow SiO_2 + 2 H_2O$$

It follows from this that illite should be more abundant in older rocks, an inference supported by field and laboratory data (Foscolos, 1984). Clays rich in montmorillonitic layers form about 60% of modern sediments but only 5% of lower Paleozoic clays. Illite, on the other hand, increases in abundance from 25% to 80% over the same interval of time. Kaolinite is minor in most mudrocks, regardless of age.

An average mudrock also contains 4–5% feldspar, 3–4% carbonate minerals, and lesser amounts of organic matter, hematite, pyrite, and accessory minerals. The feldspars and accessory minerals in mudrocks can be used for provenance studies in the same manner as is traditionally done with sandstones (Blatt, 1982, Chapter 3).

Sedimentary Structures

The coarser-grained mudrocks such as siltstones contain the same variety of sedimentary structures as sandstones: ripple marks, cross-bedding, graded bedding, and so on. As average grain size decreases and clay minerals become more abundant, however, a freshly deposited mud becomes more cohesive, so structures such as ripples and cross-bedding cannot form. In their place, fissility may appear. Fissility is a property of a mudrock that causes it to break along thinly spaced planes parallel to stratification, a property related to the percentage of platy clay minerals in the rock but dependent on other factors as well. Of greatest importance are bioturbation and flocculation. Flocculation is a process of clumping of clay flakes into randomly oriented silt-size particles that occurs in saline waters. Flocs are also produced by bottom-dwelling organisms as a byproduct of the feeding and burrowing processes. Probably all clays are aggregated into flocs in marine waters by one or both of these processes, so that fissility may be a secondary structure produced during diagenesis. The visibility of fissility in mudrocks is greatly enhanced by uplift and weathering. A mudrock at depth under a large overburden may appear massive, but on removal of the vertical stress, fissility can appear.

Lamination in mudrocks is common, but its origin is usually indeterminate without thin-section examination. Common causes of lamination in mudrocks include differences in clay mineral content, differences in silt content, and formation of varves resulting from seasonal variation in sediment supply to lakes. In many areas, burrowing organisms destroy any lamination formed by other processes (bioturbation). In the Mississippi Delta, regularly laminated muds are confined to parts of the shallow sea floor adjacent to active distributaries, where deposition is moderately rapid and there are few bottom-dwelling organisms. Further seaward, the muds become irregularly layered and then mottled or homogeneous, usually because of bioturbation (Fig. 2-14). Laminae in these shallow marine environments are produced by

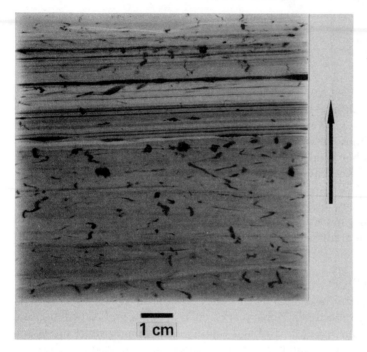

FIG. 2-14 • Radiograph of section of subsurface core in Selmier Shale (Devonian-Mississippian), Illinois, showing interbedded, thickly laminated black shale (upper half) and indistinctly bedded gray shale (uppermost centimeter and lower half). Lower contact of laminated shale is sharp, but upper contact is gradational and several burrows extend into the top of the laminated bed. Burrows and many of the laminations cannot be seen without radiography. Width of core is 10 cm. (Reprinted by permission of the publisher from Cluff, *Journal of Sedimentary Petrology*, v. 50, Society of Economic Paleontologists and Mineralogists, 1980.)

seasonal fluctuations in sediment supply or periodic stirring of the bottom by wave action. Mud may, however, be deposited so rapidly by flocculation at the mouth of a delta distributary that the deposit shows no lamination. Muds deposited in freshwater lakes commonly show good lamination because little flocculation of the clay takes place and there are comparatively few burrowing organisms.

Limestones

The textures of limestones are exceedingly complex, in part because of the biologic origin of most primary carbonate grains and in part because of the relatively soluble nature of calcium carbonate and the ease with which it recrystallizes (James and Choquette, 1983). Nevertheless, most aspects of the original carbonate texture are well enough preserved in most limestones to permit environmental interpretations in the field setting, if the particles are coarse enough to be seen and identified on the outcrop. Typically, however, many of the critical observations that depend on particle identification must be postponed until polished slabs or thin sections of the rock are available. Most carbonate particles are so small, and the outcrop surface so rough, that accurate identification and commonly even recognition are impossible.

Three types of carbonate particles are normally present in limestones: (1) gravel-

and sand-size grains, usually polycrystalline (allochemical particles), (2) calcite mud, and (3) calcite cement. The gravel- and sand-size grains, although all composed of calcite in most rocks, are of four types: fossils, peloids, ooliths, and limeclasts. Although they are usually all intrabasinal in origin, each type carries distinctive information about the depositional basin. **Fossils** carry temporal, oceanographic, ecologic, and environmental information. Fossils can answer questions such as

Was the water saline or fresh?

Were the currents strong or weak (fragmented shells)?

Was the depositional environment intertidal or subtidal?

It is clear that even a small amount of paleontologic knowledge can be of immense help in stratigraphic work.

Peloids are aggregates of microcrystalline carbonate that lack internal structure. The most common origins of such grains are as fecal pellets or as the product of metabolic processes of endolithic (boring) algae. Endolithic algae attach to the exterior surface of clastic particles, grow inward into the particle, destroy its internal structure, and precipitate microcrystalline carbonate as a replacement. The relative proportions of fecal and algal-produced peloids are not known.

Ooliths are formed almost exclusively in highly agitated marine waters, such as between sand bars where water is funnelled and moves with high velocities. Ooliths are one of the few types of primary calcium carbonate particles produced inorganically. They are of sand size, form around a nucleus of some sort, such as a quartz grain or fossil, and have either a concentric or a radial internal structure. Both the nucleus and the internal structure can be seen by using a hand lens. Oolitic limestones typically are cross-bedded because of the type of environment in which they form.

Limeclasts are fragments of earlier-formed carbonate rock. Most are **intraclasts**, pieces of penecontemporaneous lithified carbonate torn up from either the margin or the interior of the depositional basin. Those from the basin margin may be dolomitic because of their origin on a supratidal flat (see Chapter 9) and tend to be platy in shape. Descriptions of "flat-pebble conglomerates" (Fig. 2-15) are common in stratigraphic sections of limestones. Some limeclasts are derived from outside the basin of deposition, particularly limeclasts in regions of active tectonism during deposition, as in the Alps and the Paleozoic Marathon Basin in west Texas. The presence of these externally derived limeclasts is recognized not so much by their individual characteristics (e.g., reworked fossils) but by their association with large amounts of other terrigenous debris such as quartz, feldspar, and lithic fragments. That is, limestones rich in terrigenous limeclasts are notably impure and contain abundant sand-size silicate detritus.

Orthochemical particles are the calcium carbonate matrix (micrite) and clearly secondary cement (spar) that bind the allochems to lithify the sediment. The difference in crystal size and degree of translucency in hand-specimen or thin section

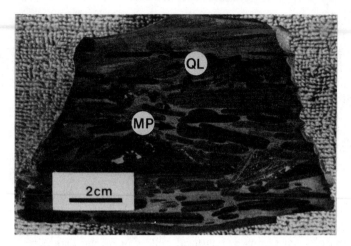

FIG. 2-15 • Acid-etched slab of flat-pebble limestone conglomerate from Conococheague Limestone (Upper Cambrian, Virginia). Abundant dark, micritic, peloidal intraclasts (MP) commonly contain quartz silt laminae (QL). The upper part of the slab is cross-laminated peloidal sand and silt. The rock is a storm deposit from a mid- to outer shelf environment. (Reprinted by permission of the publisher from Whisonant, *Journal of Sedimentary Petrology*, v. 57, Society of Economic Paleontologists and Mineralogists, 1987.)

serve to separate matrix from secondary cement. Matrix is aphanitic in crystal size and appears opaque; secondary cement is conspicuously coarser and appears transluscent, much like quartz.

Macroscopically visible chert is common in carbonate rocks and is most commonly of intrabasinal origin, generated by dissolution and crystallization of opaline skeletal parts. Either diatoms, radiolaria, or siliceous sponges can supply the silica needed for the chert nodules.

Classification of Limestones

Limestones can be classified by using some of the same criteria that are used for sandstones: the presence or absence of matrix and the types of allochemical grains (Fig. 2-16). Two classifications are in wide use, those of Folk (1959) and Dunham (*in* Ham, 1962). In the Folk system of descriptive terms the allochem "fossil" becomes *bio-*, peloid becomes *pel-*, oolith is shortened to *oo-*, and intraclast becomes *intra-*. Microcrystalline carbonate matrix is shortened to *mic-*, and coarse, clear sparry calcite becomes *spar-*. Thus we recognize biosparites, pelmicrites, and intrasparudites (gravel-size intraclasts). The main part of the name is based on the major allochem and orthochem; appropriate modifiers may precede the main part of the name.

Dunham's system of limestone classification centers on the concept of grain support. When deposited, did the sediment consist of a self-supporting framework of allochems, or do the allochems float in a micrite matrix? Although different in concept, Dunham's scheme (Table 2-6) has many similarities with Folk's. Dunham's mud-

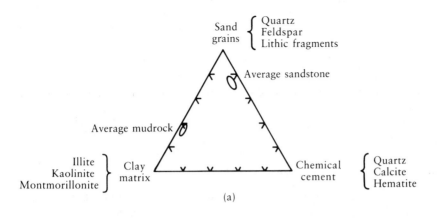

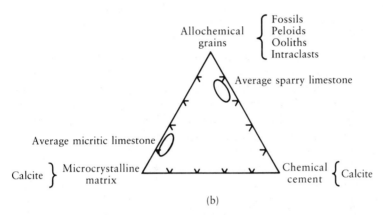

FIG. 2-16 • Triangles showing analogy between components of (a) sandstones and (b) limestones. (Reprinted by permission of the publisher from Blatt, *Sedimentary Petrology*, W. H. Freeman & Co., 1982.)

TABLE 2-6 • Dunham's classification of limestones

Original components not bound together during deposition				Original components were bound together during deposition . . . as shown by intergrown skeletal matter, lamination contrary to gravity, or sediment-floored cavities that are roofed over by organic or questionably organic matter and are too large to be interstices.
Contains carbonate mud (particles of clay and fine silt size)		Lacks carbonate mud		
Mud-supported		Grain-supported		
Less than 10% grains	More than 10% grains			
Mudstone	*Wackestone*	*Packstone*	*Grainstone*	*Boundstone*

(Reprinted with permission of American Association of Petroleum Geologists from Dunham, *in* Ham, *Classification of Carbonate Rocks*. American Association of Petroleum Geologists, Mem. No. 1 1962.)

stone and wackestone are micritic rocks, and grainstones are sparry rocks. Packstone, however, might be either a micrite or a sparite. The choice between the two systems is arbitrary.

Interpretation of Limestone Textures

The interpretation of the textures of limestones is more difficult than that for sandstones because of biologic influences and common recrystallization effects. For example, the fossils in a limestone can be whole ostracods of a particular species that were buried in the carbonate mud in which they lived. The fact that their size distribution is "well sorted" is unrelated to current strengths in the environment of deposition, as is reflected by the muddy matrix in which they were buried. Similarly, fragmented shells do not necessarily indicate high-velocity currents. Many predators live in the marine environment and obtain their food by boring through and/or cracking open the shells of benthonic organisms. Organisms supply abundant highly rounded fecal pellets to the environment, and certainly this rounding is unrelated to current strength. As an example of fecal production, a healthy adult *Callianasa major* shrimp produces about 450 fecal pellets each day.

The energy level of the environment of carbonate deposition is evaluated mostly from the presence or absence of calcium carbonate mud. It is assumed that microcrystalline ooze is always available in carbonate environments. Therefore if a limestone lacks these aphanitic particles, it means that current strengths were high enough to remove them. If the limestone is rich in microcrystalline carbonate, we interpret the depositional environment as having been of low kinetic energy. It must be remembered, however, that the "source area" of most carbonate particles is very close to the site of final deposition; allochems may be produced rapidly and in large numbers, thus forming a rock with many allochems and little mud in an environment of low kinetic energy.

Other factors used to evaluate the kinetic energy level of the environment are:

1. Evidence of mechanical abrasion during transport
2. Presence of ooliths
3. Current structures such as cross-bedding.

Insoluble Residues

Although most limestones contain less than 5% of mineral matter that is not calcite or aragonite, the non-calcium carbonate material can be isolated and concentrated very rapidly in a laboratory and can supply valuable information about the origin and diagenesis of the rock. These insoluble grains can be used to determine the nature of the rocks surrounding the carbonate basin (quartz, feldspar, and rock fragment content) and the abundance of siliceous organisms in the vicinity of the depositional site (chert content) and can also be used for correlating widely separated outcrops or subsurface sections of limestone.

Sedimentary Structures

Accumulations of calcitic or aragonitic sediment are of two types, biostromes and bioherms. **Biostromes** are areally extensive, bedded accumulations of carbonate sediment, concentrations of shell debris analogous to concentrations of quartz grains or plant fragments. Biostromes can have the same sedimentary structures as quartz sandstones, although the structures in carbonate rocks may be more difficult to see because mineralogic differences are not present and because calcite is much more soluble than silicate debris. **Bioherms** are in-place, constructional features produced by attached or colonial organisms with aragonitic or calcitic skeletal parts (Fig. 2-17). Because they are growth structures, biohermal accumulations are topographically higher than their surroundings and have a regional fabric, a preferred orientation (Shaver, 1977; Smith and Legault, 1985; Fagerstrom, 1987). These prominences, reefs, can strongly influence surrounding sedimentation and sedimentary structures. Within the reef itself may be a diverse and extremely complex group of ecologic structures interspersed with clastic carbonate debris (intraclasts) and perhaps overlaid with a growth of laminated, soft-bodied algae. Micrite and assorted types of allochems may have settled into crevices in the reef structure, further increasing its textural and structural complexity. Adjacent to the reef a talus apron is formed by the impact of ocean waves on the organic growth, and the areal extent of this apron may be greater than that of the reef core itself. The grossly intraclastic talus unit

FIG. 2-17 • Algal bioherm, Point Peak member, Wilberns Formation (Dresbachian), central Texas, one of thousands at this stratigraphic level in this area. The reef is about 35 m long and 15 m thick in this cross-section view. (Reprinted by permission of the publisher from Ahr, *Journal of Sedimentary Petrology*, v. 41, Society of Economic Paleontologists and Mineralogists, 1971.)

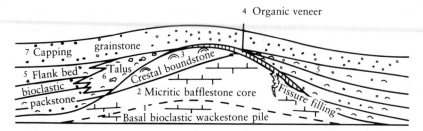

4 Organic veneer

7 Capping   grainstone

5 Flank bed   Talus   3 Crestal boundstone

bioclastic   6   5

packstone   2 Micritic bafflestone core   Fissure filling

1

Basal bioclastic wackestone pile

FIG. 2-18 ● Ideal carbonate mound with seven commonly developed facies. The sequence of facies develops when the accumulation grows into wave base and is controlled by varying rates of sediment production, rates of subsidence, and hydrographic factors. (Adapted from Wilson, *Carbonate Facies in Geologic History*, Springer-Verlag New York, p. 367, 1975.)

usually is very poorly sorted but can contain current structures such as cross-bedding if the allochems are of sand size. A crude size grading from very coarse gravel to fine sand and silt will be present, representing proximal and distal facies of the forereef.

In-place organic growth is not the only cause of constructional highs in carbonate environments. In many ancient limestones there exist **lime-mud mounds**, constructional highs formed of accumulations of bioclastic debris or mucilaginous pellets swept together by current activity (Fig. 2-18) (Toomey et al., 1988). Such accumulations are particularly common in rocks of Paleozoic age but occur throughout the Phanerozoic column and can be observed forming today in Florida Bay. These mounds may display a variety of current-generated sedimentary structures and can be interleaved with sponges, algae, or bryozoans that used the micritic substrate as an attachment surface.

Fenestrae are primary or penecontemporaneously formed gaps in a carbonate rock fabric, larger than grain-supported cavities. Fenestrae occur in rocks ranging in texture from mudstone to grainstone, are generally less than 10 mm in longest dimension, and in ancient rocks are usually filled by either secondarily introduced geopetal sediment or calcite cement. Their distinguishing characteristic is that they have no apparent support in the framework of nonfossil primary grains composing the sediment. Terms with approximately the same meaning as fenestrae include *stromatactis* (Fig. 2-19), *birdseyes* (Fig. 2-20), and *loferites* (Shinn, 1983). Field studies reveal that these structures form in supratidal, peritidal, and shallow subtidal settings, and therefore they need to be distinguished from animal burrow structures and root tubes, which can have similar appearances but form in a much wider range of depositional environments. Fenestrae form either as shelter porosity in coarse-grained fragmental limestones (for example, under a sheltering concave-downward clam shell), by differential internal shrinkage of a carbonate matrix during desiccation, or as bubble-escape structures formed penecontemporaneously with sedimentation. Preservation of these early cavities requires either preburial cementation or early infilling of the vugs by calcite or evaporite minerals.

Solution seams are very common in limestones because of the relatively high

FIG. 2-19 • Stromatactis structure in micritic mud mound, Ireland. Lighter-colored filling of structures is sparry calcite. Length of the specimen is 15 cm. (Photo courtesy R. C. Murray.) (Blatt, Middleton, and Murray, *Origin of Sedimentary Rocks*, 2nd ed., © 1980. Reprinted by permission of Prentice-Hall, Inc., Englewood Cliffs, NJ.)

FIG. 2-20 • Birdseye structures in Ordovician Valley View Limestone at Whiterock Quarry near Pleasant Gap, Pennsylvania. Note variability of shape and size. (Reprinted by permission of the publisher from Shinn, *Journal of Sedimentary Petrology*, v. 38, Society of Economic Paleontologists and Mineralogists, 1968.)

FIG. 2-21 • Diagenetic microlamination in Upper Maastrichtian White Chalk from Stevns Klint, Højerup, Denmark. Note that the solution seams crosscut, and thus postdate, the totally bioturbated ichnofabric. Scale bar = 4 mm. (Reprinted by permission of the publisher from Ekdale and Bromley, *Journal of Sedimentary Petrology,* v. 58, Society of Economic Paleontologists and Mineralogists, 1988.)

solubility of calcium carbonate. Commonly, the seams are laterally extensive and easily seen in outcrop, occurring either in the fitted teeth-and-socket patterns called stylolites or in the more subtle nonstylolitic seams that can mimic normal bedding (Fig. 2-21). The explanation for the variation in character of solution seams lies in the structural resistance to deformation of the limestone and to the proportion of less soluble sediment (quartz and clay) it contains (Wanless, 1979). Most commonly, the stress that produced the seam resulted from overburden, and the stylolites or seams formed are parallel with and typically coincident with bedding surfaces. In areas of deformed rocks, however, the maximum stress direction may not have been normal to bedding (for example, during the formation of an anticline), and any solution surface formed during deformation will be at an angle to bedding.

Dolostones

Field studies have documented that carbonate beds tend to be composed either entirely of calcite or entirely of dolomite. Less than 20% of carbonate beds contain between 20% and 80% dolomite; one-third contain 80–100% dolomite.

Most dolostone beds are clearly replacements of limestones, as shown by cross-cutting relationships, selective replacement of fossils or parts of other allochems, and geochemical studies. As a result, textural descriptions of dolostones use the same terminology as is used for limestones, giving rise to descriptions such as dolomitic biosparite, oolitic dolostone, or fossiliferous dolomicrite. Always keep in mind that the purpose of terminology is communication, not rigorous adherence to a set of published terms that may be inadequate for some rocks.

Summary

Analysis of sedimentary textures, mineralogy, and structures can supply important information in stratigraphic work. In siliciclastic rocks, textures can suggest depositional environments as well as current strengths and directions. Mineral composition reflects tectonic setting and provenance. Sedimentary structures may also reflect current strengths and directions, degrees of continuity of sediment deposition, and depositing mechanism (for example, wind, water, or turbidity current).

Carbonate rocks are largely biologic in origin and are sensitive indicators of climate and the properties of the water in which the organisms lived. Current strengths, water depth, water temperature, and water salinity can all be reflected in the character of the limestone. In addition, the biologic diversity of carbonate fossils reflects environmental and evolutionary changes through time.

A good understanding of sedimentology and petrology are essential for good stratigraphic work.

References

Ahr, W. M., 1971. Paleoenvironment, algal structure, and fossil algae in the Upper Cambrian of central Texas. J. Sed. Petrology, v. 41, pp. 205–216.

Allen, J. R. L., 1970. Physical Processes of Sedimentation: An Introduction. London, George Allen & Unwin, 248 pp.

Atkinson, D. J., 1962. Tectonic control of sedimentation and the interpretation of sediment alternation in the Tertiary of Prince Charles Foreland, Spitsbergen. Geol. Soc. Amer. Bull., v. 73, pp. 343–364.

Barrett, P. J., 1964. Residual seams and cementation in Oligocene shell calcarenites: The Kuiti Group. J. Sed. Petrology, v. 34, pp. 524–531.

Blatt, H., 1982. Sedimentary Petrology. New York, W. H. Freeman & Co., 564 pp.

Blatt, H., Middleton, G. V., and Murray, R. C., 1980. Origin of Sedimentary Rocks, 2nd ed. Englewood Cliffs, N.J., Prentice-Hall, 782 pp.

Blatt, H. and Totten, M. W., 1981. Detrital quartz as an indicator of distance from shore in marine mudrocks. J. Sed. Petrology, v. 51, pp. 1259–1266.

Bouma, A. H., 1962. Sedimentology of Some Flysch Deposits: A Graphic Approach to Facies Interpretation. Amsterdam, Elsevier Science Publishers, 168 pp.

Bouma, A. H., 1969. Methods for the Study of Sedimentary Structures. New York, Wiley-Interscience, 458 pp.

Busby-Spera, C., 1985. A sand-rich submarine fan in the lower Mesozoic Mineral King caldera complex. Sierra Nevada, California. J. Sed. Petrology, v. 55, pp. 376–391.

Cluff, R. M., 1980. Paleoenvironment of the New Albany Shale Group (Devonian-Mississippian) of Illinois. J. Sed. Petrology, v. 50, pp. 767–780.

Collinson, J. D. and Thompson, D. B., 1988. Sedimentary Structures, 2nd ed. London, Allen & Unwin, 208 pp.

Compton, R. R., 1985. Geology in the Field. New York, John Wiley & Sons, 398 pp.

Conybeare, C. E. B. and Crook, K. A. W., 1968. Manual of Sedimentary Structures. Canberra Australia Department of Natural Development, Bureau of Mineral Resources, Geology and Geophysics Bulletin 102, 327 pp.

Curran, H. A. (ed.), 1985. Biogenic Structures: Their Use in Interpreting Depositional Environments. Tulsa, Society of Economic Paleontologists and Mineralogists, Spec. Publ. No. 35, 347 pp.

Dobkins, J. E., Jr. and Folk, R. L., 1970. Shape development on Tahiti-Nui. J. Sed. Petrology, v. 40, pp. 1167–1203.

Duke, W. L., 1985. Hummocky cross-stratification, tropical hurricanes, and intense winter storms. Sedimentology, v. 32, pp. 167–194.

Ekdale, A. A. and Bromley, R. G., 1988. Diagenetic microlamination in chalk. J. Sed. Petrology, v. 58, pp. 857–861.

Fagerstrom, J. A., 1987. The Evolution of Reef Communities. New York, John Wiley & Sons, 600 pp.

Feiznia, S. and Carozzi, A. V., 1987. Tidal and deltaic controls on carbonate platforms: Glen Dean Formation (Upper Mississippian) of Illinois Basin, USA. Sedimentary Geol., v. 54, pp. 201–243.

Folk, R. L., 1951. Stages of textural maturity in sedimentary rocks. J. Sed. Petrology, v. 21, pp. 127–130.

Folk, R. L., 1959. Practical petrographic classification of limestones. Amer. Assoc. Petroleum Geol. Bull., v. 43, pp. 1–38.

Folk, R. L., 1960. Petrography and origin of the Tuscarora, Rose Hill, and Keefer formations, Lower and Middle Silurian of eastern West Virginia. J. Sed. Petrology, v. 30, pp. 1–58.

Foscolos, A. E., 1984. Catagenesis of argillaceous sedimentary rocks. Geoscience Canada, v. 11, pp. 67–75.

Frey, R. W. (ed.), 1975. The Study of Trace Fossils: A Synthesis of Principles. New York, Springer-Verlag, 562 pp.

Frey, R. W. and Pemberton, S. G., 1984. Trace fossil facies models. In R. G. Walker (ed.), Facies Models, 2nd ed. Toronto, Geological Association of Canada, pp. 189–207.

Ham, W. E. (ed.), 1962. Classification of Carbonate Rocks. Tulsa, American Association of Petroleum Geologists, Mem. No. 1.

Hutcheon, I., 1983. Diagenesis, 3: Aspects of the diagenesis of coarse-grained siliciclastic rocks. Geoscience Canada, v. 10, pp. 4–14.

Ingram, R. L., 1954. Terminology for the thickness of stratification and parting units in sedimentary rocks. Geol. Soc. Amer. Bull., v. 65, pp. 937–938.

James, N. P. and Choquette, P., 1983. Diagenesis, 5: Introduction: Limestones. Geoscience Canada, v. 10, pp. 159–161, pp. 162–179.

Klein, G. deV., 1977. Clastic Tidal Facies. Urbana, Ill., Continuing Education Publishing Co., 149 pp.

Krinsley, D. B. and Doornkamp, J. C., 1973. Atlas of Quartz Sand Surface Textures. Cambridge, England, Cambridge University Press, 91 pp.

Marsaglia, K. M. and Klein, G. deV., 1983. The paleogeography of Paleozoic and Mesozoic storm systems. J. Geol., v. 91, pp. 117–142.

McBride, E. F., 1963. A classification of common sandstones. J. Sed. Petrology, v. 33, pp. 664–669.

McCall, P. L. and Tavesz, M. J. S. (eds.), 1982. Animal-Sediment Relations: The Biogenic Alteration of Sediments. New York, Plenum Press, 336 pp.

O'Brien, N. R., Nakazawa, K., and Tokuhashi, S., 1980. Use of clay fabric to distinguish turbiditic and hemipelagic siltstones and silts. Sedimentology, v. 27, pp. 47–61.

Pelletier, B. R., 1958. Pocono paleocurrents in Pennsylvania and Maryland. Geol. Soc. Amer. Bull., v. 69, pp. 1033–1064.

Pettijohn, F. J. and Potter, P. E., 1964. Atlas and Glossary of Primary Sedimentary Structures. New York, Springer-Verlag, 370 pp.

Pettijohn, F. J., Potter, P. E., and Siever, R., 1987. Sand and Sandstone, 2nd ed. New York, Springer-Verlag, 553 pp.

Reineck, H-E. and Singh, I. B., 1980. Depositional Sedimentary Environments, 2nd ed. New York, Springer-Verlag, 549 pp.

Scoffin, T. P., 1987. An Introduction to Carbonate Sediments and Rocks. New York, Chapman and Hall, 274 pp.

Shaver, R. H., 1977. Silurian reef geometry—New dimensions to explore. J. Sed. Petrology, v. 47, pp. 1409–1424.

Shinn, E. A., 1968. Practical significance of birdseye structures in carbonate rocks. J. Sed. Petrology, v. 38, pp. 215–233.

Shinn, E. A., 1983. Birdseyes, fenestrae, shrinkage pores, and loferites: A reevaluation. J. Sed. Petrology, v. 53, pp. 619–628.

Smith, A. L. and Legault, J. A., 1985. Preferred orientations of Middle Silurian Guelph-Amabel reefs of southern Ontario. Bull. Canadian Petroleum Geol., v. 33, pp. 421–426.

Stow, D. A. V. and Piper, D. J. W. (eds.), 1984. Fine-Grained Sediments: Deep-Water Processes and Facies. Boston, Blackwell Scientific Publications, 664 pp.

Toomey, D. F., Mitchell, R. W., and Lowenstein, T. K., 1988. "Algal biscuits" from the Lower Permian Herrington/Krider limestones of southern Kansas-northern Oklahoma: Paleoecology and paleodepositional setting. Palaios, v. 3, pp. 285–297.

Walker, R. G., 1966. Shale Grit and Grindslow Shales: Transition from turbidite to shallow water sediments in the Upper Carboniferous of northern England. J. Sed. Petrology, v. 36, pp. 90–114.

Wanless, H. R., 1979. Limestone response to stress: Pressure solution and dolomitization. J. Sed. Petrology, v. 49, pp. 437–462.

Whisonant, R. C., 1987. Paleocurrent and petrographic analysis of imbricate intraclasts in shallow-marine carbonates, Upper Cambrian, southwestern Virginia. J. Sed. Petrology, v. 57, pp. 983–994.

Wilson, J. L., 1975. Carbonate Facies in Geologic History. New York, Springer-Verlag, 471 pp.

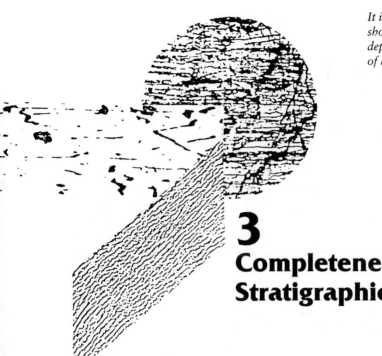

It is most important that a clear distinction should be made between the temporary deposition and the permanent accumulation of deposits.

Henry Sorby

3
Completeness of the Stratigraphic Record

Time is continuous; it "flows" and is not naturally compartmentalized into discrete units bounded by surfaces or zones in which time stops or does not exist. It is always present. Unfortunately, however, time cannot be seen, smelled, heard, tasted, or touched, and we are aware of its existence or passage only because of events that occur in sequence. The events are the measuring stick of time. For the long spans of geologic time the physical events that we use include organic evolution (for example, the increasing complexity of suture patterns on ammonite shells during Mesozoic time) and the superposition of sedimentary layers as seen in outcrop. But in recent years the accuracy and even the validity of these measuring sticks have been seriously questioned. Does 1 m of black shale imply the same span of time in different areas? What are we to make of the fact that 0.3 m of sediment in Sicily contain 30 ammonite zones, while 5000 m represent a single zone in Oregon (Ager, 1981)? What can be the relationship between the passage of time and sedimentation rate when we find in the Apennines of Italy Late Jurassic through Early Oligocene time (133 m.y.) represented by an apparently unbroken stratigraphic sequence only 70 m thick, while the rest of Oligocene time (6 m.y.) is represented by 3000 m of sediment?

Uniformitarianism

The concept termed uniformitarianism was originated by James Hutton in the latter part of the 18th century and was popularized by Charles Lyell in the many editions of his historical geology text a few decades later. Lyell's concept was that the physical, chemical, and biological laws that operate today are those that operated in the geologic past and, in addition, that the rates and intensities of these processes have been constant throughout geologic history. The current view of uniformitarianism is that although the laws have indeed been constant, rates and intensities have varied greatly from time to time. For example, the rate of weathering is influenced by the ratio of oxygen to carbon dioxide in the atmosphere, and there is abundant evidence that this ratio has increased greatly since Archean time.

Uniformitarian processes are typically thought of as those that are uninterrupted, such as weathering. However, uniformitarian processes may also be intermittent. For example, earthquakes are intermittent, floods may be annual, bolide impacts occur on the order of 10^6 years, and events that may depend on the position of the solar system in the Milky Way have an even longer recurrence interval. This chapter is concerned with the effects of such intermittent uniformitarian events on the stratigraphic record.

Distribution and Mass of Sedimentary Rocks

A rapid glance at a geologic map of any large area such as the United States, Europe, or Africa reveals that sedimentary rocks are not distributed uniformly. Some extensive areas have no sedimentary cover (excluding Quaternary soils and glacial debris), while other regions contain thousands of meters of ancient sediment. Geophysical data indicate sediment thicknesses of more than 20,000 m in some basinal areas. How much sedimentary rock is there in the Earth's crust?

The most direct way to approach this question is by examining published geologic maps to determine the area of outcrop of sedimentary rocks and then geophysical data to establish the thickness of sediments in the depositional basins. The results of such studies (Blatt, 1970; Blatt and Jones, 1975) indicate the volume to be 4.1×10^8 km^3 and to have an average thickness of 900 m. About 80% of the sedimentary mass lies on the continental blocks; 20% lies in the deep ocean basins that cover 65% of the Earth's surface. Thus the average thickness on the continental blocks is almost 1400 m. Clearly, the amount of sediment on the Earth is remarkably small, considering the great length of geologic time. An average sediment thickness of 900 m representing 4.6×10^9 years is only 0.2 μm/year. At such a rate it would require 300 years of continuous deposition to accumulate a sediment layer as thick as a single sand grain! Even ignoring rocks of Precambrian age, which are almost all metamorphic and igneous, the apparent rate of deposition is only 1 μm/year.

In fact, of course, there is no simple or direct relationship between the existing mass of sediments and rates of deposition because of erosion accompanying epeirogeny and orogeny. Rates of sediment accumulation range between highly positive (rapid accumulation and burial) and highly negative (rapid erosion). In addition, sediment is continually lost through metamorphism and transport into the mantle at convergent plate margins. The preserved stratigraphic record is well described as an immense blank roll of paper (time) that is very occasionally partly traversed at an

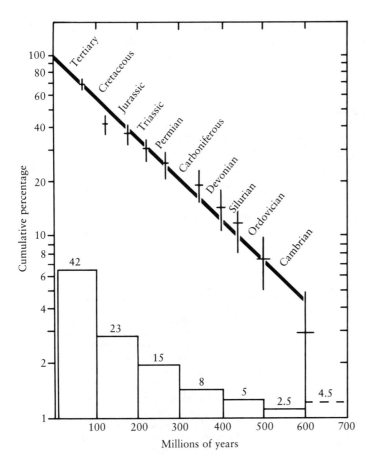

FIG. 3-1 • Relationship between sedimentary rock age and amount of outcrop area. Vertical error bars indicate precision at the 95% confidence level based on number of data points in each age category. Horizontal error bars indicate precision of radioactive dating procedures. (Does not include Quaternary data points.) The histogram shows the percentage of outcrop within each 10^8-year interval (arbitrary vertical scale). (Adapted from Blatt and Jones, 1975.)

angle by thin pieces of string. In this regard the standard stratigraphic columns that we see in published stratigraphic studies are terribly misleading. They convey the impression of rock sections as a long record of sedimentation with occasional gaps. A far more accurate picture of the stratigraphic record is one of long gaps with only very occasional records of sedimentation. Interpretation of the stratigraphic record is an excellent illustration of Mark Twain's aphorism about the wonder of science—one gets such wholesale returns of conjecture from such a trifling investment of fact.

Another aspect of the sedimentary mass that is of stratigraphic importance is the relative areas of outcrop of rocks as a function of age (Fig. 3-1). As we would anticipate, younger rocks are exposed over larger areas than older rocks. Paleogeographic interpretations for the last 100×10^6 years of geologic time will be about twice as accurate as those for the preceding 100×10^6 years and about eight times as accurate as interpretations for Ordovician-Silurian time. Precambrian paleogeography will forever be particularly obscure, because about 4×10^9 years are represented by less than 5% of the sedimentary mass. The decline in abundance of sedimentary rocks with increasing age results from the fact that older rocks have had more opportunity to be destroyed by subsequent uplift and erosion.

Rates of Sediment Accumulation

The relative abundances of preserved sedimentary rocks available for examination are a direct function of geologic age. The more recent the deposition, the more abundant the rocks. Geologic age, however, is not the only factor controlling the amount of sedimentary rock available for study. Another important control is the intermittent character of the processes that cause sediment to be deposited. For example, a spring flood may deposit 1 m or more of sediment within a few hours, followed by weeks of nondeposition. In the marine environment a turbidity current

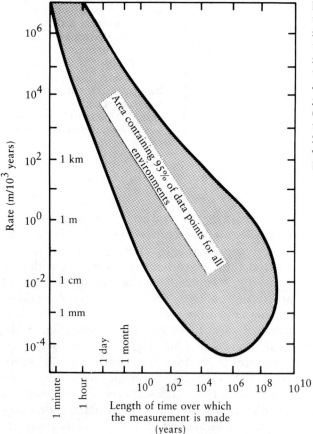

FIG. 3-2 • Relationship between rate of deposition of sediment and the length of time over which the measurement is made. Lengths of time longer than a few years are calibrated by biostratigraphic or radiometric data. (Reprinted by permission of the publisher and author from Sadler, *Journal of Geology,* v. 89, University of Chicago Press, 1981.)

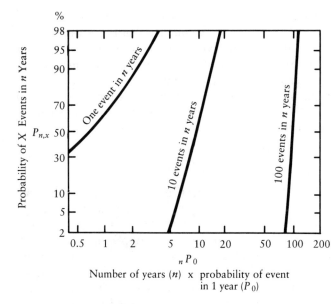

FIG. 3-3 • Probability (p_{nx}) of a rare event to occur at least x times in n years; p_n is the probability that the event will take place in a single year. (Reprinted by permission of the American Association of Petroleum Geologists from Gretener, *American Association of Petroleum Geologists Bulletin*, v. 51, 1967.)

may deposit 1 km^3 of detritus in a few days, which will then be surficially toyed with by bottom currents for thousands of years with no net change in sediment thickness. Thus sedimentary sequences, even those lacking obvious disconformities, record the passage of geological time as an alternating set of sedimentary increments and gaps. The ratio of these two components is the completeness of the section.

The question of incompleteness in apparently uninterrupted stratigraphic sections has been considered in some detail by Sadler (1981). In a comprehensive compilation of published data involving nearly 25,000 rates of sediment accumulation in a wide range of depositional environments, he found that rates are extremely variable, spanning at least 11 orders of magnitude (Fig. 3-2). There is a systematic trend of falling mean rate with increasing time span. For example, in a 10-year study of sediment accumulation rate in a river, a typical result is 10^2 m/10^3 years. But when we examine an ancient fluvial deposit accumulated over a period of 10 million years (assume accurate radiometric dating of the top and bottom of the deposit), the rate is only 10^{-3} m/10^3 years. The longer the time span over which the measurement is made, the lower the apparent rate of sediment accumulation.

How can this result be explained? The answer may be found in a fact often overlooked by geologists. On the geologic scale of time, that which seems impossible by human standards becomes possible, and the improbable becomes inevitable (Fig. 3-3). We see from the $x = 1$ curve that if there is a 60% chance of one event in 1 year, then there is an 85% chance of that event occurring once in a 2-year span and a virtual certainty (98%) of the event occurring within a 4-year span. For a much rarer event with a one-in-a-million chance of occurring once in 1 year, there is nonetheless near certainty of the event occurring five times in 10 million years. Applying this reasoning to Sadler's (1981) data, we would say that the reason the apparent accumulation rate decreased markedly (exponentially) is because events that seem to us very uncommon or rare or improbable and that thin a sedimentary accumulation are not at all rare when millions or tens of millions of years are involved. These infrequent events carry the sediment they remove out of the local depositional system and into the ocean basins, where it may ultimately disappear into the mantle. What are these relatively unusual events?

Floods

For the dominant types of deposits above base level, such as alluvial fans and fluvial environments, cataclysmic floods seem to be the answer. The extreme example known from ancient deposits is the Scablands flood of Pleistocene age in eastern Washington, described most recently by Baker (1973). Giant ripples up to 10 m high were formed in gravel; coarse gravel behaved like sand grains in ordinary stream flows. Baker estimated peak flow velocities of 100–130 km/hr, based on maximum diameters of boulders moved by the surging waters. Such velocities are perhaps 20 times that of an average stream flow rate and might be equal to a 1,000-year flood or perhaps a 10,000-year flood. Earlier cataclysmic floods are known from fragmentary evidence to have occurred in the same area of eastern Washington during Pleistocene time, but their magnitudes cannot be determined. Because of the recency of the Scablands flood, 22,000 years ago, its geomorphic results are still clearly seen, but in more ancient sedimentary sections there would be no evidence of such an occurrence. These "rare" floods could easily remove many millions of years of earlier fluvial deposits.

Hurricanes

Hurricane effects have received considerable attention in recent decades, and it seems clear that they are geologically significant. But like floods on the land surface, they are infrequent on a human time scale. Hurricanes (also termed typhoons and cyclones) originate only over tropical oceans and are quite frequent in the North Atlantic Ocean. By definition, winds of at least 120 km/hr are required to be classified as being of hurricane force; speeds of 240 km/hr have been recorded, with gusts estimated as high as 400 km/hr. As the storm approaches a coast, the piling up of water by strong winds may produce a storm surge with extreme erosive effects in the near-shore region. In the United States the coastline of the Gulf of Mexico is the area most frequently affected, and data indicate that any local area of the coast can expect about one hurricane per 100 years. On a geological time scale it is questionable whether tropical hurricanes should even be considered as uncommon, much less "rare." As an example of the effect of a hurricane, the 1961 storm that hit the central Texas Gulf Coast is often cited. Carla, as it was named, raised the sea level 4 m above normal, and as the storm passed, retreat of the surge and associated currents carried much sediment seaward and deposited it as a conspicuous graded layer up to 6 cm thick. A recent examination of the area, however, found the originally graded deposit so thoroughly homogenized by burrowing organisms that it is no longer recognizable. Carla is a good example of the lack of preservation of the results of a major rare (on a human scale) event.

Many examples of storm or hurricane deposits have been described from the geologic record. Marsaglia and Klein (1983) tabulated a total of 69 such deposits. Since their listing, other deposits have been found, one example being the Baraboo Formation (Cambrian) in Wisconsin described by Dott (1983). The effects of the hurricane include truncation of the upper parts of animal burrows, homogenization of successive burrowed intervals (**amalgamation**) to produce an exceptionally thick and thoroughly burrowed sequence, hummocky cross-bedding, and shelly layers attributed to massive suffocation of organisms by an unusual storm-induced cloud of mud.

Aigner (1985) has interpreted the entire Upper Muschelkalk Limestone (Triassic) of West Germany as a sequence of storm deposits. The sequence is up to 100 m thick and consists of a series of coarsening-upward cycles showing mixing of depth-dependent shallow-water faunas, bipolar sole marks, and large abraded carbonate clasts. Estimates of storm erosion depth were made by using abundances of vertical trace fossil burrows.

Earthquakes

Earthquakes may cause substantial alteration of local or regional stratigraphic sections by initiating landslides, rockfalls, avalanches, soil slumps and slides, or other mass movements. The degree of disturbance depends in part upon the amount of energy released, focal depth, type of faulting, and characteristics of the local geology, such as degree of slope and rock type. In a study of landslides caused by earthquakes, Keefer (1984) illustrated the positive correspondence between earthquake magnitude (which is a function of energy released) and area affected (Fig. 3-4). For earthquakes of the largest magnitude ever recorded (9.2 for the Alaskan earthquake of March 28, 1964) the area potentially affected may approach 500,000 km^2. Earthquakes are quite common in the vicinity of convergent plate margins, such as the Pacific Ocean rim (Fig. 3-5), because of the abundant faults and subduction that occur there. Both types of movements produce quakes, but the number and size of slumps produced are difficult to determine because most of the tremors occur offshore in deep water. Historical records indicate that the correlation between subduction movement and earthquakes can be poor when measured over short time intervals.

Perhaps the best documented and described example of the effect of a large submarine earthquake is the slump and turbidity current produced by the 1929 quake on the Grand Banks of Newfoundland (Piper et al., in Clifton, 1988). The current is estimated to have reached a peak velocity of 67 km/hr, and the sediment was transported hundreds of kilometers across the abyssal plain and deposited as a

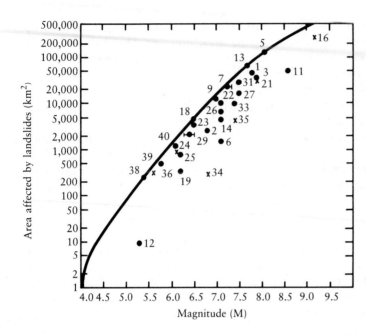

FIG. 3-4 • Area affected by landslides in earthquakes of different magnitudes. Dots indicate onshore earthquakes; *x* indicates offshore earthquakes. Horizontal bars indicate range on reported magnitudes. The solid curve is the upper bound enclosing all the data. (Reprinted by permission of the author from Keefer, *Geological Society of America Bulletin*, v. 95, 1984.)

widespread graded bed more than 1 m thick. Rare, unusually thick ancient turbidite beds, commonly seen within otherwise thinly bedded distal or basin-plain mudstone sequences, probably reflect similar exceptionally large flows triggered by either major earthquakes or tsunamis.

Detection of the effects of earthquakes may be easier in shallow marine settings than in deeper locations because shallower deposits are more likely to be exposed in outcrop in the stratigraphic record. On the other hand, the effect may be harder to detect because of increased subtlety; sea bottom slopes are lower in shallower water, and therefore the amount of sediment produced by the quake will be less. Atwater (1987) postulated that vegetated lowlands just high enough to avoid inundation by the sea most of the time would sink far enough to be submerged regularly by down-toward-the-coast faulting and become barren tidal mud flats. In intervals between great earthquakes, enough mud could fill the tidal flats to raise them once more to the level that supports extensive vegetation. Thus repeated earthquakes should produce alternating layers of lowland soil and tidal flat mud. Such alternation was found in five estuaries along a 220-km stretch of coastline in Washington State. In each of the estuaries, Atwater found one to six or more layers of peaty mud, which he took to be the soil of long-dead, vegetated lowlands, overlain by sharply defined layers of gray mud, the tidal flat deposits. Atwater (1987) inferred at least six major quakes during the past 7000 years, but the accurate ^{14}C dating required to document the contemporaneity of the peat layers has not yet been accomplished. Particularly noteworthy from Atwater's study is the fact that an earthquake considered "large" by human standards may leave only a subtle (or no?) record in the stratigraphic

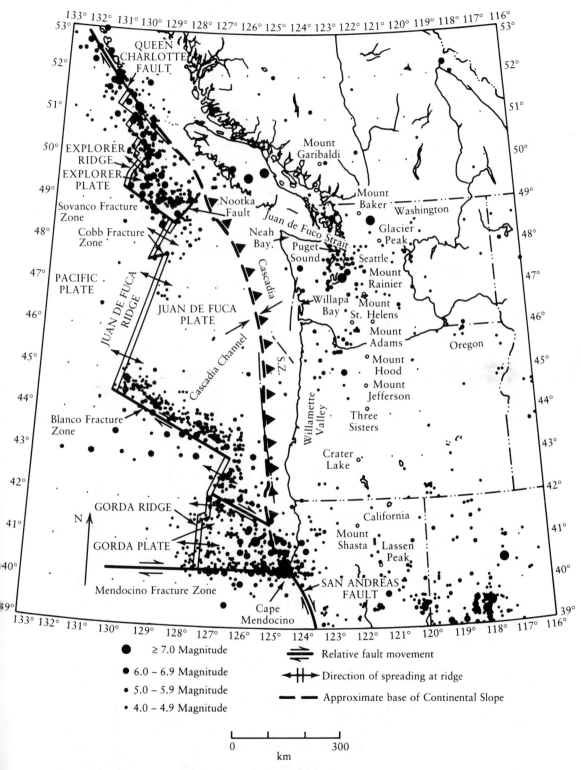

G. 3-5 ● Seismicity and plate tectonics of the Pacific Northwest. (Reprinted by permission of the author and e publisher from Heaton and Hartzell, *Science*, v. 236, p. 163, 10 April 1987.)

column. Many of the thin layers of sediment along coastlines or in ancient coastal deposits may have been generated by faulting and the earthquakes they generated.

The potential effects of earthquakes upon carbonate platform sequences have been modeled by Cisne (1986), who noted that if carbonate basin margins that are fringed with reefs are subjected to periodic stick-slip faulting, causing the basin to subside and the platform to rise in relation to sea level, an unconformity will be produced as erosion removes the now subaerial sediments. Furthermore, repeated basin margin faulting in this setting will cause hemicyclic carbonate sequences bounded by platform unconformities that die out within short distances from the platform's margin. Such a pattern is found in the Löfer cyclothems, numbering more than 300, preserved in late Triassic sediments of the Northern Calcareous Alps (Cisne, 1986). If each cyclothem represents the occurrence of one earthquake (there may have been more of lesser energy, of course, that went unrecognized in the stratigraphic section), then during the approximately 12 m.y. represented by the section, the average recurrence interval per earthquake is 40,000 years. Although this recurrence interval is less by several orders of magnitude than rates typical at plate margins, Cisne notes that it is "consistent with the (carbonate) platform's more quiescent tectonic setting."

Tsunamis

Tsunamis are gravitational sea waves produced by any large-scale, short-duration disturbance of the ocean floor, principally by a shallow submarine earthquake, but also by submarine earth movement, subsidence, or volcanic eruption. They have wave lengths up to 200 km and maximum recorded speeds of about 950 km/hr, and they may travel thousands of kilometers in the ocean basin before striking bordering coastlines. Tsunami waves may pile up to heights of 30 m or more along exposed coasts, causing severe erosion. Because they have such long wave lengths, tsunamis touch bottom everywhere in the ocean and can cause intense agitation there as well. Examples of the sedimentologic effects of tsunami waves are scarce, but a possible example has been described from 3500-year-old sediments on the eastern Mediterranean Sea floor (Kastens and Cita, 1981). The tsunami was generated by the eruption of the Santorini volcano in the Aegean Sea and appears to have caused homogenization of a mass of stratified marly sediments at a depth of about 3500 m. These remobilized sediments were deposited and preserved in deeper basins downcurrent of the eruption.

Another possible tsunami deposit has been described from the Cretaceous-Tertiary boundary sequence in eastern Texas (Bourgeois et al., 1988). The deposit is a coarse-grained sandstone bed, 30–130 cm thick, that contains clasts of microfossiliferous mudstone up to 100 cm in length. The bed has an erosional base with up to 70 cm of local relief and grades upward to wave-ripple laminated, very fine-grained sandstone. Underlying the bed is more than 10 m of Upper Cretaceous shelf mudstone containing no distinct sandstone layers; above the bed is 1–4 m of similar mudstone of Paleocene age. The mudstone clasts within the sandstone originated at depths of 75–200 m, based on the paleoecology of the benthic foraminifera they

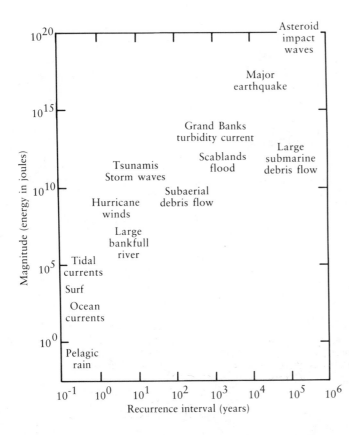

FIG. 3-6 • Graph of magnitude versus characteristic recurrence intervals for twelve important sedimentary processes. (Reprinted by permission of the publisher from Dott, *Journal of Sedimentary Petrology*, v. 53, Society of Economic Paleontologists and Mineralogists, 1983.)

contain. Immediately above the bed is an iridium anomaly widely thought to indicate a bolide impact because of its geochemical similarities with meteoritic material.

Based on calculations of bolide sizes, impact velocities, and the shear stresses that would be generated on the sea floor, Bourgeois et al. (1988) suggested that the tsunami wave was 50–100 m high and could have been generated by a bolide 10 km in diameter hitting the deep ocean about 5000 km distant from the depositional site of the sandstone bed. They also suggested that other coarse-grained sandstone beds near the Cretaceous-Tertiary boundary that have been described from continental margin deposits around the Gulf of Mexico and elsewhere may be tsunami deposits.

Magnitudes and Recurrence Intervals

As was noted by Dott (1983), the range of kinetic energies represented by sedimentary processes is enormous (Fig. 3-6). Most day-to-day currents and surf have energies that are in the range of 10^1–10^5 joules. The more familiar high-intensity episodic processes such as floods, hurricanes, and tsunamis have magnitudes in the range of 10^5-10^{10} joules. The energy generated by the impact on the ocean surface of an asteroid 10 km in diameter might be as much as 10^{20} joules. It has been estimated that such an impact in the ocean can be expected to occur every 30 million years, on average, suggesting that perhaps 20 of them have occurred during Phanerozoic time.

Evidences of Interruptions in Deposition

There are numerous features in stratigraphic sequences that indicate breaks in the continuity of deposition. Among the more abundant and important of these are hardgrounds, condensed sections, and bedding surfaces.

Hardgrounds

A modern **hardground** is a zone at a carbonate sediment-water interface, usually a few centimeters thick, that is lithified (Fig. 3-7). The hardground commonly is en-

FIG. 3-7 • Burrowed and mineralized hardground in shallow-water skeletal wackestone to mudstone of the Deschambault Formation near Quebec City, Canada. The sinuous, tubular features are typical of the ichnogenus *Thalassinoides* and are formed of sediment that differs from that immediately above and below. Overlying unit is the Neuville Formation, also of Middle Ordovician age, which is abiotic and a deeper water facies. (Photo courtesy of S. C. Ruppel.)

crusted, discolored, and bored by subtidal marine organisms and contains many solution features. Commonly associated with ancient hardgrounds are crusts of, or impregnations by, glauconite, phosphorite, iron, and manganese salts (Fig. 3-8), materials that form very slowly and in the absence of sediment influx. Frequently, the upper surface of the hardground coincides with a gap in the expected faunal succession in ancient examples. All these qualities show that the hardground was lithified before deposition of the overlying sediment; it is a surface of nondeposition that has been lithified by submarine cementation. Based on faunal discontinuities, hardgrounds typically represent hiatuses of several hundred thousand years. The exact nature of the chemical conditions that cause hardgrounds to form in some shallow water areas and not in others is uncertain. It may be that the extreme slowness of sediment deposition allows colonization and boring by organisms to an unusual degree (Fig. 3-9), and resulting biochemical interactions between the organisms and the carbonate sediment result in cementation and hardground formation.

Condensed Sections

A condensed section is a relatively thin but apparently uninterrupted stratigraphic succession representing a considerable length of time (Kidwell, 1989). The deposit accumulated very slowly and is generally represented by a time-equivalent thick succession elsewhere in the depositional basin. One method of producing a condensed section is storm activity in shallow seas. For example, storms agitate the sea floor to unusually great depths and winnow finer carbonate sediment from coarser benthonic shell debris. The shells are thus broken, abraded, and concentrated into a coquina (Fig. 3-10). Storms of less intensity will winnow less effectively, so a complete range of possibilities exists from undisturbed to highly disturbed sediment layers. The most condensed sections are those most highly disturbed and winnowed. Many examples of carbonate beds condensed by storm activity are given by Einsele and Seilacher (1982).

The vertebrate equivalent of a highly winnowed shell concentrate is a **bone bed**, a thin bed in which fossil bones or bone fragments are abundant. Commonly, other phosphatic remains such as teeth, scales, and coprolites are also present. Bone beds can be either of shallow marine or terrestrial origin and appear abruptly in sections that are otherwise poor in vertebrate remains. Often, bone beds occur in series containing 2–20 layers within a single outcrop, with individual beds up to 20 cm thick and with lateral extents up to 50,000 km^2. As with shell coquinas, the clues to origin are provided by the fractured nature of the remains, their good sorting, their gravel sizes, and the characteristics of the beds with which they are associated. In both shallow marine and terrestrial environments, unusually high kinetic energies produce the concentrates (Antia, 1979; Reif, *in* Einsele and Seilacher, 1982)—floods on the land surface or nearshore storms that stir up the shallow sea floor.

Condensed stratigraphic sections are not formed only by winnowing of originally thicker deposits. Some depositional sites are typically characterized by slow accumulation over long periods of time, examples being abyssal deposits far from terrigenous sediment sources and shallow depocenters protected in some manner

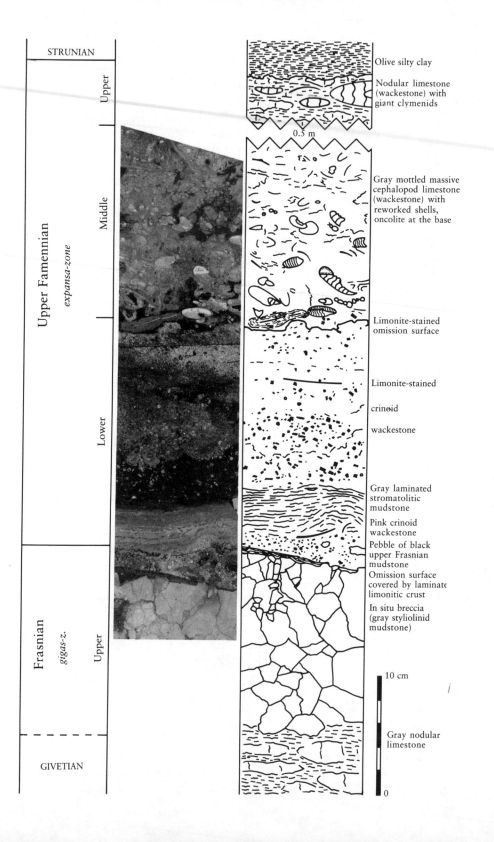

STRUNIAN

Upper

Olive silty clay

Nodular limestone
(wackestone) with
giant clymenids

0.5 m

Upper Famennian

expansa-zone

Middle

Gray mottled massive
cephalopod limestone
(wackestone) with
reworked shells,
oncolite at the base

Limonite-stained
omission surface

Limonite-stained

crinoid

wackestone

Lower

Gray laminated
stromatolitic
mudstone

Pink crinoid
wackestone

Pebble of black
upper Frasnian
mudstone

Omission surface
covered by laminate
limonitic crust

In situ breccia
(gray styliolinid
mudstone)

Frasnian

gigas-z.

Upper

10 cm

Gray nodular
limestone

GIVETIAN

0

FIG. 3-9 • Encrusting crinoid holdfasts attached to near-planar hardground in lower left of photo. The unit above the hardground is a crinoidal wackestone. Note the irregular, partly branched shape of holdfasts formed by repeated encrustation. Jebel Mech Ir-dane, Tifilalt Platform, eastern Anti-Atlas, Morocco. Scale is in centimeters. (Photo courtesy J. Wendt; reprinted by permission of the publisher from Wendt, *Eclogae Geol. Helv.*, Swiss Geological Association, 1988.)

FIG. 3-10 • Kenwood Beach shell layers (Miocene) at Calvert Cliffs, Maryland. Physical stratigraphic evidence reveals that each coquina is an amalgam and not an accumulation swept together by marine currents at a single point in time. The amount of time represented by the coquinas is too small for evolutionary changes to be detectable in the faunas. The arrow points to burrowed disconformity at base of a shell deposit. (Reprinted by permission of the publisher from Kidwell, *Journal of Geology*, v. 97, University of Chicago Press, 1989.)

◄ **FIG. 3-8 •** Core and descriptive sketch of condensed Upper Devonian section near Amelane road pass, Tafilalt Platform, eastern Anti-Atlas, Morocco. (Photo courtesy J. Wendt; reprinted by permission of the publisher from Wendt, *Eclogae Geol. Helv.*, Swiss Geological Association, 1988.)

from terrigenous influx and limestone formation. The thin but areally extensive black shales of Devonian-Mississippian age in southeastern and south-central United States are an example of a condensed section formed by slow deposition. Based on Sadler's (1981) trends of accumulation rate versus the length of time during which accumulation occurred, the Chattanooga Shale and its variously named equivalent black shales are perhaps an order of magnitude thinner than would be expected. In central Tennessee, for example, Conant and Swanson (1961) report an average thickness of only 10 m accumulated over a period of 14 million years, although Sadler's data predict that the expected thickness is 90 m. The Chattanooga Shale is typical of many black shales in the geologic record, not only because of its slow accumulation rate and black color, but also because of its high degree of fissility.

Bedding and Fissility

Bedding planes indicate breaks in sedimentation. If there had been no interruption in sediment accumulation, there would be no discontinuity between successive layers of the particles that form the bed. The discontinuity with no apparent erosion signifies a period of time when the marine currents were, on the basis of net effect, simply swishing the bottom sediment around without significantly adding to or subtracting from the amount present. There is no way of knowing how long this may continue, perhaps only for 1 year, perhaps for 10,000 years, but the very large number of bedding surfaces that we observe indicates the total time during which no sediment is added can be very large. The same statement can probably be made for fissility surfaces, although the development of fissility may be in part diagenetic. Bedding surfaces are mini-unconformities termed **diastems**. As defined by Bates and Jackson (1987), a diastem is

> . . . a relatively short interruption in sedimentation, involving only a brief interval of time, with little or no erosion before deposition is resumed; a depositional break of lesser magnitude than a *paraconformity*, or a paraconformity of very small time value. Diastems are not ordinarily susceptible of individual measurement, even qualitatively, because the lost intervals are too short. . . .

Sometimes there is not even a bedding surface or fissility surface to suggest that a diastem is present, as in the case described by van Andel (1981):

> I was much influenced early in my career by the recognition that two thin coal seams in Venezuela, separated by a foot of grey clay and deposited in a coastal swamp, were respectively of Lower Palaeocene and Upper Eocene age. The outcrops were excellent but even the closest inspection failed to turn up the precise position of that 15 Myr gap.

Newer radiometric and magnetostratigraphic data indicate the span from Early Paleocene to Late Eocene to be between 24 m.y. and 29 m.y., a gap even greater than suggested by van Andel.

In the deep ocean basin, sometimes thought of as a region of fairly uninterrupted sedimentation, albeit at a very slow rate in pelagic areas, many areally extensive hiatuses are present, some of which appear to be global in extent (Keller et al., 1987). As an example, Figure 3-11 illustrates the temporal pattern of hiatus distribution for

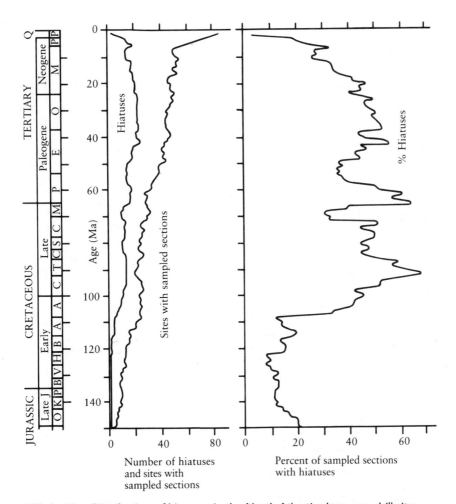

FIG. 3-11 ● Distribution of hiatuses in the North Atlantic deep-sea drill sites. Hiatus distributions are given in absolute and relative figures. (Reprinted by permission of the publisher from Ehrmann and Thiede, *History of Mesozoic and Cenozoic Fluxes to the North Atlantic Ocean*, E. Schweizerbart'sche Publishers, 1985.)

the North Atlantic Ocean (Ehrmann and Thiede, 1985). It shows a number of well-defined maxima and minima located at irregular intervals. Minima occur between 140 Ma and 110 Ma, close to 70 Ma, at 57–50 Ma, and since approximately 15 Ma ago. Maxima of hiatus development occur between the minima and peak at 91 Ma, 65 Ma, and 42–18 Ma. It is noteworthy that most of the maxima and minima are separated from each other by transitional zones, indicating relatively slow development of specific oceanographic scenarios over millions to tens of millions of years.

At least three factors are involved in the origin of ocean basin hiatuses:

1. The presence of deep or bottom water capable of transport and erosion. Currents are ubiquitous along the deep ocean floor because of the Earth's rotation and because of temperature and salinity gradients. Velocities as high as 50 cm/sec (2 km/hr) have been measured by current meters, velocities high enough to move quartz sand.

2. Differential rates of plankton production, which give rise to regional variations in sedimentation rates. These variations are controlled by water temperatures, ecologic factors, and evolutionary trends in shell composition and architecture.

FIG. 3-12 • Idealized diagram of the sequences preserved in solution hollows in Jurassic limestones of west Sicily. (Reprinted by permission of John Wiley & Sons, Inc., from Ager, *Nature of the Stratigraphic Record*, copyright © 1981.)

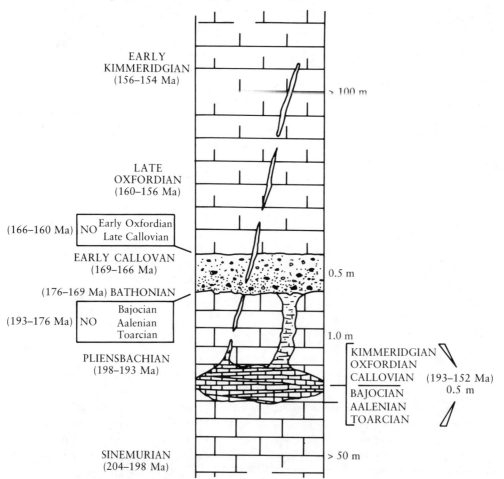

3. The coincidence of regions of low rates of plankton production and sedimentation with regions of high rates of dissolution at the sediment-water interface.
   These coincidences are produced either by the presence of a cold corrosive current or by the buildup of carbon dioxide due to slow mixing and ponding of
   oceanic deep water.

Paraconformities and Disconformities

As the duration of time increases for which there is no sedimentary rock, diastems
grade into surfaces termed **paraconformities**. They are discontinuity surfaces defined
by detectable paleontologic breaks that are not easily attributable to environmental
changes or obviously the result of uplift and erosion (Newell, 1967). Paraconformities are defined by paleontologic criteria, not lithologic ones. A paraconformity may
be laterally coextensive with a disconformity (Fig. 3-12) and may, on close examination, be found to coincide with a disconformity. A paraconformity may even be
confused with a condensed sequence resulting from very slow net sediment accumulation. Because these "surfaces" are defined by biologic rather than lithologic
criteria, they do not necessarily have to coincide with a conspicuous bedding plane,
although they commonly do.

    Disconformities are lithostratigraphic breaks in parallel sequences. They may or
may not coincide with paleontologic discontinuities, depending on the duration of
time represented by the hiatus. Disconformities are based on physical evidence and
display features attributed to subaerial weathering and erosion, such as paleosols,
lag gravel concentrates, channeling, or paleokarst surfaces.

Summary

The essential point of this chapter is the gross incompleteness of the stratigraphic
record. Sedimentary geologists are seriously limited in what they can do with the
first 3.6 billion years of earth history because most of the rocks are either metamorphic or igneous. Fortunately, many Proterozoic rocks and even some Archean rocks
are altered to a small enough degree that standard stratigraphic and sedimentologic
techniques can still be applied and paleogeographic interpretations made.

    The sedimentary record is much clearer for the last billion years of Earth history,
but even this part of the rock record is difficult to interpret. The difficulty is not
metamorphism but a combination of (1) "rare events" that are really not rare,
(2) erosion resulting from orogeny and epeirogeny, (3) the approximate balance between sediment accumulation and removal that is typical of depositional environments for extended periods of time, and (4) transgressions and regressions resulting
from oceanic ridge formation and sea floor spreading. Sedimentary geologists, for
unknown reasons, are enraptured with sedimentary rocks. As a natural consequence, they unconsciously have the mental set that the stratigraphic record is reasonably complete (except for the obvious nonconformities and angular unconfor-

mities), and therefore, if we are astute and hard-working, we will someday be able to print a daily geologic newspaper of Earth events. Unfortunately, this is demonstrably untrue.

The stratigraphic record consists of a few scattered short filaments set at varying angles on a blank canvas of nearly infinite size. Even the most "complete" sections are almost all "empty space" when long periods of geologic time are considered. The problem is that the empty spaces seem fully occupied to human eyes viewing the outcrop. Discontinuities in the rock record need not even be marked by visible features such as bedding. There is a complete gradation between "continuous" and "uninterrupted" stratigraphic sections, sections with observable diastems, and sections with paraconformities and disconformities. There is a complete gradation between nondeposition, reworking of bottom sediment, and erosion (visible loss of sediment). None of these events necessarily implies obvious tectonic activity. These facts should keep us humble and less certain of our stratigraphic and paleogeographic accuracy.

References

Ager, D. V., 1981. The Nature of the Stratigraphical Record, 2nd ed. New York, John Wiley & Sons, 122 pp.

Aigner, T., 1985. Storm Depositional Systems. New York, Springer-Verlag, 174 pp.

Antia, D. D. J., 1979. Bone-beds: A review of their classification, occurrence, genesis, diagenesis, geochemistry, paleoecology, weathering, and microbiotas. Mercian Geol., v. 7, pp. 93–174.

Atwater, B. F., 1987. Evidence for great Holocene earthquakes along the outer coast of Washington State. Science, v. 236, pp. 942–944.

Baker, V. R., 1973. Paleohydrology and Sedimentology of Lake Missoula Flooding in Eastern Washington. Boulder, Geological Society of America, Spec. Paper 144, 79 pp.

Baker, V. R., 1982. The Channels of Mars. Austin, Texas, University of Texas Press, 198 pp.

Bates, R. L. and Jackson, J. A. (eds.), 1987. Glossary of Geology, 3rd ed. Falls Church, Va., American Geological Institute, 788 pp.

Blatt, H., 1970. Determination of mean sediment thickness in the crust: A sedimentologic method. Geol. Soc. Amer. Bull., v. 81, pp. 255–262.

Blatt, H. and Jones, R. L., 1975. Proportions of exposed igneous, metamorphic, and sedimentary rocks. Geol. Soc. Amer. Bull., v. 86, pp. 1085–1088.

Bourgeois, J., Hansen, T. A., Wiberg, P. L., and Kauffman, E. G., 1988. A tsunami deposit at the Cretaceous-Tertiary boundary. Science, v. 241, pp. 567–570.

Byers, C. W., 1982. Stratigraphy—the fall of continuity. J. Geol. Educ., v. 30, pp. 215–221.

Cisne, J. L., 1986. Earthquakes recorded stratigraphically on carbonate platforms. Nature, v. 323, pp. 320–322.

Clifton, H. E. (ed.), 1988. Sedimentologic Consequences of Convulsive Geologic Events. Boulder, Geological Society of America, Spec. Paper 229, 157 pp.

Conant, L. C. and Swanson, V. E., 1961. Chattanooga Shale and Related Rocks of Central Tennessee and Nearby Areas. Washington, D.C., U.S. Geological Survey, Prof. Paper 357, 91 pp.

Dott, R. H., Jr., 1983. Episodic sedimentation—How normal is average? J. Sed. Petrology, v. 53, pp. 5–23.

Ehrmann, W. U. and Thiede, J., 1985. History of Mesozoic and Cenozoic Sediment Fluxes to the North Atlantic Ocean. Stuttgart, Schweitzerbart'sche, 109 pp.

Einsele, G. and Seilacher, A. (eds.), 1982. Cyclic and Event Stratification. New York, Springer-Verlag, 536 pp.

Gretener, P. E., 1967. Significance of the rare event in geology. Amer. Assoc. Petroleum Geol. Bull., v. 51, pp. 2197–2206.

Heaton, T. H. and Hartzell, S. H., 1987. Earthquake hazards on the Cascadia subduction zone. Science, v. 236, pp. 162–168.

Kastens, K. A. and Cita, M. B., 1981. Tsunami-induced sediment transport in the abyssal Mediterranean Sea. Geol. Soc. Amer. Bull., v. 92, pp. 845–857.

Kauffman, E. G., 1988. Concepts and methods of high-resolution event stratigraphy. Ann. Rev. Earth Planet Sci., v. 16, pp. 605–654.

Keefer, D. K., 1984. Landslides caused by earthquakes. Geol. Soc. Amer. Bull., v. 95, pp. 406–421.

Keller, G., Herbert, T., Dorsey, R., D'Hondt, S., Johnsson, M., and Chi, W. R., 1987. Global distribution of late Paleogene hiatuses. Geology, v. 15, pp. 199–203.

Kidwell, S., 1989. Stratigraphic condensation of marine transgressive records: Origin of major shell deposits in the Miocene of Maryland. J. Geol., v. 97, pp. 1–24.

Kreisa, R. D., 1981. Storm-generated sedimentary structures in subtidal marine facies with examples from the Middle and Upper Ordovician of south-western Virginia. J. Sed. Petrology, v. 51, pp. 823–848.

Lualdi, A., 1986. Early Sinemurian hardgrounds in the Ligurian Alps, northwestern Italy (Prepiemontese domain, Arnasco-Castelbianco unit). Geol. Rundschau, v. 79, pp. 365–384.

Marsaglia, K. M. and Klein, G. deV., 1983. The paleogeography of Paleozoic and Mesozoic storm depositional systems. J. Geol., v. 91, pp. 117–142.

Newell, N. D., 1967. Paraconformities. In C. Teichert and E. L. Yochelson (eds.), Essays in Paleontology and Stratigraphy. Lawrence, University of Kansas Press, pp. 349–367.

Retallack, G. J., 1984. Completeness of the rock and fossil record: Some estimates using fossil soils. Paleobiology, v. 10, pp. 59–78.

Rundberg, Y. and Smalley, P. C., 1989. High-resolution dating of Cenozoic sediments from northern North Sea using $^{87}Sr/^{86}Sr$ stratigraphy. Amer. Assoc. Petroleum Geol. Bull., v. 73, pp. 298–308.

Sadler, P. M., 1981. Sediment accumulation rates and the completeness of stratigraphic sections. J. Geol., v. 89, pp. 569–584.

Sadler, P. M. and Dingus, L. W., 1982. Expected completeness of sedimentary sections: Estimating a time-scale dependent, limiting factor in the resolution of the fossil record. In Third North American Paleontology Convention, Proceedings, v. 2, pp. 461–464.

Silver, L. T. and Schultz, P. H. (eds.), 1982. Geological Implications of Impacts of Large Asteroids and Comets on the Earth. Boulder, Geological Society of America, Spec. Paper 190, 528 pp.

van Andel, T. H., 1981. Consider the incompleteness of the geological record. Nature, v. 294, pp. 397–398.

Wendt, J., 1988. Condensed carbonate sedimentation in the late Devonian of the eastern Anti-Atlas (Morocco). Eclogae Geol. Helv., v. 81, pp. 155–173.

4
Geologic Time and
Correlation

One of the permanent problems facing stratigraphers in dealing with the existing mass of sedimentary rocks is estimating their relative and numerical ages. The importance to a stratigrapher of both relative dating and numerical dating is that these establish when geologic events occurred, which makes correlation possible. Ideally, when all events recorded in sedimentary rocks are identified and correlated, we will have as complete a "daily newspaper" of Earth history as is obtainable, given the inherent incompleteness of the stratigraphic record (Chapter 3).

About 80 years ago, the famous geologist T. C. Chamberlain remarked that diastrophism is the ultimate basis of correlation. The truth of this statement is even clearer today than it was then, as our understanding of lithospheric tectonics has grown. Mantle convection instigates crustal fragmentation, which causes the formation of midoceanic ridges, which displaces ocean water, which causes flooding of continental margins, which results in environmental and faunal changes. The land surface is partially covered with seawater, and marine sediment is deposited, creating an unconformity. As the continental fragments plow their way slowly through the lithosphere, mountain ranges are created at their frontal edges, for example, at the western margin of North and South America. Should continental fragments collide, as during the Paleozoic along the Atlantic seaboard of the United States, more mountains will be created. Without diastrophism the land surface would quickly be leveled by erosion, the sediment would be transformed to clay minerals and quartz silt, and surficial sedimentologic processes would essentially grind to a halt. Fortunately for those of us who enjoy playing with sedimentary rocks, very little of the Earth's heat has escaped from the interior during the past 4.6 billion years, so the mantle convection that results eventually in the sustenance of our professional lives is in no danger of ending.

Recognition of the great length of time recorded and legible in the Earth's crust is one of humankind's most important scientific achievements. It is arguably the only new fact in basic science discovered by geoscientists. But the technological ability to determine numerical ages of minerals and rocks was preceded by a long history of investigation centering on *relative* ages of events (Berry, 1987). The concepts now termed the "Law of Superposition" and the "Law of Crosscutting Relationships" can be traced back to Old Testament writers and the Greek and Roman scholars who followed them, although official credit is usually awarded to the Danish scientist Nicholas Steno, who stated them explicitly in the 17th century.

As valuable and useful as Steno's concepts are, they cannot establish the temporal relationships among widely scattered areas. Is a vertical sequence of sandstone, shale, and limestone in one region of the same age as a similar sequence in an area 100 km distant? Such sequences have been deposited repeatedly throughout the world, so it is clear that lithology does not necessarily indicate time equivalence. The key to establishing such an equivalence is fossils, as was first recognized by William Smith, an English surveyor at the beginning of the 19th century. He discovered that he was able to recognize a rock unit by its fossil content and also to determine a particular layer's position in a succession of layers by its fossils, even where the relationships were otherwise obscured. Smith established the principal of faunal (and floral) succession. His work showed that there is a series of unique fossil assemblages that can be recognized when the fossils are collected in precise superpositional order. Smith's insights were used subsequently by Darwin in the development of his ideas about evolution.

Zones and Stages

Smith's work was formalized in the concept of *zones* and *stages*, developed in the middle of the 19th century by the German geologist Friedrich Quenstedt and his student Albert Oppel. They established the fundamental biostratigraphic unit as the fossil **zone**, a faunal association that is different from those immediately above and below. To establish faunal zones, Oppel said, geologists need to explore, on the scale of centimeters,

> the vertical stratigraphic range of each separate species in the most diverse localities, while ignoring the lithological development of the beds; by this means will be brought into prominence those zones which, through constant and exclusive occurrence of certain species, mark themselves off from their neighbors as distinct horizons. In this way the ideal profile is obtained in which layers of the same age in different areas are characterized by the same association of species.

Oppel named each zone after a prominent species in the zone, and this is the way it is still done today. A faunally or stratigraphically distinctive group of zones is termed a **stage**, a term devised by D'Orbigny, a French contemporary of Quenstedt and Oppel. Stages are formally tied to a **stratotype** or type section, a body of rock that contains the upper and lower limits of the stage.

The most useful types of fossils for establishing faunal zones depend on geologic age, depositional environment, and rock lithology. Among Paleozoic units, graptolites, goniatitic ammonites, conodonts, and fusulinids have proven most useful. Ammonites have been most useful in Mesozoic sequences, while Cenozoic rocks have been zoned most commonly on the basis of planktonic foraminifera, coccolithophorids, and radiolaria. The zonal sequence established by using the method of Oppel is as fundamental to erecting a biostratigraphic or chronostratigraphic frame of reference as the formation is to the establishment of a lithostratigraphic frame of reference.

Modern technological developments such as subsurface coring on the continents and in the ocean basins have greatly expanded the number and detail of stratigraphic sequences that can be studied in great detail. For example, the recovery of cores of oceanic sediment has provided in some areas almost continuous sequences of richly fossiliferous materials that represent tens of millions of years of sediment accumulation. These core materials have yielded tens of thousands of specimens in stratigraphic succession and permit study of the development and spread of minute morphologic change seen in thousands of individuals over very long intervals of time. The record of fine-scale morphologic change over a continuous record of millions of years of accumulation can lead to a very refined scale of interpretation of the history of the ocean basins and of life in them. The ocean core materials permit delineation of zones using Oppel's methodology based upon fine-scale understanding of the ranges of species, including their originations and extinctions.

Fossil-based time correlations involve setting up a matrix for correlations among stratigraphic sections. The steps in developing a correlation matrix include the following.

1. Set up a sequence of zones and perhaps stages for the interval of geologic time and the area under study. Alternatively, review of the literature may show that an existing sequence of zones and stages is available for use in the area and for the time interval being studied. If so, then examination of existing fossil collections from the rocks being studied as well as making new collections is involved. Then the species and genera identified in these collections should be compared with the fossils used to recognize the existing zones and stages.
2. Delineate a set of zones and stages for each major environment in the area and time interval studied to find out the details of geologic history.
3. Establish time equivalencies or time correlations among the sets of zones and stages used within each major environment. Such features as debris flows that carry fossils from one set of environments into the rocks bearing the record of another environment are useful in making these correlations. For example, a turbidite or debris flow containing fossils such as near-shore clams and snail shells may be found within a set of planktonic foram-bearing pelagites or hemipelagites deposited on a slope or in a basin. Spores and pollen will be blown

from a terrestrial set of environments into a marine setting and so permit temporal correlation of nonmarine and marine units.

4. Add radiometric dates, if available, to the biostratigraphic column based on zones and stages. The numerical ages will aid in making time correlations not only in one area but also with other areas.

5. Determine, if possible, the magnetic polarity of iron particles for all rocks present (see below). Similarities in the "thickness" and sequence of normal or reversed signatures and their radiometric ages will be useful in establishing time synchroneities among strata in the area as well as synchroneities with other areas.

After the basic matrix of zones and stages has been developed such that all environments represented in the area may be linked, then a correlation chart can be made. First, the basic correlation matrix, the sets of zones and stages for each environment and any radiometric and magnetostratigraphic information, is plotted on the left side of the chart. Then fossil information from each stratigraphic section under study is examined. Matches are made, if possible, between each fossil collection made at every stratigraphic position up the stratigraphic section with the fossil association that typifies each of the zones and stages. The set of strata bearing the fossils that are found to match with a set in the correlation matrix is plotted on the chart. Step by step, each collection found in each stratigraphic section is matched with the groups of fossils that characterize each zone and stage. The time span represented by each stratigraphic section examined is plotted on the correlation chart. Each column on the chart represents one stratigraphic section. After the time synchroneities have been worked out by making matches between fossils collected in each section with those in the correlation matrix on the left side of the chart, then the lithologic features of the rocks containing the fossils can be studied, area by area and time division by time division. Also, the environments represented in each stratigraphic section can be analyzed, time interval by time interval. The correlation chart of information plotted using the correlation matrix becomes the fundamental tool in the analytical steps toward understanding tectonic, geomorphic, and environmental histories of an area for any time interval.

Zonation Using Spores

Spores are valuable in oil exploration for both biostratigraphic and paleoenvironmental interpretations. The following zonation example involving Siluro-Devonian spores illustrates some of the steps in developing a correlation matrix. As may be seen in the example, zones based on spores were developed first. Then the spore zones were equated with those based on marine organisms that floated (graptolites), swam (ammonoids), or were allied to various environments on the shelf (conodonts).

Siluro-Devonian spores have been analyzed from near-shore marine and terrestrial settings from many parts of North America and Western Europe by Richardson and McGregor (1986). The spores found in many stratigraphic sequences in eastern North America and Western Europe were examined closely, and Richardson and

McGregor (1986) pointed out "widespread, apparently synchronous 'first appearances' of certain taxa and form features of spores that occur repeatedly in the geological column . . . such occurrences are assumed to reflect evolution, and morphological antecedents can be suggested for many taxa." Richardson and McGregor (1986) went on to note that "Whatever their evolutionary antecedents, certain (but not all) species and structural features seem to have attained widespread geographical distribution instantaneously in terms of geologic time. Consequently, we have accorded considerable biochronological significance to first appearances of selected species and form features" (Fig. 4-1). To be noted in the figure is the number of zonal boundaries drawn at the appearance of a new morphological feature among spores.

Richardson and McGregor (1986) used unique associations of species based upon the overlapping ranges of spore species as the basis for their spore zones. They stated that

> the earliest records of one or two selected species, observed in a variety of facies and irrespective of relative abundance (which may, however, be significant for local correlations), are used to define the base of each zone in a reference section. Assemblages occurring in consistent order of succession, each characterized by several species, provide the basis for correlations on a global scale.

Richardson and McGregor (1986) used the names of two widely found and characteristic spore taxa in each zonal assemblage as the name of each zone. They emphasized that the base of each zone was drawn at the first occurrence of at least one of the taxa used to give the name to the zone. Most of the species used to give their names to zones are widespread geographically. The presence of a number of the characteristic species of each zonal assemblage is needed to recognize a zone. The name-giving species may range out of the zone stratigraphically. The boundaries of some zones are drawn at the first appearance of new morphological features in the spore taxa. Zonal boundaries were defined in a specific stratigraphic section, and an attempt was made to draw each boundary at a stratigraphic level in which no change in lithological aspect or structural deformation could be observed.

For each of the zones the following features were cited:

1. The age in terms of a more generalized time scale
2. The reference section for the base of the zone
3. Characteristic species
4. General description of morphological features exhibited by spores making up the assemblage
5. Geographic distribution of the zone assemblage
6. Additional remarks, such as possibilities for division of the zone with additional study

Because many of the spore associations have been found in strata with other types of fossils, Richardson and McGregor (1986) could describe correlation of their spore zones with zonal sequences based upon graptolites, conodonts, ammonoids, tentaculites, and land plants as well as with the stage divisions for the Siluro-Devonian

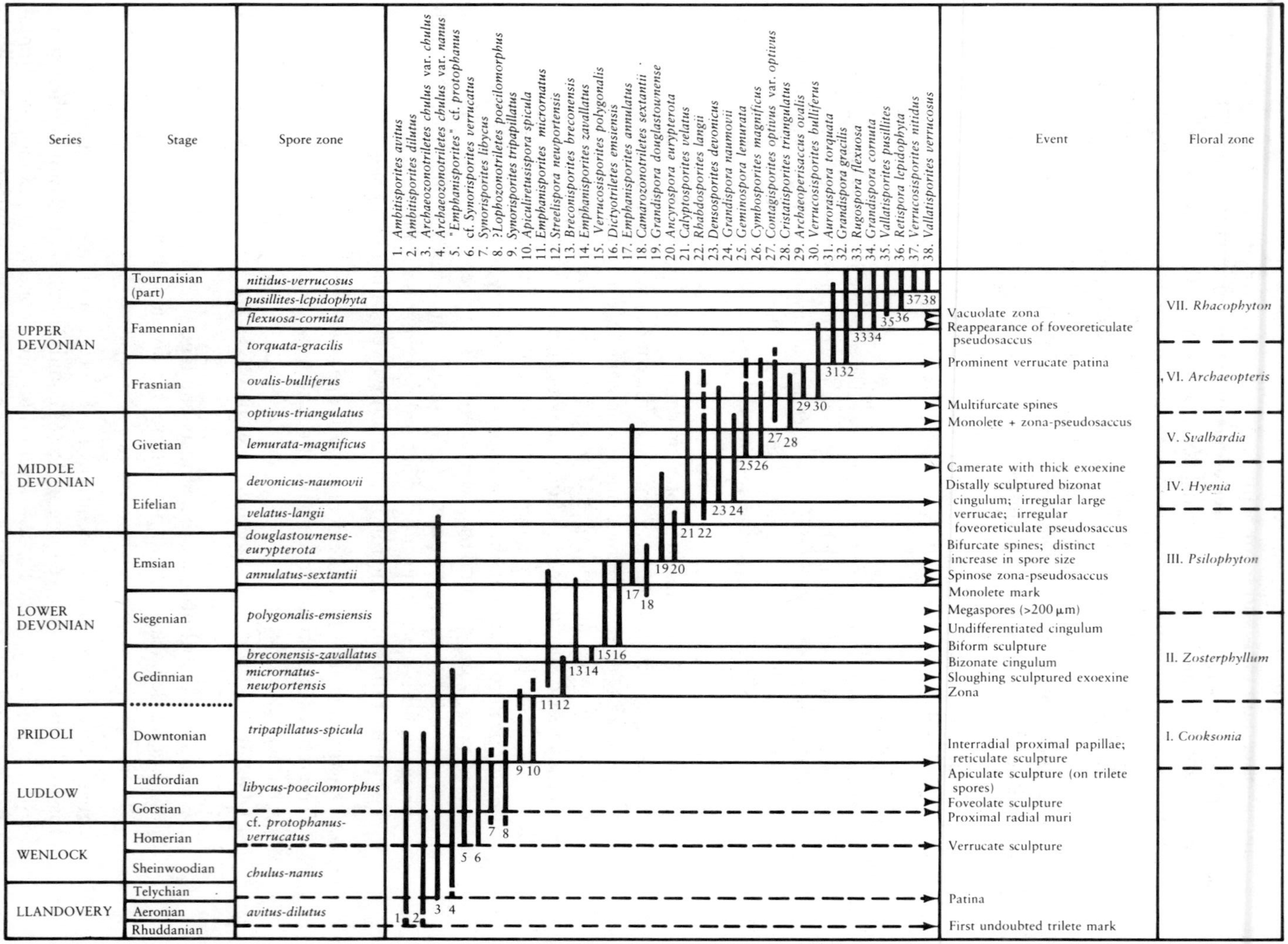

FIG. 4-1 ● Zonation of Siluro-Devonian spores in North America and western Europe and its relationship to floral zones (Richardson and McGregor, 1986; reproduced with permission of the Minister of Supply and Services, Canada.)

94

that are based upon associations of a number of "shelly fossils," including corals and brachiopods. Spores can be transported to a broad spectrum of environments by both wind and water. Accordingly, they are very useful tools in establishing time synchroneities among zones based upon organisms that grew on land and those that lived in diverse marine environments as well. The correlations or time synchroneities were established with zonal sequences that represent the full spectrum of shelf sea settings (tentaculites, conodonts, ammonoids) as well as environments on the shelf margin (ammonoids and graptolites). Because they may be transported so widely, spore associations are perhaps the most effective means of establishing time synchroneities among zonal sequences based upon organisms that lived in a range of different marine environments and those that lived on land.

Neogene Time Scale

Because continental areas were extensive in the Neogene (Miocene to Recent), land mammals have proven valuable in delineating the geographic positions and shapes of continental areas during this youngest interval in the time scale. Accordingly, land mammals have been used as a basis for stages. Land mammal stages have been developed for many of the major continental regions of the Cenozoic. The drilling program into the sediments of the deep oceans has yielded a wealth of fossil information. To make use of that information, fossil-based time scales based on the evolution of the fossils found in the cores, as discussed on earlier pages, have been developed. Traditionally, marine shelf sea clams and snails have been used to make stages as divisions of the Cenozoic. These molluscan-based stages are shown on the accompanying chart as "Standard Ages." Each of them has a type section to which anyone may go to look at the stratigraphic positions at which the fossils that characterize the "Standard Age" occur.

The correlation matrix shown (Fig. 4-2) indicates the open ocean zones based on the evolution of different organisms as well as the land mammal stage/age divisions recognized in North America and Europe. The different sets of zones and stages have been equated in time by using radiometric age and magnetic reversal evidence. The information available from the Neogene is more comprehensive and more detailed than for other, older parts of the geologic record. The matrix chart indicates the high degree of detailed information available for making Neogene correlation charts.

Fossils and Isochroneity

How precise are the time equivalences established by biostratigraphic correlation? Modern evolutionary theory, whether based on gradualism or punctuated equilibria, states that a new species originates at a specific place at a specific instant. The organism must then radiate from its point of origin to populate the ecologic niches to which it can adapt. The amount of time this requires will vary for different types of

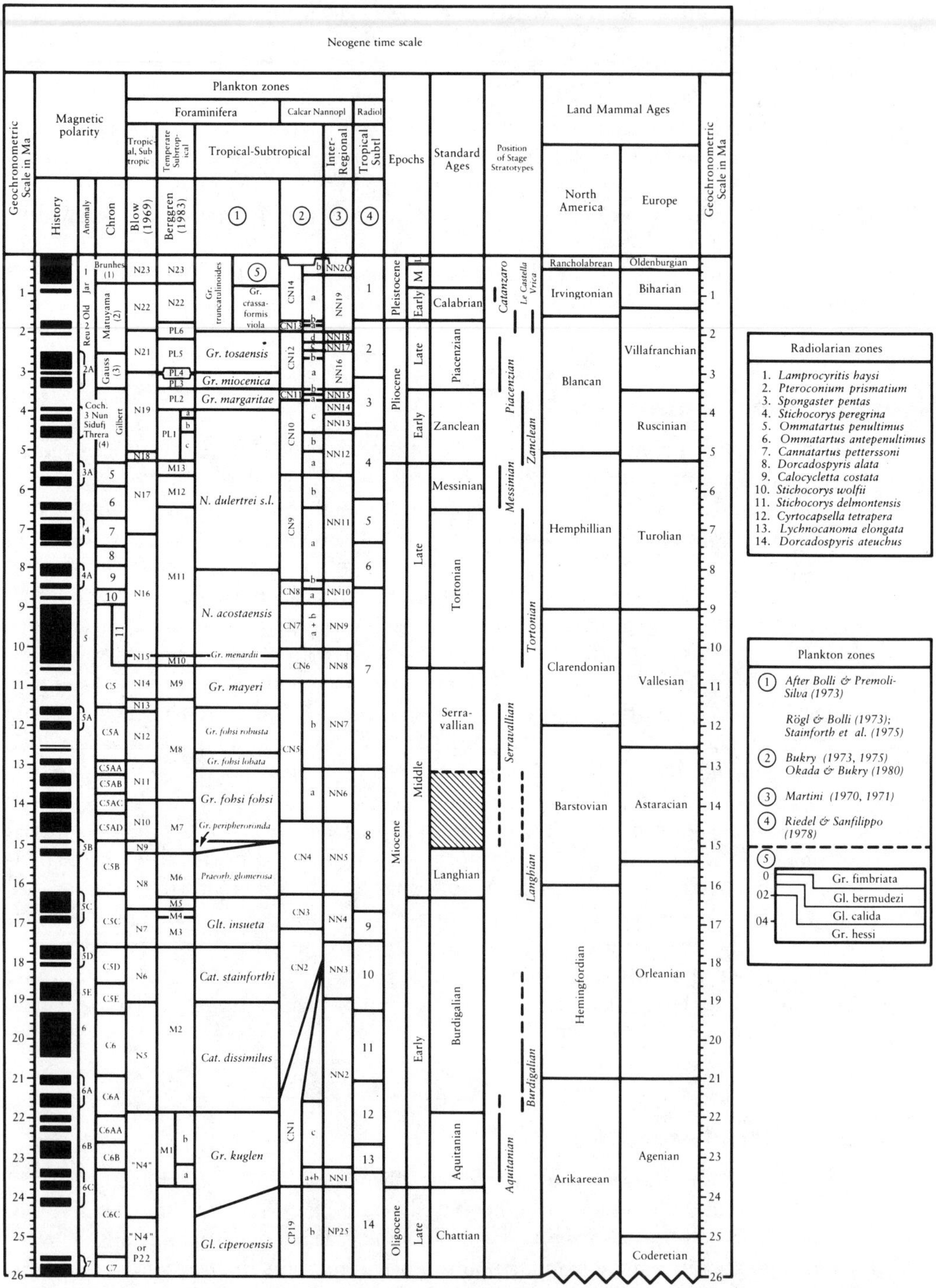

Neogene time scale

Geochronometric Scale in Ma	Magnetic polarity			Plankton zones							Epochs	Standard Ages	Position of Stage Stratotypes	Land Mammal Ages		Geochronometric Scale in Ma

Radiolarian zones

1. *Lamprocyritis haysi*
2. *Pteroconium prismatium*
3. *Spongaster pentas*
4. *Stichocorys peregrina*
5. *Ommatartus penultimus*
6. *Ommatartus antepenultimus*
7. *Cannatartus pettersoni*
8. *Dorcadospyris alata*
9. *Calocycletta costata*
10. *Stichocorys wolfii*
11. *Stichocorys delmontensis*
12. *Cyrtocapsella tetrapera*
13. *Lychnocanoma elongata*
14. *Dorcadospyris ateuchus*

Plankton zones

① After Bolli & Premoli-Silva (1973)

Rögl & Bolli (1973); Stainforth et al. (1975)

② Bukry (1973, 1975) Okada & Bukry (1980)

③ Martini (1970, 1971)

④ Riedel & Sanfilippo (1978)

⑤

	Gr. fimbriata
	Gl. bermudezi
	Gl. calida
	Gr. hessi

organisms, so the amount of time transgressed by the surface marking the first appearance of an organism in rocks will vary.

Two factors control migration time, biology of the organism and barriers to migration.

Biology of the Organism

Suppose a new species of snail evolves in New York City, avoids being crushed by pedestrians, and heads for Los Angeles. At a nonstop rate of crawl of 1 m/hr it would require about 700 years for the snail to arrive in the City of Angels (assuming that friends along the way supplied it with an overcoat for crossing the Rockies and a water-filled canteen for the desert areas and neglecting the fact that the life span of a land snail is only a few years). Or consider a new species of marine organism with planktonic larvae (Fig. 4-3), whose larvae are transported by oceanic currents (Scheltema, *in* Kauffman and Hazel, 1977). The velocity of the current might be on the order of 10 km/day, and the spread of the new planktonic species around the world would require at least tens of years. An analogous consideration of species extinctions makes it apparent that not all individuals in a species will die simultaneously. Thus it is clear that even in the best of circumstances paleontologic isochroneity does not exist. Normally, however, it is an acceptable approximation because of the relatively large imprecisions in numerical dating methods now available.

Barriers to Migration

An even more serious influence on migration time for organisms is the formation and persistence of physical or chemical barriers. Well-known examples of physical barriers to migration include the oceanic barrier to terrestrial organisms formed by the breakup of Pangaea and the separation of land areas and the tectonic raising of the Central American isthmus, which separated Atlantic and Pacific faunas. It is relatively easy to establish the approximate time of formation of such barriers by using standard tectonic and stratigraphic methods and, from this, to explain faunal provincialities (Chapter 5) and differing times of first appearances of organisms in different localities. More difficult types of barriers to recognize and evaluate are those that exist within a water mass and are not preserved physically in the geologic record. The Gulf Stream in the Atlantic Ocean is an example of such a feature. At what time after the formation of the mid-Atlantic Ridge did this warm current come

---

◀ **FIG. 4-2** • Neogene geochronology. The geochronologic scale at the margins of the figure is derived from the magnetic polarity chronology. This chronology is in turn derived from radiometrically dated calibration points controlled by fossils and/or paleomagnetic zones. The position of the calcareous plankton zonal boundaries is based, for the most part, upon direct (first-order) correlation between biostratigraphic datum levels and paleomagnetic polarity stratigraphy as determined in deep-sea cores or continental marine sediments. In this way a true magnetobiochronology is possible. The extent (duration) of standard time-stratigraphic units and their boundaries and the position of stage stratotypes are estimated on the basis of their relationship to standard plankton biostratigraphic zones. (Reproduced by permission of the Geological Society from Berggren et al., *in* Snelling, *The Chronology of the Geological Record,* 1985.)

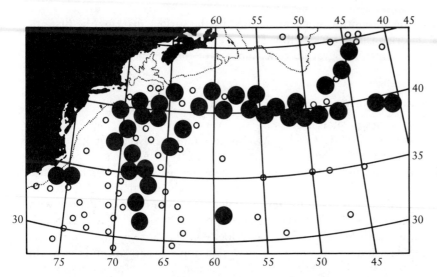

FIG. 4-3 • Dispersal of inarticulate brachiopod larvae of *Glottidia pyramidata* off the Atlantic coast of the United States. Larvae of this Caribbean species are carried northward on coastal currents and the Gulf Stream; those found off the Grand Banks (upper right) will not survive to settlement because larval development is too brief for transport to a region where the postlarvae can survive. Large circles are locations where brachiopod larvae were taken; open small circles are locations without brachiopod larvae. (Reprinted by permission of the publisher from Scheltema, *in* Kauffman and Hazel, *Concepts and Methods of Biostratigraphy*, Van Nostrand Reinhold, 1977.)

into existence? Another example is the mass of cold oceanic bottom water generated by the melting of large ice caps in polar regions. During interglacial times the bottom water was necessarily warmer, with attendant but uncertain effects on the evolution and migration patterns of benthonic organisms.

Lithostratigraphic Correlation

All ancient lithologic units are time-transgressive. For example, the basal Cambrian sand is early Cambrian in southern California and southwestern Nevada but becomes progressively younger toward the transcontinental arch, so it is late Cambrian—perhaps 35 m.y. younger—at the arch in central Colorado. However, in some cases the amount of time transgressed by a unit is so small that no significant error in geologic interpretation is incurred if isochroneity is assumed although no fossils are present. For example, bentonite beds are generally used as time-parallel surfaces, and a numerical date obtained in one part of a bed is used as the age for all other parts of the bed. Is this a valid procedure? Bentonite is a rock formed by the alteration of volcanic tuff or ash and is composed mostly of clay minerals of the smectite (montmorillonite) group with subordinate amounts of opaline or microcrystalline silica and accessory minerals of volcanic origin. The fragments that were the parent material of the bentonite were blown out of one or more volcanoes into the atmosphere and settled to the ground in the surrounding area. Obviously, it takes time for a fragment to move away from the volcanic vent, so the age of the base of the sediment layer formed is older nearer the vent than farther away. The difference in age is small, no greater than a few days at most for a bentonite bed. But nevertheless, the base of the bed is time-transgressive. We can consider a bentonite bed to be isochronous because the precision of radiometric dating for ancient rocks is perhaps 10^9 poorer than the 3 days of time transgressed by the base of the bentonite. As the accuracy and precision of radiometric dating improve, however, the assumption of isochroneity for lithologic surfaces becomes less acceptable. For example, consider the upper surface of glacial sediments that marks the Pleistocene-Holocene boundary. In Minnesota this surface is 10,000 years old, but in Greenland the upper surface of glacial sediment has not yet been deposited. It is still "Pleistocene time" in Greenland. The reason we cannot ignore this fact in interpreting relatively recent Earth history is that the precision of the dating method for recent times, ^{14}C disintegration, is much better than 10,000 years; it is only a few hundred years.

All attempts at lithologic correlation presuppose that something is known about the geologic ages of the rocks being correlated. The fact that the sequence sandstone, shale, gypsum is found in one area does not support the hypothesis that a disconnected sequence of sandstone, shale, gypsum 50 km distant is lithostratigraphically correlatable unless there is some basis for believing that the two sequences are not of vastly differing ages. Such a belief may arise from several types of field observations at ground level.

1. The rocks in the intervening 50 km are flat-lying or have a uniform dip, no significant faults are present, and "topographic tracing" is possible in cross-section. When correlating rock sequences (or individual beds) in this way, however, always keep in mind that rock types change laterally (facies change), so sandstone in the first area may be temporally equivalent to shale in the second and the shale in the first area may be equivalent to both shale and gypsum in the second.

2. The macroscopically visible fossils in the first area seem to be the same as those in the second. In this case the ranges of the fossils are critical. For example, the fact that both sequences contain pelecypods and fish scales is not too helpful in age determination because both taxa have been present since Cambrian time. Commonality of trilobites and rugose corals in the two sections is a bit more useful, because both taxa became extinct by the close of the Paleozoic Era. Better still would be correspondence in fossil content at the generic or specific level, but such specificity is difficult to establish without laboratory work.

3. A distinctive lithologic marker bed occurs in the several separated stratigraphic sections, such as a distinctive ash or bentonite bed, a single turbidite bed or a carbonate hardground. Each of these is formed instantaneously (in the geologic sense, not in an absolute sense) and can be so unique in field appearance that miscorrelation among scattered outcrops is unlikely. One can also use a series of

less distinctive geologic features as lithostratigraphic markers. For example, an angular unconformity overlain by a fossil soil, a coal bed, and a gypsiferous dolomite may be present in the same order in scattered sections of a thin stratigraphic sequence. In this case, lithostratigraphic correlation is hypothesized on the basis that a sequence of such events is not likely to occur twice within a geographically limited area whose regional geologic character is roughly known. Clearly, the higher the number of unusual characteristics the two sections have in common, the more secure the lithostratigraphic correlation. The principle is the same as occurs in the form of entertainment called craps. If a player throws 14 sevens in a row, there is a reasonable inference that the phenomenon is correlated with the event known as cheating.

Frequently in a geographic area, topographic relief is adequate for significant outcrops to be present, but the climate is sufficiently temperate and humid that soil cover is extensive. Rock outcrops are scarce. In such cases the character of the soil can be informative. Soils developed on quartz sandstones are different from those on carbonate rocks; soils on granite differ from those on shale. A knowledge of elementary soil science can be very useful in such situations. Also useful in areas of extensive soil cover is some knowledge of cultural geography. For example, along the Texas Gulf coast the Tertiary shales and fine-grained sandstones strike parallel to the present coastline, and lithologies are fairly constant within units along strike. These coastal areas were settled for farming in the 19th century by Czech, German, and English immigrants in sequence. A Czech would settle on one shale band, and succeeding Czechs would settle adjacent to their countrymen on the shale. A German would arrive but settle on a stratigraphically different shale band, ignoring the intervening band of fine sand because of its relative infertility, and succeeding German immigrants would settle nearby on the shale. Each nationality thus came to occupy a series of farms along geologic strike and was separated from neighboring nationalities by relatively poor land underlain by sand. The pattern (lithostratigraphy) is easily seen today in the names on rural mailboxes and in the architecture of the old churches in this part of southeast Texas.

Lithostratigraphic correlations can also be made by using aerial photographs and various types of remotely sensed data from satellites (Chapter 1). Satellite images and very high altitude aerial photography have not been of significant use in stratigraphic studies because of the small scales of such photos that are available to the public. The potential of computer-guided photographic and image-generating equipment is almost beyond imagining, but as of 1990, standard aerial photographs are still the most useful above-ground pictures for lithostratigraphic work. When used in stereo pairs, lithologic variations of rather subtle character may be detectable on the basis of their stratification and resistance to weathering (topographic irregularities).

Correlation of Varves

Many lacustrine and evaporite sequences are composed in part of thinly laminated couplets repeated without interruption through a considerable stratigraphic thick-

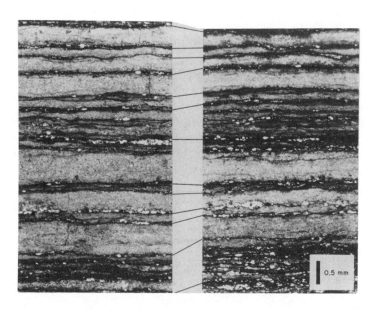

FIG. 4-4 • Correlative thin sections of calcite-sapropel laminae of Todilto limestone from two localities separated by 2.3 km. Light layers are calcite; dark layers are sapropel. (Reprinted by permission of the author from Anderson and Kirkland, *Geological Society of America Bulletin*, v. 77, 1966.)

ness. Commonly, these varved sequences are laterally extensive as well, but because outcrops of specific stratigraphic horizons are discontinuous, correlation problems invariably exist. The key to solving the question of correlation in scattered outcrops of thick, varved units is statistical comparison of laminae thicknesses (Anderson and Kirkland, 1966; Anderson et al., 1972). The technique used is essentially the same as that of correlation of electric logs, in which the two log traces are placed side by side and slid past each other until the best match of log characteristics is found for the length of the log. The goodness of fit for any position of the two logs can be tested statistically at different horizons, and the position that gives the highest correlation coefficient is taken as the correlation position. Figure 4-4 illustrates an excellent match (correlation) between two varved cores of the Todilto Formation (Jurassic, New Mexico). In this case, statistical manipulation is hardly needed, as the match is visually acceptable only in the position shown in the photograph. Anderson and co-workers have obtained correlation coefficients consistently exceeding 0.70 (out of a mathematically possible 1.0) over lateral distances of up to 290 km. Problems with this technique of correlation are that local disturbances of the lake or sea bottom during sediment deposition can destroy correlations and that in some formations the degree of correlation decreases with increasing lateral distance and the rate of decrease may vary among outcrops and basins.

Correlation Using Organic Matter

Most fine-grained rocks contain significant amounts of organic matter, and it may be possible to use various characteristics of such material for correlation purposes. The most obvious effect of the presence of organic matter in a limestone or mudrock is a dark color; 1% in a limestone and 2% in a mudrock causes the rock to be black.

The amount of organic matter in rocks can vary widely with variation in depositional environment, so great care must be taken in using color in studies of rock equivalence.

More subtle analytic methods can be used on organic matter isolated from sedimentary rocks to determine the amounts of particular organic constituents, such as lignins or carbohydrates, and these can be used to determine the depositional environment of the sediment. Lignins, for example, characterize woody land plants. It also is possible to determine the dominant functional groups in organic matter, and these have been used by Schnitzer and Schwab (1975) in correlation and facies analysis of otherwise cryptic mudrock units ranging in age from Precambrian to Quaternary.

Radiometric Dating

The only practical method available for determining the numerical age of a lithic unit is through the use of unstable nuclides in minerals. These nuclides decay to "daughter products" (other nuclides, plus particles, rays, neutrons and protons, and energy) at a rate unique to the particular parent nuclide, and the rate is unaffected by changes in either the temperature or pressure at which the decay occurs. Although the idea of decay seems straightforward, there are, unfortunately, several problems with sampling and analysis whose additive nature make the accuracy and precision of radioactive dating less than we might wish (Odin, 1982).

1. **Decay Constants.** The fundamental question of the rate of decay of each radioactive nuclide is solved today by use of a set of International Conventional Constants (ICC) that were adopted in 1976 (Steiger and Jäger, *in* Cohee et al., 1978; Table 4-1). Measurements made before the ICC were established used a variety of decay rates determined in different laboratories, adding imprecisions to radiogenic ages in addition to those inherent in the various methods of analysis of parent/daughter ratios in mineral and rock samples.

   All analytical instruments have limitations on the degree of precision they can attain. As better instruments are devised, the precision improves, but perfection can never be achieved. With regard to determination of radiometric ages the reproducibility of apparatus at present is commonly better than 1%, but the calibration among laboratories doing the determination on a different split from the same mineral sample may differ by 5–10%. This latter imprecision is often ignored in published radiometric ages.

2. **Stratigraphic Uncertainties.** Commonly, a radioactive nuclide such as ^{235}U is used to determine the date of crystallization of a plutonic rock. If the pluton is intrusive and conformably overlain by sedimentary rocks that extend upward to the surface, the date of intrusion establishes only a minimum age for the sedimentary section. If the pluton is truncated by an unconformity, only the maximum possible age of the overlying sediments is known. If the sedimentary rocks are sandwiched between two datable intrusions, the age of the sediments is

TABLE 4-1 • International Conventional Constants for the most commonly used radiogenic nuclides. Whole-rock ages are commonly obtained in addition to or in place of mineral ages in crystalline rocks.

Nuclide	Daughter Products	Half-Life (yrs)	Common Minerals Used
^{235}U	^{207}Pb	7.0381×10^8	Zircon, uranium minerals, rare-earth phosphates
^{40}K	^{40}Ar	1.250×10^9	Micas, feldspars, glauconite
^{238}U	^{206}Pb	4.4683×10^9	Same as for ^{235}U
^{232}Th	^{208}Pb	1.401×10^{10}	Same as for ^{235}U
^{87}Rb	^{87}Sr	4.89×10^{10}	Feldspars, micas, glauconite

(Reprinted by permission of the American Association of Petroleum Geologists from Steiger and Jäger, *in* Cohee et al., *Studies in Geology*, No. 6, 1978)

bracketed but is not accurately determined. Similar statements apply to sedimentary rocks underlain by dated metamorphic rocks or sandwiched between dated volcanic rocks.

3. **Geochemical Uncertainties.** Two genetic problems are present when minerals separated from magmatic rocks are used to bracket the age of sedimentary rocks. First, the closure temperature (time zero) is established a very long time after the intrusion and varies with the cooling rate. For example, radiogenic argon and strontium are mobile in some minerals at temperatures well below that of crystallization. The closure temperatures of the Rb-Sr system in muscovite and biotite are $500 \pm 50°C$ and $300 \pm 50°C$, respectively, and the closure temperatures of the K-Ar system are $350 \pm 50°C$ for muscovite and $300 \pm 50°C$ for biotite. In plagioclase the closure temperature for radiogenic argon is about $260°C$; in orthoclase it is about $160°C$. Thus it is clear that the apparent age of an igneous or metamorphic rock may differ depending on the isotopes measured and the mineral used. One must also keep in mind that mineral ages are easily reset when closure temperatures are low. Second, significant amounts of daughter product (such as argon) may be inherited and result in the determined radiogenic age being greater than the actual age.

Another problem in radiogenic dating is that a mineral extracted for dating purposes from a volcanic unit may have been inherited from an older volcanic unit remobilized by a new volcanic event (zircons). A related difficulty occurs when the volcanism occurs in the marine environment during conversion of an ash into a bentonite. The age of the mineral in the bentonite can be different from the time of ash deposition and *can be either much younger or much older* than it, depending on the type of alteration of the dated mineral.

A related type of genetic uncertainty is present when glauconite in sedimentary deposits is used for dating (Odin, 1982). Glauconite is a diagenetic product and therefore must be younger than the age of the rock unit in which it is located. Also, glauconite grains (ovoid, greenish particles) tend to lose some of the radiogenic argon generated by decay of the ^{40}K in them, depending on the details of mineralogic composition of the glauconite. This uncertainty is of the order of 10^6-10^7 years.

Strontium Isotopic Dating

As was noted previously, ^{87}Rb, which makes up 28% of the total rubidium in the crust, undergoes radioactive decay to form ^{87}Sr. Natural strontium consists of three other stable isotopes, ^{84}Sr, ^{86}Sr, and ^{88}Sr, and the addition of ^{87}Sr by the continual decay of unstable ^{87}Rb causes the $^{87}Sr/^{86}Sr$ ratio to change through time. If the $^{87}Sr/^{86}Sr$ ratio changes in a regular or predictable manner through geologic time, it may be possible to obtain numerical ages of strontium-bearing sediments by means of strontium isotopic analyses. Burke et al. (1982) made precise measurements of 786 apparently unaltered marine carbonate, evaporite, and phosphate samples of known geologic age to obtain a curve of seawater $^{87}Sr/^{86}Sr$ variation through the

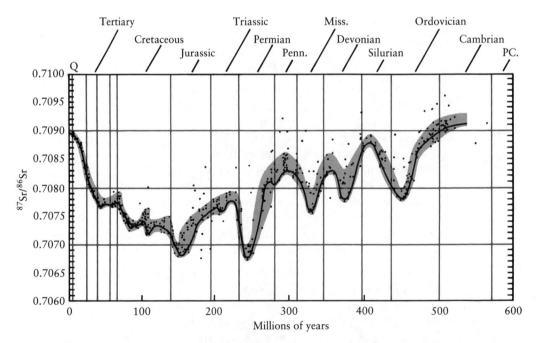

FIG. 4-5 • Variation of $^{87}Sr/^{86}Sr$ ratio through Phanerozoic time. An updated, detailed set of data points for the past 100 m.y. is provided by Elderfield, (1986). (Reprinted by permission of the author from Burke et al., *Geology*, v. 10, 1982.)

Phanerozoic (Fig. 4-5). The gray band around the line through the points includes 93% of the data points. Points outside the band probably represent either nonmarine samples that did not obtain the $^{87}Sr/^{86}Sr$ of contemporaneous seawater or diagenetically altered marine samples whose strontium was contaminated by foreign strontium with a different ratio.

The explanation for the temporal variation in $^{87}Sr/^{86}Sr$ ratio lies in the varying contribution through time of strontium isotopes from isotopically distinct crustal sources. The two prominent sources are the weathering of old sialic rocks of continental interiors ($^{87}Sr/^{86}Sr$ average of 0.720), mafic volcanic and intrusive rocks from active plate margins that average 0.704, and marine carbonates and sulfates that average 0.708. High $^{87}Sr/^{86}Sr$ ratios, then, indicate a dominance of old sialic sources in determining the marine isotopic ratio, and low values indicate a dominance of mafic sources. Thus, the shape of the $^{87}Sr/^{86}Sr$ curve can be influenced significantly by the history of plate interactions and by rates of sea floor spreading throughout the Phanerozoic.

Magnetostratigraphy

Magnetostratigraphy is one of the most recently developed techniques of use to the stratigrapher for correlation purposes (Prothero, 1988). The principle on which the technique is based is that during cooling of hot rock or during deposition of sediments, the magnetic poles of iron oxide minerals align themselves with the existing magnetic field. During crystallization of an igneous or metamorphic rock the direction of magnetization of magnetite will usually be the same as the direction of grain elongation, if any. In the sedimentary environment, moving currents will rotate such unfractured elongate grains of magnetite into alignment with the existing magnetic field. If a grain has been fractured so that its long dimension is no longer coincident with its magnetic lineation, however, the water or wind currents may not produce a good magnetic fabric in the sediment. Fortunately, the deposited grains will be rotated into magnetic parallelism by the strength of the magnetic field as they rest at the sediment-water interface. Equant grains will also be physically rotated into alignment by the strength of the magnetic field as they settle. The result of these grain alignments is a **remanent magnetism**.

Studies of lava flows have revealed that all flows younger than 690,000 years have **normal polarity**, with the north-seeking magnetic pole pointing in the direction of the present magnetic north pole. Lavas between 690,000 and 890,000 years in age have **reversed polarity**; the north-seeking pole in the magnetic minerals of the rock point in the direction of the present south magnetic pole. It is now generally accepted that these differences are due to the Earth's magnetic field having reversed itself in the past, and on the basis of results from many paired radiometric and polarity analyses of lavas and deep-sea sediment cores, it has been found that the frequency of polarity reversals is irregular but ranges between 10^4 and 10^7 years. A geomagnetic time scale has been established that correlates in detail times of nor-

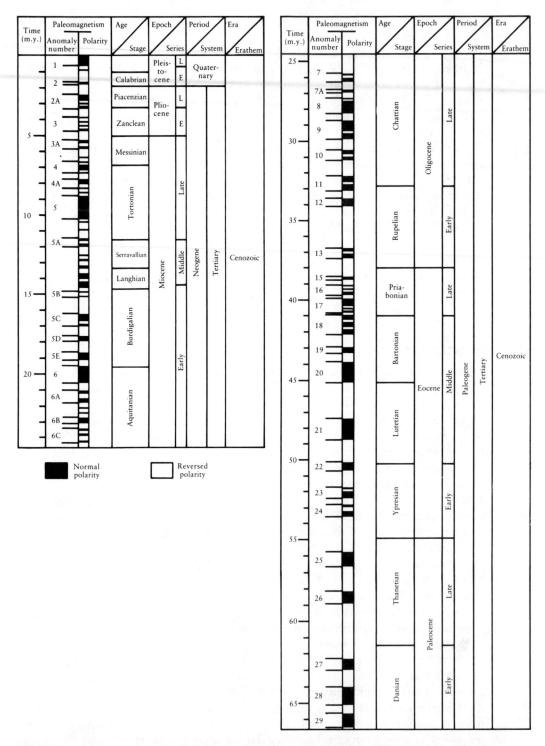

FIG. 4-6 • Magnetic polarity time scale for Tertiary and Cretaceous rocks based on correlations between polarity reversals and radiometric dates; data from oceanographic studies. Tarling (1983) gives analogous data for Jurassic rocks. (Reprinted by permission of the author from Lowrie and Alvarez, *Geology*, v. 9, 1981.)

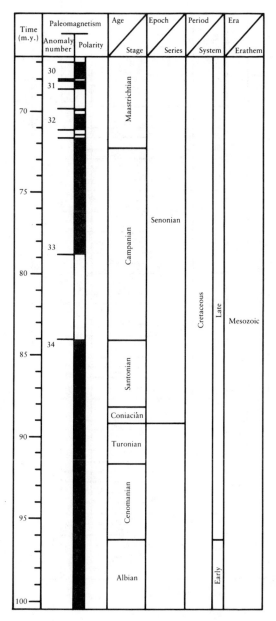

FIG. 4-6 • (*continued*)

mal and reversed polarity for the past 10^8 years with the K-Ar ages of the rocks (Fig. 4-6). It should be noted, however, that magnetic polarity in minerals is established at the moment of crystallization, but radioactivity clocks are set somewhat later, at the time the closure temperature is reached during cooling of the rock. In instances in which the same normal or reversed interval that is dated radiometrically can also be recognized in fossil-bearing rock, the numerical age of a fossil zone or

Left panel

Eonothem	Erathem	System	Series	Stage	Geochronometry Ma BP	Magneto-strat.
PHANEROZOIC	PALEOZOIC / Upper Paleozoic	Permian	Upper	Tatarian	255	
				Kazanian	260	
				Kungurian	270	
			Lower	Artinskian		
				Sakmarian	280	
				Asselian	290	
		Carboniferous / Upper Subsystem	Stephanian	Barruelian		Mainly Reverse
				Cantabrian	300	
			Westphalian	Bolsovian		
				Duckmantian		
				Langsettian	310	
			Namurian	Yeadonian		
				Marsdenian		
				Kinderscoutian		
				Alportian		
				Chokierian		
				Arnsbergian		
				* Pendleian	325	
		Lower Subsystem	Visean	Brigantian		
				Asbian		Mixed
				Holkerian		
				Arundian		
				Chadian		
			Tournaisian	Ivorian		
				Hastarian	355	
		Devonian	Upper	Famennian		
				Frasnian	375	
			Middle	Givetian		
				Eifelian		
			Lower	Emsian	390	Mainly Reverse
				Pragian		
				Lochkovian	410	
		Silurian	Pridoli			Mixed
			Ludlow	Ludfordian		
				Gorstian	424	Reverse
			Wenlock	Homerian		
				Sheinwoodian	428	
			Llandovery	Telychian		
				Aeronian		
				Rhuddanian	438	
	Lower Paleozoic	Ordovician	Upper / Cincinnatian (Ashgill)	Hirnantian		
				Rawtheyan		
				Cautleyan		
				Pusgillian	446	Mainly Normal
			Champlainian (Caradoc)	Onnian		
				Actonian		
				Marshbrookian		
				Longvillian		
				Soudleyan		
				Harnagian		
				Costonian	455	
			Lower	Llandeilo-Llanvirn	455 / 470	
			Arenig (Can.)	Fennian-Moridunian	490	Mixed
			Tremadoc		510	
		Cambrian	Upper	Trempealeauan		Mainly Reverse
				Franconian		
				Dresbachian		
			Middle	Mayaian		
				Amgaian		
			Lower	Toyonian		Normal
				Botomian	530	
				Atdabanian	550	Mainly Reverse
				Tommotian	570	

Right panel

Eonothem	Erathem	System	Series	Stage	Geochronometry Ma BP	Magneto-strat.
PHANEROZOIC	CENOZOIC	Quaternary	Holocene			Brunhes (Magnetochron. 1)
			Pleistocene	Upper		Matuyama (Magnetochron. 2)
				Middle		
				Lower	1.6	
		Neogene	Pliocene	Upper Piacenzian	3.3	Gauss (3)
				Lower Zanclean	5.3	Gilbert (4)
			Miocene	Upper Messinian	6.5	
				Tortonian	11	
				Middle Serravallian		
				Langhian		
				Lower Burdigalian	19	
				Aquitanian	23	
		Paleogene	Oligocene	Chattian	27	
				Rupelian	(36.5)	Mainly Mixed (Magnetochrons. 5-34)
			Eocene	Priabonian	34	
				Bartonian	39	
				Lutetian	45	
				Ypresian	53	
			Paleocene	Thanetian	59	
				Danian	65	
	MESOZOIC	Cretaceous	Upper	Maastrichtian	65	
				Campanian	83	
				Santonian	86	Predominantly Normal
				Coniacian	88	
				Turonian	91	
				Cenomanian	95	
				Albian	107	
			Lower	Aptian	114	
				Barremian	116	
				Hauterivian	120	Mixed (Anomalies M1-M29)
				Valanginian	128	
				Berriasian	135	
		Jurassic	Upper	Tithonian	139	
				Kimmeridgian	144	
				Oxfordian	152	
				Callovian	159	Mainly Normal
			Middle	Bathonian	170	
				Bajocian	176	
				Aalenian	180	Mainly Mixed
			Lower	Toarcian	188	
				Pliensbachian	195	
				Sinemurian	201	
				Hettangian	205	
		Triassic	Upper	Rhaetian	210	Mainly Normal
				Norian	220	
				Carnian	230	
			Middle	Ladinian	235	Reverse
				Anisian	240	
			Lower	Scythian	250	

FIG. 4-7 • Geologic time scale. Newer data that shift the dates of some boundaries by a few million years are given by Carr et al., 1984, Geology, v. 12, p. 276; and by Odin, 1986, Chem. Geol., v. 59, p. 107. The most significant change suggests the base of the Cambrian to be at 530–550 m.y. (Reprinted by permission from *Geology*, 1983, v. 11, p. 504.)

stage may be determined (Fig. 4-7). Zones and stages developed in deep ocean core studies must be equated with zones and stages erected for use in shelf sea deposits and those developed in examination of terrestrial deposits. Pollen, spores, and mammalian bones and teeth commonly have been used to erect time divisions of Cenozoic terrestrial deposits. Benthonic foraminifers and molluscs have proven the most useful in fossil-based time division of Cenozoic shelf sea deposits. Planktonic foraminifers, coccolithophorans, diatoms, and radiolarians are used in developing zones and stages of open ocean sequences. Time synchroneities among these different fossil-based time divisions commonly are most easily established by using radiometric dates. Snelling (1985) shows the relationships between several fossil-based zones and stages and radiometric ages as well as the magnetic field reversal chrons.

Studies of the strength of magnetization of rocks formed during a reversal in polarity indicate that the field decreases in intensity by a factor of ten during a reversal and that the switching of the poles takes about 4000–5000 years to be completed. Thus the stratigraphic resolution of the technique is better than the best paleontologic data. In addition, the appearance of a polarity reversal at the Earth's surface following its generation in the core may be nearly synchronous at all locations because of its origin in the center of a near-sphere.

Changes in the polarity of the magnetization of rocks are relatively easy to measure, as high accuracy is not required (Tarling, 1983). If two stratigraphic sequences accumulated at similar rates, matching of polarity reversal sequences is relatively easy and reliable. But different rates of sedimentation can create apparent increases or decreases in the lengths of individual polarity intervals, rendering correlation more difficult unless some other marker horizon is present and the interruptions in the sequences are not so great that entire polarity zones are omitted. The presence of other marker horizons, such as a specific biomarker, may also mean that numerical dating is possible, thus making possible determination of rates of deposition.

At present the polarity reversal scale is well developed only as far back in time as the late Middle Jurassic and is based largely on analysis of deep-sea pelagic sediments. These sediments have the advantages that (1) they contain relatively few discontinuities and these are mostly of short duration; (2) rates of deposition are relatively constant; (3) they have not been remagnetized by subsequent metamorphism; and (4) they are intruded by sheeted dikes that can be radiometrically dated. In contrast, the hemipelagic, near-shore marine, and continental sediments exposed on continental blocks are dominated by rock sequences containing numerous, temporally significant unconformities; were deposited at vastly differing rates; and commonly have a magnetic fabric that has been overprinted by later events. Nevertheless, some polarity reversal stratigraphy is available for pre-Jurassic rocks, one example being for rocks of Carboniferous age (Palmer et al., 1985).

In 1972 a subcommission on a Magnetic Polarity Time Scale was established as part of the International Commission on Stratigraphy. The deliberations of this group, plus those of the Subcommission on Stratigraphic Classification, have resulted in a formal nomenclature for magnetostratigraphy, analogous to rock- and time-stratigraphic nomenclature (Table 4-2).

TABLE 4-2 • Nomenclature of magnetostratigraphy

Magnetostratigraphic Units	Geochronologic Units	Approximate Durations
Polarity subzone	Subchron	$10^4 - 10^5$ years
Polarity zone	Chron	$10^5 - 10^6$ years
Polarity superzone	Superchron	$10^6 - 10^7$ years

A wide variety of igneous and sedimentary rocks record the direction of the geomagnetic field at the time of their formation, during sedimentation or igneous cooling. Polarity stratigraphy can be used as a means of correlation not only between sedimentary sections with different biozonations, but also between radiometrically dated lava sequences and paleontologically dated sedimentary sequences. Magnetic stratigraphy provides a means of correlating different biozonations with each other and with radiometrically determined numerical ages. For example, Ward et al. (1983) examined outcrops of Late Cretaceous age on opposite sides of the Great Valley in northern California. Biostratigraphic correlation between the two areas, about 100 km apart, had been imprecise because of their very different facies and different fossils. The eastern outcrops are composed of cross-bedded coarse clastics deposited on the continental shelf and were zoned on the basis of ammonites, gastropods, and bivalves. In contrast, outcrops on the western side of the valley are largely shale and turbidites that rarely contain molluscs but are rich in foraminifera and radiolaria. Biostratigraphic zonation of these western deposits is based on the microfossils. Ward et al. discovered an interval of reversed magnetic polarity on both sides of the valley that is associated with the same ammonite local range zone. The zone is known from other areas to be of Campanian age, and the Campanian Stage is known to contain only one polarity reversal. The reversed interval thus can serve as a marker horizon linking the different faunal assemblages and sedimentary facies of the Great Valley Sequence. It can also be used to extend the correlation to parts of the Upper Cretaceous section with relatively little biostratigraphic control and to tie these on-land sections to marine magnetic anomalies found during deep-sea drilling in the world ocean.

In passing, we might note that magnetostratigraphy may be a method of resolving the perennial stratigraphic problem of defining the Precambrian-Cambrian boundary. Preliminary studies of polarity sequences across this boundary in both central Australia and Nevada reveal that an uninterrupted reversed polarity zone is present immediately below the possible boundary and is followed above the boundary by a zone of rapidly changing polarities. Because of the very rapid and globally synchronous character of magnetic polarity reversals, they can be used to resolve many of the ubiquitous debates over whether a rock unit is "uppermost Reaganian" or "lowermost Bushian."

Fission-Track Dating

The radioactive isotopes of uranium decay spontaneously, and as the fragments produced travel through a solid medium, they leave a trail of crystallographic damage in the host mineral. These disordered channels within the crystal are termed fission tracks. When the crystal is suitably etched and examined using a petrographic microscope, the tracks are seen to be about 20 μm in length and can occur in amounts exceeding 10^6 tracks/cm^2. The number of tracks is a function of the uranium concentration and the temperature history of the mineral. Once the uranium content is determined, the thermal age of the mineral is determined by counting the number of tracks per unit area. Fission tracks are, like any radiation damage, very sensitive to elevated temperatures, and above a certain temperature characteristic of each mineral the tracks are "healed" by annealing.

The most important mineral for stratigraphic applications of fission track dating is zircon, whose relative abundance in sediments and relatively high uranium content allow age determination of single grains. Studies on borehole rocks indicate that temperatures of about 200°C prevailing for 10^8 years can significantly lower the fission track age of zircon. For this reason, estimates of maximum burial depth of the rock unit and geothermal gradient must be made before an apparent age from zircon fission tracks can be evaluated. Stratigraphic studies using fission tracks have been carried out on sedimentary and volcanic rocks of various Phanerozoic ages in many parts of the world. For example, Hurford et al. (1984) used modes in the frequency distribution of fission track ages from detrital zircon grains in two Lower Cretaceous sandstone units in England to identify changes in provenance from southerly to northwesterly.

Because of the known relationship between track fading, temperature, and burial depth, the fission track technique may also be useful in determining thermal histories of large basins of deposition, such as the Williston Basin (Crowley et al., 1985) in the north-central United States and south-central Canada.

Seismic Stratigraphy and Eustatic Sea Level Changes

One of the major advances in our understanding of temporally correlatable stratigraphic relationships has resulted from seismic analysis. In a series of now-classic papers, P. R. Vail and associates (*in* Payton, 1977) used the concept that seismic reflections follow chronostratigraphic boundaries to establish cycles of worldwide (eustatic) changes of sea level during Phanerozoic time. They observed that many regional cycles determined on different continental margins are simultaneous (within the limits of resolution of the technique), and the relative magnitudes of the changes are similar.

On the global curve of sea level changes detectable during Phanerozoic time, three major orders of cycles are present. First-order cycles have durations of

200–300 million years; second-order cycles, 10–80 million years; and third-order cycles, 1–10 million years. Two first-order cycles, at least 31 of the second order, and approximately 80 of the third order were distinguished, not counting Late Paleozoic cyclothems (Fig. 4-8) (see also Chapter 9). Documentation of these is, of course, better for Tertiary and Cretaceous time than for earlier periods because much of the data used for interpretations of the cycles come from the ocean basins. Vail et al. estimated that the sea level reached a high point near the end of Cenomanian time (Late Cretaceous) about 200 m above present sea level and reached a pre-Holocene low point about 50 m below the present level during Permo-Triassic time with a total range in sea level of 250 m. The cause of the first-order cycles and most of those of second order is believed to be change in the shape and volume of the ocean basins resulting from the evolution of midocean ridges and changes in rate of sea floor spreading. For example, the sea level has fallen steadily but at varying rates since the Late Cretaceous, apparently as a result of contraction in size of oceanic ridges related to decreased rates of sea floor spreading. At the same time, passive continental margins of the Atlantic and other ocean basins have subsided tectonically at decreasing rates, following a predictable thermal cooling curve. Numerical calculations are consistent with this interpretation. Some second-order and most third-order cycles may result from waxing and waning of glaciation, the causes of which are not well understood.

The broad outlines of the "Vail curves" and the recent modifications by Haq et al. (1987) are now generally accepted by stratigraphers as valid, but much of the data on which they are based are unpublished and remain in the files of the various oil companies as proprietary. Given the numerical uncertainties of both radiometric dating and biostratigraphic zonation, as well as local problems resulting from flexural subsidence and facies shifts, it is not always possible to use the Vail curves without modification (Burton et al., 1987). An example of the level of uncertainty in the Vail curve is shown in Figure 4-9.

Summary

Two types of geologic ages are obtained by stratigraphers, relative age and numerical age. When dealing with fossils the basic biostratigraphic unit is the fossil zone, a faunal or floral association that is different from those immediately above and below. A biostratigraphically distinctive group of zones is termed a stage. The most useful types of fossils for zonation are short-ranging species with great lateral extent, those not influenced by facies changes. The best temporal resolution of fossil zones is on the order of 10^5 years.

Radiometric dating has a precision of 5% to 10%, so for Neogene rocks is approximately the same as the best resolution possible for fossils, 10^5 years. For

▶ **FIG. 4-8** • Mesozoic-Cenozoic chronostratigraphy and cycles of sea-level change. (Reprinted by permission from the author and the publisher from Haq et al., *Science*, v. 235, p. 1159, 6 March 1987. Copyright 1987 by the AAAS.) (*continues*)

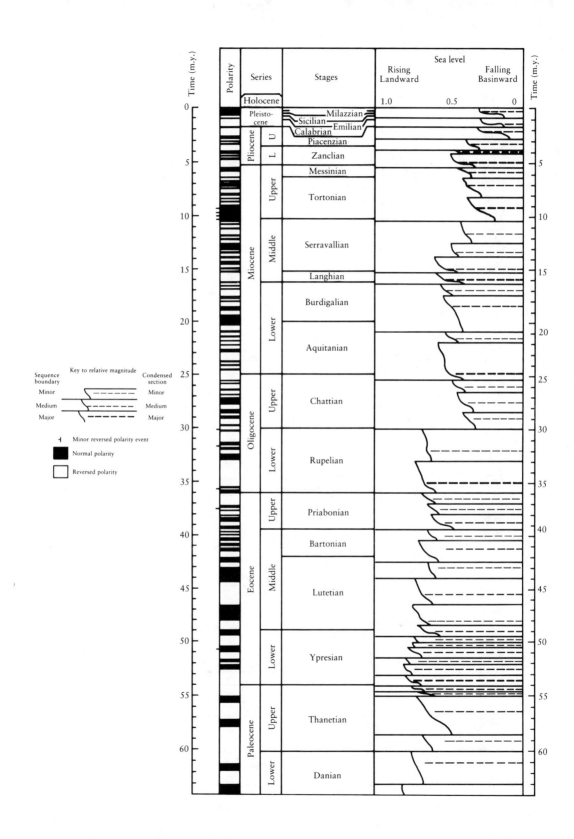

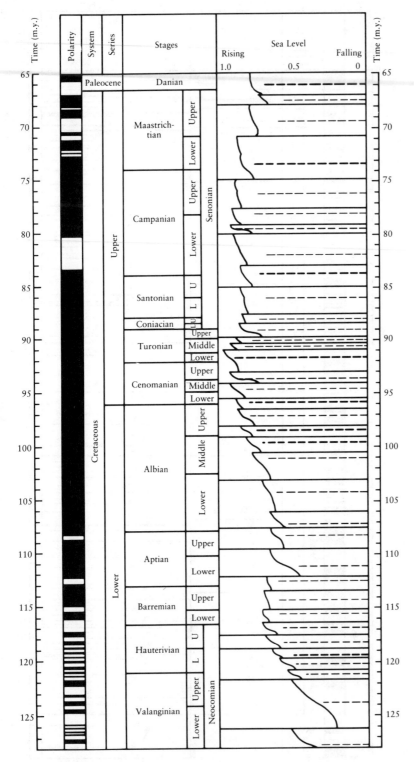

FIG. 4.8 • (*continued*)

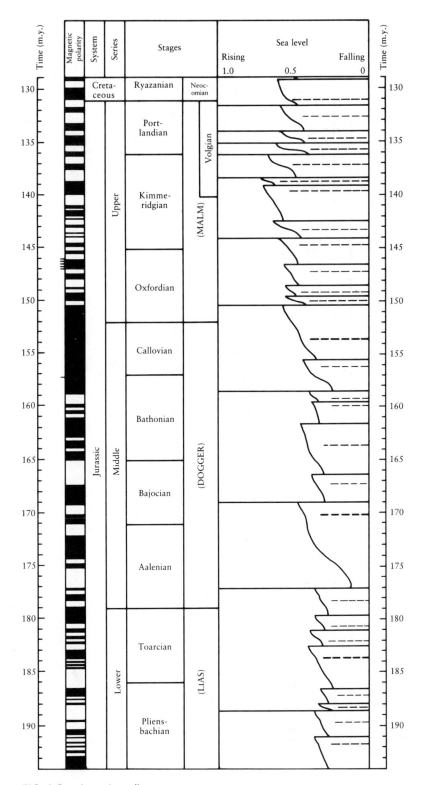

FIG. 4.8 • *(continued)*

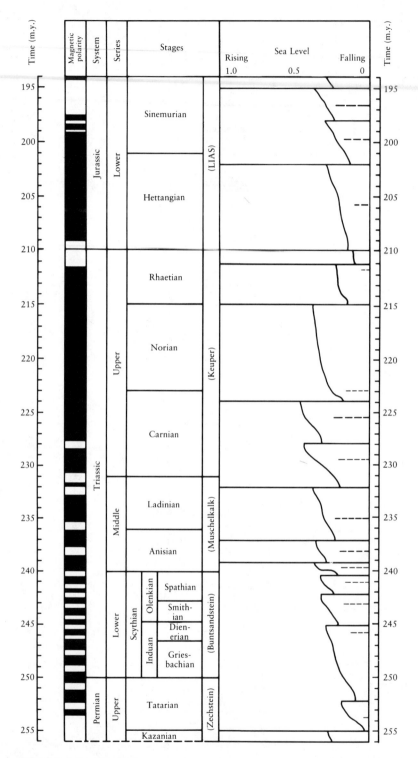

FIG. 4.8 • *(continued)*

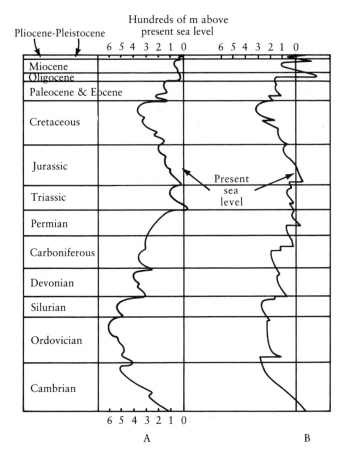

Hundreds of m above
present sea level

Pliocene-Pleistocene

6 5 4 3 2 1 0   6 5 4 3 2 1 0

Miocene
Oligocene
Paleocene & Eocene

Cretaceous

Jurassic

Present
sea
level

Triassic

Permian

Carboniferous

Devonian

Silurian

Ordovician

Cambrian

6 5 4 3 2 1 0

A   B

FIG. 4-9 ● Eustatic curves for the Phanerozoic. (A: Hallam, 1984. B: Vail et al., *in* Payton, 1977.)

older rocks, which form the bulk of the sedimentary stratigraphic record, the precision of radiometric dates is on the order of 10^6 or 10^7 years.

Numerical dating of sedimentary rocks using magnetic polarity reversals is used commonly in Cretaceous and Tertiary continuous marine sections and can have a resolution of 10^4 years in the best circumstances.

References

Anderson, R. Y., Dean, W. E., Jr., Kirkland, D. W., and Snider, H. I., 1972. Permian Castile varved evaporite sequence, west Texas and New Mexico. Geol. Soc. Amer. Bull., v. 83, pp. 59–86.

Anderson, R. Y. and Kirkland, D. W., 1966. Intrabasin varve correlation. Geol. Soc. Amer. Bull., v. 77, pp. 241–256.

Bayer, U., 1987. Chronometric calibration of a comparative time scale for the Mesozoic and Cenozoic. Geol. Rundschau, v. 76, pp. 485–503.

Berry, W. B. N., 1987. Growth of a Prehistoric Time Scale. Boston, Blackwell Scientific Publications, 202 pp.

Burke, W. H., Denison, R. E., Hetherington, E. A., Koepnick, R. B., Nelson, H. F., and Otto, J. B., 1982. Variation of seawater $^{87}Sr/^{86}Sr$ throughout Phanerozoic time. Geology, v. 10, pp. 516–519.

Burton, R., Kendall, C. G. S. C., and Lerche, I., 1987. Out of our depth: On the impossibility of fathoming eustasy from the stratigraphic record. Earth-Sci. Rev., v. 24, pp. 237–277.

Cohee, G. V., Glaessner, M. F., and Hedberg, H.D. (eds.), 1978. Contributions to the Geologic Time Scale. Tulsa, American Association of Petroleum Geologists, Studies in Geology No. 6, 388 pp.

Crowley, K. D., Ahern, J. L., and Naeser, C. W., 1985. Origin and epeirogenic history of the Williston Basin: Evidence from fission-track analysis of apatite. Geology, v. 13, pp. 620 623.

Dalrymple, G. B. and Lanphere, M. A., 1969. Potassium-Argon Dating. New York, W. H. Freeman & Co., 258 pp.

Elderfield, H., 1986. Strontium isotope stratigraphy. Paleogeog., Paleoclim., Paleoecol., v. 57, pp. 71–90.

Faure, G., 1986. Principles of Isotope Geology, 2nd ed. New York, John Wiley & Sons, 589 pp.

Hallam, A., 1984. Pre-Quaternary sea-level changes. Ann. Rev. Earth Planet. Sci., v. 12, pp. 205–243.

Haq, B. U., Hardenbol, J., and Vail, P. R., 1987. Chronology of fluctuating sea levels since the Triassic. Science, v. 235, pp. 1156–1157.

Haq, B. U., Hardenbol, J., and Vail, P. R., 1988. Mesozoic and Cenozoic chronostratigraphy and cycles of sea-level changes. In C. K. Wilgus et al. (eds.), Sea Level Changes—An Integrated Approach. Tulsa, Society of Economic Paleontologists and Mineralogists, Spec. Pub. No. 42, 400 pp.

Hurford, A. J., Fitch, F. J., and Clarke, A., 1984. Resolution of the age structure of the detrital zircon populations of two Lower Cretaceous sandstones from the Weald of England by fission track dating. Geol. Mag., v. 121, pp. 269–277.

Kauffman, E. G. and Hazel, J. E. (eds.), 1977. Concepts and Methods of Biostratigraphy. Stroudsburg, Pa., Dowden, Hutchinson and Ross, 658 pp.

Lowrie, W. and Alvarez, W., 1981. One hundred million years of geomagnetic polarity history. Geology, v. 9, pp. 392–397.

Mahaney, W. C. (ed.), 1984. Quaternary Dating Methods. New York, Elsevier Science Publishers, 431 pp.

Margaritz, M., 1985. The carbon isotope record of dolostones as a stratigraphic tool: A case study from the Upper Cretaceous shelf sequence, Israel. Sedimentary Geol., v. 45, pp. 115–123.

Odin, G. S., 1982. Numerical Dating in Stratigraphy. Part I. New York, John Wiley & Sons, 630 pp.

Palmer, J. A., Perry, S. P. G., and Tarling, D. H., 1985. Carboniferous magnetostratigraphy. J. Geol. Soc., v. 142, pp. 945–955.

Palmer, M. R. and Elderfield, H., 1985. Sr isotope composition of sea water over the past 75 Myr. Nature, v. 317, pp. 526–528.

Payton, C. E. (ed.), 1977. Seismic Stratigraphy—Applications to Hydrocarbon Exploration. Tulsa, American Association of Petroleum Geologists, Mem. 26, 516 pp.

Prothero, D. R., 1988. Mammals and magnetostratigraphy. J. Geol. Educ., v. 36, pp. 227–236.

Richardson, J. B. and McGregor, D. C., 1986. Silurian and Devonian spore zones of the Old Red Sandstone continent and adjacent regions. Geol. Assoc. Canada Bull., v. 364, pp. 1–79.

Sarna-Wojcicki, A. M., Morrison, S. D., Meyer, C. E., and Hillhouse, J. W., 1987. Correlation of Upper Cenozoic tephra layers between sediments of the western United States and eastern Pacific Ocean and comparison with biostratigraphic and magnetostratigraphic age data. Geol. Soc. Amer. Bull., v. 98, pp. 207–223.

Schnitzer, W. A. and Schwab, R. G., 1975. Neue Möglichkeiten erdgeschichtlicher Forschung mit Hilfe des Palaogeruches. Erlanger Geol. Abhandlungen, No. 101, pp. 1–19.

Scholle, P. A. and Arthur, M. A., 1980. Carbon isotope fluctuations in Cretaceous pelagic limestones: Potential stratigraphic and petroleum exploration tool. Amer. Assoc. Petroleum Geol. Bull., v. 64, pp. 67–87.

Snelling, N. J. (ed.), 1985. The Chronology of the Geological Record. Geological Society of London, Mem. 10, 343 pp.

Tarling, D. H., 1983. Paleomagnetism. New York, Chapman and Hall, 379 pp.

Ward, P. D., Verosub, K. L., and Haggart, J. W., 1983. Marine magnetic anomaly 33–34 identified in the Upper Cretaceous of the Great Valley Sequence of California. Geol., v. 11, pp. 90–93.

Watson, R. A., 1983. A critique of chronostratigraphy. Amer. J. Sci., v. 283, pp. 173–177.

Williams, D. F., Lerche, I., and Full, W. E., 1988. Isotope Chronostratigraphy: Theory and Methods. New York, Academic Press, 345 pp.

Wyatt, A. R., 1986. Post-Triassic continental hypsometry and sea level. J. Geol. Soc., v. 143, pp. 205–243.

Wyatt, A. R., 1987. Shallow water areas in space and time. J. Geol. Soc., v. 144, pp. 115–120.

Biologists work very close to the frontier between bewilderment and understanding.

Peter Medawar

5
Patterns in the Distribution of Organisms

The distribution and dispersal patterns of organisms in time and space are the subject matter of biogeography, a synthetic subject that depends on an understanding of topics as diverse as ecology, meteorology, oceanography, paleogeography, and plate tectonics at least. For example, most corals have in their tissues a symbiotic alga, which restricts these corals to very shallow depths where solar radiation is highest. Many land animals are cold-blooded and thus restricted in habitat to tropical and warm temperate areas of the Earth. Many planktonic and benthonic marine communities are localized by the distribution of a water mass with specific characteristics of temperature and salinity. Species of trees are limited by elevation; leafy trees occur in lowlands, pines dominate at higher elevations. And, of course, the ability of land animals to migrate is limited by the size and latitudinal distribution of land masses, which change in accordance with plate tectonics. Most organisms have rather narrow tolerances for environmental change.

The obvious difficulties in the interpretation of modern biogeographies are increased greatly when dealing with the ancestors of existing species. Were soft-bodied symbionts that are never preserved important in limiting the geographic spread of a fossil group? Has selective preservation of mineralogically different hard parts affected our ability to define the range of an extinct organism? Does a species appear to become extinct before its actual demise only because of a permanent change in the environment seen at an outcrop? Has there been biochemical evolution in a lineage that affected its distribution but that caused no apparent change in the shell architecture used by paleontologists to define the species? How did the post-Jurassic

121

origination of the Gulf Stream affect the distribution of marine organisms? Perhaps the most difficult factor to evaluate is the knowledge that the fossils found in a bed, or even along a single bedding plane, may not have coexisted. As we have seen (Chapter 3), even an infinitely thin surface such as a bedding surface represents a condensation of sediment accumulated over an extended period of time. Thus fossil collections, no matter how well collected, include organisms that lived in the area over a period of several years or possibly several centuries or millennia. So even if the shells have not been transported after death, there may be a greater variety of plants and animals on a square meter of bedding plane than existed at any instant while that surface was exposed on the sea floor. All fossil assemblages are the result of condensations of assemblages accumulated over a considerable span of time. Biostratigraphic boundaries are inherently fuzzy.

Biogeography and Biostratigraphy

Biostratigraphy is essentially biogeography through geologic time, biogeography with the additions of changes in astronomical factors and the latitudinal location of land masses. With regard to astronomical factors we are limited largely to speculations (Chapters 3 and 10). We are aware of geologically short-term variations (Milankovitch cycles) and the possible effect of bolide impacts on fauna and flora (the Cretaceous-Tertiary extinction controversy), but it is reasonable to suppose that many other astronomical influences on life remain undetected so far. For example, radiolarian extinctions have been successfully correlated with reversals in the Earth's magnetic field (Hays, 1971), times when the field intensity was very low. Perhaps other organisms were affected as well but in a less obvious way. The next few decades will see an explosion of interest in viewing the Earth as part of a larger astronomical framework: magnetic field, gravitational field, cosmic rays, and so on. The benefits of such an approach may be as revolutionary to our thinking about biostratigraphic problems as was the introduction of the theory of plate tectonics.

Plate movements have been a part of the Earth's history since at least Proterozoic time, and the changes that plate movements have wrought on biogeography can only be described as revolutionary. On the grossest scale (Fig. 5-1) the theory states that about 200 m.y. ago, all the major land masses were joined, and land was continuous in a north-south direction from pole to pole (Pangaea). By 180 m.y. ago, Pangaea had split into two major land areas, termed Laurasia (Northern Hemisphere) and Gondwana (Southern Hemisphere), and Gondwana had begun to fragment into three segments (South America–Africa, Antarctica-Australia, and India). By the end of Jurassic time, 135 m.y. ago, South America and Africa were separating, and by the end of the Cretaceous Period, 65 m.y. ago, North America and Europe were parting, and Australia was soon to split from Antarctica. Simultaneously, Africa, India, and Australia were drifting northward. And, of course, islands, isthmuses, and other small and now undetectable units of land were no doubt coming and going during this time span. An added complication is that numerous epicontinental sea-

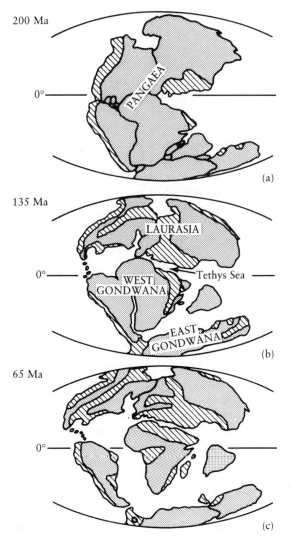

FIG. 5-1 • Gross paleogeography of the Earth showing the breakup of Pangaea and the positions of epicontinental seas. Within the accuracy of available data, land was continuous between the poles 200 million years ago. Mobile terrestrial organisms could migrate freely, restricted only by their environmental tolerances. By the close of Triassic time, the Tethys seaway was present just north of the equator, separating land faunas. Shallow-water marine organisms could colonize freely between "south China" and "Mexico." Sixty-five million years ago, at the close of Cretaceous time, "Canada" and "Scandanavia" were still joined, as were Antarctica and Australia. During Tertiary time, movements of land masses increasingly localized less mobile land faunas. (Reprinted by permission of John Wiley & Sons, Inc., from Pielou, *Biogeography*, copyright © 1979.)

ways were present on the continental blocks during the drifting process, and some of them totally bisected the land area of some plates.

Different kinds of barriers differ from each other in their imperviousness, that is, in the extent to which they prevent an interchange of biota across them; also, they affect different kinds of organisms differently. Even a narrow salt water strait is an impassable barrier to fresh water fishes, but plants with wind-dispersed seeds can cross easily. A stretch of frozen sea is a barrier to small mammals but not to large ones. The degree of imperviousness does, of course, change with time. For example, the lowering of the sea level that accompanied Pleistocene glaciations (and all others in the geological past) resulted in movements of land organisms across previously impassable zones, as in the Bering Strait area or among many of the islands in Indonesia.

Within the mass of the world ocean, changes in the variables that affect biogeography were equally great (Schopf, 1980; Hsü and Weissert, 1985). The relatively narrow, sima-bottomed, east-west seaway in the Northern Hemisphere between the equator and 30°N latitude (Tethys) must have had pinchouts at various longitudes from time to time during its 130-m.y. existence, just as the Mediterranean Sea had during Miocene time (the Messinian "salinity crisis"). As the South Atlantic Ocean came into existence during late Jurassic time, as South America separated from Africa, new cold water currents (Benguela, Falkland) and warm water currents (South Equatorial, Brazil) were generated, and as the North Atlantic opened later in the Mesozoic, the warm Gulf Stream formed. Schopf (1980) and Hay (1988) provide extended consideration of the more general topic of paleoceanography.

Faunal Character

Because of the restricted ecologic tolerances of organisms, climatic variations through time, orogenic activity, and plate movements, certain associations of plants and animals are unique to specific geographic areas. Koalas, kangaroos, and eucalyptus trees characterize Australia, a result of plate movements in early Tertiary time. Big-eared elephants, hippopotami, and wildebeest are present in central Africa but not in northern Africa or Europe, a result of aridity brought on by Pleistocene climatic changes in north Africa. The Amazon jungle is home to a multitude of plants and animals not found elsewhere, a result stemming partly from the drift of central South America into the humid tropics as Gondwana fragmented and partly to migrations of organisms through the Central American isthmus during Tertiary time (Stehli and Webb, 1985). For most paleobiogeographic patterns, however, the explanations are unclear or unknown (Briggs, 1987). Indeed, even the paleobiogeographic patterns themselves are typically poorly defined. But knowledge of ancient biogeographies is essential to biostratigraphy because it limits the geographic area in which a set of zones can be recognized. Obviously, if a set of zones and stages in an area is based on the evolutionary development of marsupials, ferns, or clams and such fossils are absent in an adjacent area, correlation between the two regions must be based on other types of fossils. Indeed, Oppel encountered this limitation in the mid-1800s when he tried to use in Britain his zonal sequence of Jurassic ammonites developed in Germany. The most useful time-stratigraphic zonal groupings are those based on fossils from deposits formed in many different environments, ideally with some overlap in species among adjacent environments. For example, in a marine setting with sediment supplied by turbidity currents, both shallow-water and deep-water faunas of the same age may occur together. In an analogous manner, plant spores may be carried by the wind into near-shore marine environments, providing a mixed terrestrial–shallow marine time-stratigraphic correlation.

Realms, Regions, and Provinces

Distribution patterns of modern organisms are divided by biogeographers into areas termed—from largest to smallest—**realms**, **regions** (and subregions), and **provinces** (Table 5-1). Each of these divisions is intended to outline areas with a considerable uniformity and distinctiveness of fauna and/or flora (Fig. 5-2), but because such

TABLE 5-1 • Modern terrestrial zoogeographic realms, regions, subregions, and provinces

Realms	Regions	Subregions	Provinces
Arctogean	Holarctic	Arctic	Arctic
		Nearctic	Canadian / Appalachian / West American / Sonoran
		Caribbean	Central American / West Indian
		Palearctic	European / Siberian / Manchurian / Tibetan / Mediterranean / Eremian
		Oriental	Indian / Ceylonese / Indo-Chinese / Malayan / Celebesian
	Paleotropical	Ethiopian	West African / South African
		Malagasy	Seychellian / Madagascan / Mascarene
Neogean	Neotropical	Neotropical	Amazonian / East Brazilian / Chilean
Notogean	Australian	Australian	Australian
		Papuan	Papuan
	Oceanian	New Zealandian	New Zealandian
		Oceanic	Oceanic
		Antarctic	Antarctic

(Reprinted by permission of the publisher from Schmidt, *Quarterly Review of Biology*, Stony Brook Foundation, 1954.)

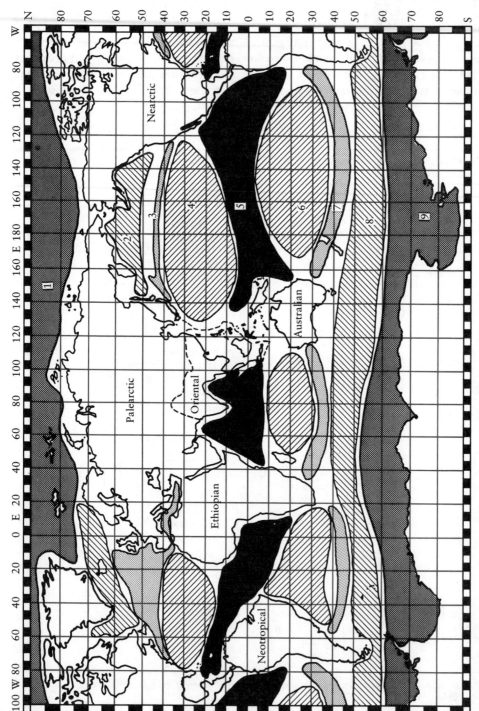

FIG. 5-2 • Biogeographic regions of the continents and world ocean; boundaries are approximate. 1 = arctic; 2 = subarctic; 3 = northern temperate; 4 = northern subtropical; 5 = tropical; 6 = southern subtropical; 7 = southern temperate; 8 = subantarctic; 9 = antarctic. (Adapted from Backus, *in* Pierrot-Bults et al., 1986.)

boundaries are almost invariably gradational, both the location and existence of many boundaries are disputed.

Clearly, as we go back in geologic time, the boundaries of all divisions shift (Fig. 5-3), but the concept of regions, subregions, and provinces is a useful one and is widely used in biostratigraphy. Biostratigraphic zones appear to be limited in extent to a single biogeographic province. Because stages are groups of two or more zones, a larger faunal aggregate is involved in characterization of a stage. The larger number of taxa in a stage means that stages generally extend into subregions or even regions.

As categorized by Pielou (1979), there are three basic methods by which organisms spread from an area in which they are indigenous. The methods are termed jump-dispersal, diffusion, and secular migration.

Jump-dispersal is the movement of individual organisms across great distances, followed by the successful establishment of a population of the original dispersers' descendants at the destination. The journey is completed in a period of time that is much shorter than the life span of an individual organism, and the journey usually takes the migrants across totally inhospitable terrain. An example is plant spores carried across a narrow seaway.

Diffusion is the gradual movement of populations across hospitable terrain for a period of many generations. Well-known examples of diffusion include the spread of *Homo sapiens* across the Bering Strait during Pleistocene glacial episodes of low sea level, the spread of many groups of land vertebrates both north and south as the Central American land bridge formed 3–3.5 m.y. ago, and the currently infamous spread northward from Brazil of originally African "killer bees." As this book goes to press, these bees have reached northern Mexico.

Secular migration is diffusion occurring so slowly that the diffusing species undergoes appreciable evolutionary change in the process. The species migrates into new environments, and if migration is sufficiently slow, the environments themselves may undergo continuous secular change. Natural selection acting on the migrants therefore causes the descendant population in a new region to differ from its ancestral population in the source region. For example, South American members of the camel family such as the llama are descended from now extinct North American ancestors that made a secular migration over the newly formed Isthmus of Panama.

As defined by Pielou (1979), the essential difference between jump-dispersal and diffusion is that "pure" diffusion consists of the crossing of a totally hospitable terrain in a series of short steps, one per generation for many generations. Jump-dispersal consists of the crossing of a totally *inhospitable* terrain in a single jump. Clearly, however, the terrain to be covered by a spreading species can have any degree of hospitableness between the two extremes, raising the question of the degree of impermeability of the ecologic or geographic barrier. The most obvious barriers are arms of the sea that completely separate land masses. These, of course, come and go during advances and retreats of epicontinental seas and as the continental blocks are moved around during plate motions. Other barriers include the emergence of orogenically produced mountain ranges, climatic zones existing as a function of lati-

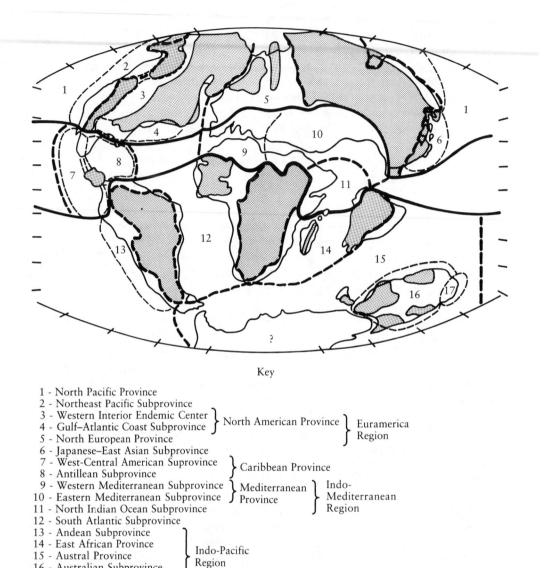

Key

1 - North Pacific Province
2 - Northeast Pacific Subprovince
3 - Western Interior Endemic Center ⎫
4 - Gulf–Atlantic Coast Subprovince ⎬ North American Province ⎫ Euramerica
5 - North European Province   ⎭   ⎬ Region
6 - Japanese–East Asian Subprovince
7 - West-Central American Suprovince ⎫ Caribbean Province
8 - Antillean Subprovince   ⎭
9 - Western Mediterranean Subprovince ⎫ Mediterranean Indo-
10 - Eastern Mediterranean Subprovince ⎭ Province   Mediterranean
11 - North Indian Ocean Subprovince   Region
12 - South Atlantic Subprovince
13 - Andean Subprovince   ⎫
14 - East African Province   ⎪
15 - Austral Province   ⎬ Indo-Pacific
16 - Australian Subprovince   ⎪ Region
17 - New Zealand Subprovince   ⎭
––– Subprovince Boundaries
▬▬ Province-Region Boundaries
▬▬ Realm Boundaries

FIG. 5-3 • Average distribution of Cretaceous marine biogeographic provinces and regions based on Bivalvia. Positions of land masses are based on paleomagnetic reconstructions. Shaded areas are a composite plot of continental areas not inundated through all Cretaceous stages. The differences between Figs. 5-2 and 5-3 result primarily from plate motions. (Reprinted by permission of the publisher from Kauffman, *in* Hallam (ed.), *Atlas of Paleobiogeography*, Elsevier Science Publishers, 1973.)

tude, and changes in ocean currents caused by plate movements. Pielou (1979) discusses these questions in detail, particularly for modern organisms, and concludes that it is commonly very difficult to explain the biogeographies of many groups. The difficulties are greatly compounded when dealing with ancient or extinct groups, as biostratigraphers do.

Disjunctions

Disjunct or discontinuous geographic ranges of species and higher taxa are quite common among modern organisms (Fig. 5-4) and have been described frequently from ancient faunas and floras as well. Separation of an initially continuous range into widely separated parts can result from geomorphological changes, climatic changes, evolutionary differentiation plus migration, or jump-dispersals. Examples of geomorphological disjunctions include the current restriction of most marsupials to Australia because of the fragmentation of Gondwana (both fossil and living marsupials occur in South America), and the separation of Caribbean from Pacific marine faunas caused by the formation of the Isthmus of Panama.

Climatic disjunctions can be caused by continental drift or crustal warping. Climatic deterioration can make the center of the range of a widespread taxon unin-

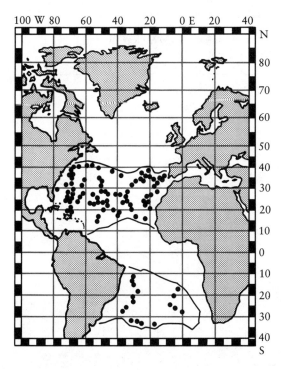

FIG. 5-4 • Disjunction in the distribution of the fish *Lepidophanes gaussi*, restricted to subtropical areas of the Atlantic Ocean. (Adapted from Backus, *in* Pierrot-Bults et al., 1986.)

habitable for it and leave the extremities as disjunct subranges. Some marine disjunctions stem from changes in the climate of the sea, its warm and cold currents. A rise in sea level over the Atlantic coastal shelf of North America during Holocene time appears to have affected the thermal stratification of the water and hence the environment of benthonic organisms. No doubt the loss of cold, dense bottom water from the ocean floor during interglacial epochs had similar effects in the deep ocean environment. Bottom water may also be heated sporadically by lava flows and the intense heat emanating from rifts on the ocean floor. No numerical data exist at present, so the effect of this potentially great heat transfer from the Earth's interior to the sea cannot now be evaluated. But its effect on marine organisms may be quite large.

Evolutionary disjunctions are generated when a pair of sister subspecies or species has become differentiated on opposite sides of the area occupied by their common ancestor and the ancestor then becomes extinct. The descendant taxa form a disjunct pair. In the marine environment this might occur because of the intrusion of a warm current into the central part of the range of a cold water species and separate evolutionary development of the disjunct populations through time.

Jump-dispersal disjunctions of terrestrial organisms are largely restricted to plant taxa. In the marine realm, such disjunctions are rare because the motile,

FIG. 5-5 • Paleobiogeographic map showing Ordovician trilobite provinces, which differ on the three major landmasses because of the extensive deep marine oceans separating them. The free-swimming larvae of trilobites apparently were unable to survive such extensive oceanic crossings. (Reprinted by permission of the publisher from Cocks and McKerrow, *in* McKerrow (ed.), *The Ecology of Fossils*, Gerald Duckworth and Co., Ltd., 1978.)

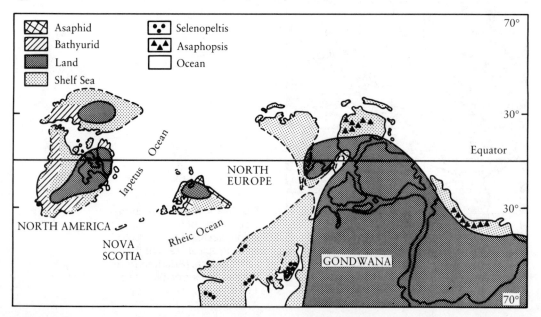

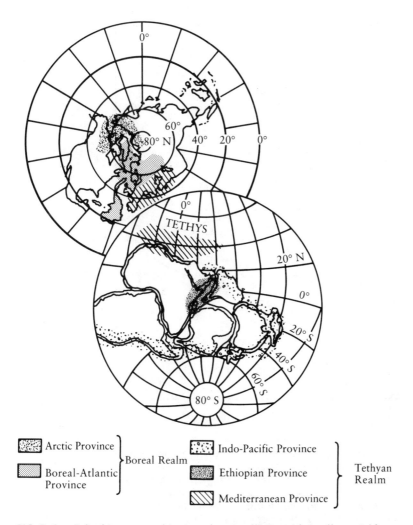

FIG. 5-6 • Paleobiogeographic map showing Kimmeridgian (Jurassic) belemnite provinces. The mid-Atlantic ridge is shown by the stippled pattern. (Reprinted by permission of the publisher from Stevens, *in* Hallam (ed.), *Atlas of Paleobiogeography*, Elsevier Science Publishers, 1973.)

migrating stages of most sedentary animals of the shelf benthos (such as molluscs and echinoderms) are immature larval stages and are much more delicate and short-lived than the spores and seeds of vascular plants. A few modern examples of trans-Atlantic jump-dispersal of mollusc genera are known, however.

Recognition of faunal and floral disjunctions in ancient rocks (Figs. 5-5, 5-6) can supply more detailed information about paleogeography and paleoclimate than are obtainable through continental relocations based on paleomagnetic data. The latter data are necessarily vague with respect to continental elevations, the locations

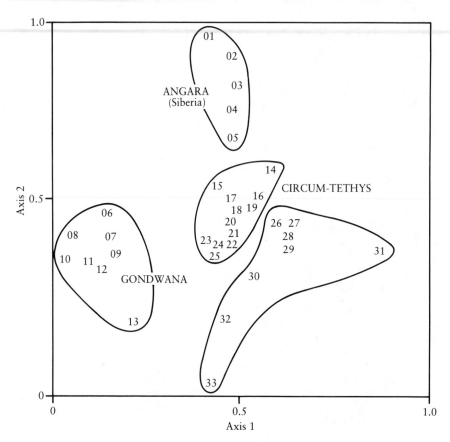

FIG. 5-7 • Polar ordination plot showing the generic relationships among Early Carboniferous plant compression assemblages from different geographic areas. Polar ordination is a statistical technique designed to reduce the number of variables being considered by making use of correlations among groups of variables. (A. M. Ziegler, et al., "Paleozoic Biogeography and Climatology," in Karl J. Niklas, ed., *Paleobotany, Paleoecology, and Evolution*, v. 2 (Praeger Publishers, New York, 1981), p. 252. Copyright © 1981 by Praeger Publishers. Reprinted with permission.)

of shorelines, and seaway pinchouts and need to be supplemented with paleobiogeographical information. Accurate information of this type is difficult to obtain, however. Distributions of ancient taxa commonly show a striking relationship to the geographic distribution of paleontologists' places of study, a function of the availability of outcrops, their accessibility, finances, and perhaps other factors. There also is a strong subjective element in some taxonomic assessments as well as an occasional lack of ecologic background on the part of some investigators. To reduce the subjective element, there is now a widespread use of statistics in biogeographical research (Fig. 5-7). As Hallam (1973) has observed, "the majority of skilled taxonomists, who alone have a mastery of relevant data, are temperamentally somewhat cautious and conservative, and strongly disinclined to indulge in what they would regard as unwarranted speculation."

Paleohypsometry and Paleobathymetry

Terrestrial organisms are distributed across the face of the Earth in patterns that reflect temperature and precipitation (Fig. 5-8), variables whose values can usually be correlated with latitude and elevation above sea level. The correlations are not perfect, however, as is evident from the equatorial glacier that caps the 6000-m

FIG. 5-8 • Distribution of (a) species of Crocodilia and (b) genera of palms (Palmae). The isopleths join points of equal taxon density. Latitudinal control is evident, for the reptiles because they are cold-blooded and for the palms because of a lack of resistance to frost. (Reprinted by permission of John Wiley & Sons, Inc., from Pielou, *Biogeography*, copyright © 1979.)

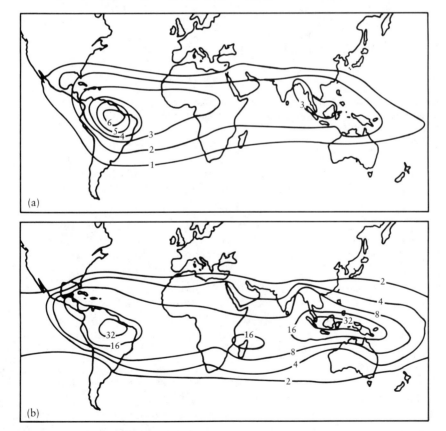

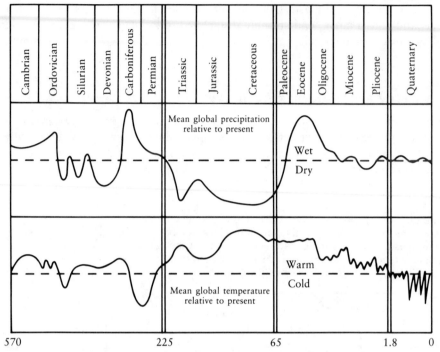

Precipitation and temperature fluctuations since the Precambrian (Ma)

FIG. 5-9 • Major fluctuations of global precipitation and temperature from the Cambrian to the present. Note progressive increase in time scale. (Adapted from Critchfield, H. J., 1983.)

summit of Mount Kilimanjaro in Kenya. Of perhaps greater importance in preventing perfect correlation between temperature and latitude are oceanic currents. For example, consider the difference in habitability between Labrador and England, both located largely between 52°N and 55°N latitude. The effect of the warm Gulf Stream water is apparent, even after it has flowed for 5000 km through cold waters between Florida and England. Figure 5-9 shows how the overall Earth's climate seems to have varied since Cambrian time, based on both biologic and lithologic evidence. The causes of these changes are largely unknown but include such imponderables as variations in solar radiation, changes in atmospheric composition (percentage of carbon dioxide, the "greenhouse effect"), variations in the Earth's albedo (reflectivity) because of changes in the relative amounts of land and sea area, and intermittent changes in the amount of volcanic dust in the atmosphere in response to plate tectonic interactions.

Temporal variations in oceanic bathymetry have received more attention from geologists than have variations in elevation of the land surface, a natural result of the fact that most preserved sediments are of marine origin. For the ocean basin as a whole, the average depth probably has remained fairly constant during Phanerozoic time at about 3800 m, with the positions of intraoceanic shallow areas determined by the temporally shifting locations of midoceanic ridges. More important to biostratigraphers than average depth is a numerically precise method of determining depth, or at least a method of determining relative depth. For most species, water temperature is the controlling factor in where the individual organism lives, although the presence of sunlight is critical for plants, and salinity or density variations may be important for a few taxa. Within the near-shore surf zone, robustness of the shell may be critical.

Georgia Coastal Region

The study of Dörjes (1972) of the distribution and zonation of macrobenthic animals off Sapelo Island, Georgia, provides an example of the existence of biozonation in the surf and nearshore zone. The environments examined included foreshore, shoreface, and offshore to a depth of about 15 m, 20 km from shore. Taxa with preservable hard parts included molluscs (90 species), crustaceans (67 species), echinoderms (13 species), coelenterates (9 species), and a few species of other types. The distribution, abundance, and density of animal species showed the following general patterns:

1. Backshore and backshore-foreshore transition composed of fine sand: Few terrestrial and marine migrant species; few individuals
2. Foreshore and shoals composed of fine sand: Few species and numerous individuals
3. Shoreface composed of fine sand (from mean low water line to water depth of 2.5 m): Increasing number of species; few individuals
4. Upper offshore formed of mud (between breaks in slope at depths of 2.5 m and 8 m): Numerous species and numerous individuals
5. Lower offshore formed of coarse sand (depth 8–15 m): Few species and few individuals

Four animal communities were recognized and named after their characteristic species; the two inshore communities were crustacean, the offshore two were based on echinoderm species. In ancient rocks the species and perhaps even the phyla might be different, but the relationship between relative numbers of species and individuals in each near-shore zone would probably remain the same. These variations depend on factors such as frequency of inundation for onshore zones, and salinity and substrate composition for the offshore zones.

Benthonic Foraminifera, Gulf of Mexico

Benthonic foraminiferan associations are remarkably sensitive to increasing depth and consequently are very useful in interpretations of paleodepth in Cretaceous to Holocene marine rocks. Poag (1981) reviewed the distribution of benthonic foraminifera in the Gulf of Mexico (Fig. 5-10), which can be viewed as a small ocean with depths up to 4300 m. The intertidal and shallow subtidal environments of the Gulf

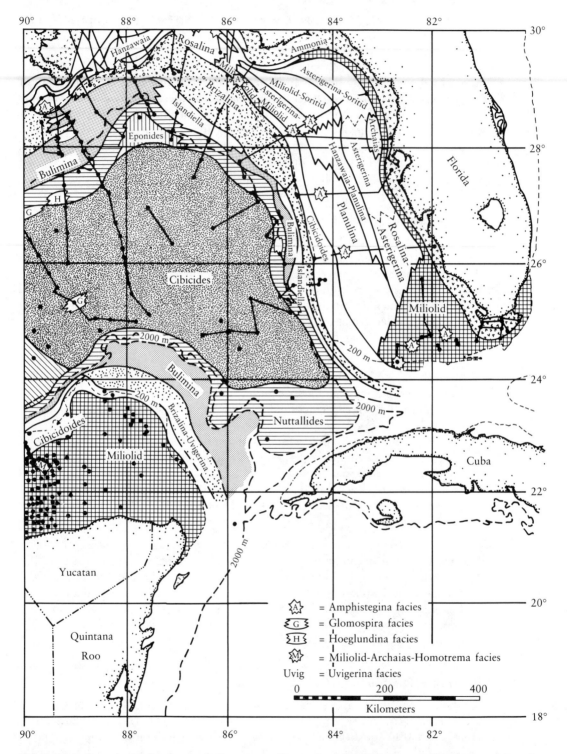

FIG. 5-10 ● Generic predominance facies among the marine benthonic foraminiferal community of the eastern Gulf of Mexico based on 3400 data points. (Reprinted by permission of the publisher from Poag, *Ecologic Atlas of Benthic Foraminifera of Gulf of Mexico*, Van Nostrand Reinhold, 1981.)

include a cluster of bays, barrier islands, estuaries, mangrove swamps, and deltas. Shelf depths range from the intertidal to as much as 200 m, though locally the shelf edge may be as shallow as 60 m. The relatively steeply sloping continental slope is found seaward from the edge of the shelf, which is about 200 m, to a depth of about 2000 m. The slope off Campeche and off Texas and Louisiana includes numerous ridges, domes, and basins, formed for the most part by subsurface salt masses, and the outer margin of the continental slope is cut by four major submarine canyons. The west-central part of the Gulf at depths greater than 2000 m is the relatively flat Sigsbee Plain. Sediment-capped knolls occur in the Sigsbee Plain and elsewhere to the east, in the eastern part of the Gulf.

Examination of bottom type, salinity, temperature, and water depth among other factors has resulted in the recognition of several associations living in the near-shore bays, lagoons, and estuaries. Each association was named after a common taxon in the association. Four to as many as ten associations of benthonic taxa have been traced in bands that parallel the shoreline across the shelf, and these parallel bands of benthonic foraminiferan associations may be traced around the shelves and slopes, each lying at about the same depth range. An association characterized by *Nutallides decorata* is found at the base of the slope. The deepest part of the Gulf, the Sigsbee Plain on the west, is inhabited by an association characterized by tiny forms of *Eponides* and *Nutallides*. The eastern deep-water sea bottom is inhabited by large forms of *Cibicides wuellerstorfi*.

Poag (1981) considered the distribution of benthonic foraminiferans in relationship to prominent bottom type. His study of the continental shelf off the Yucatan Peninsula indicated that the shelf sediments there are primarily calcium carbonate. The benthonic foraminiferan associations living on and in those bottoms are predominantly miliolids, foraminiferans different from those living on shelves to the north in the Gulf of Mexico floored by primarily siliciclastic sediments. Many other investigations into the calcium carbonate sediments on shelves and their contained benthonic foraminiferan fauna have suggested that calcium carbonate bottom sediment has a profound influence on organisms. The benthonic organisms appear to be very different from those living at similar depths and in similar environmental conditions in areas in which siliciclastic material is the dominant bottom sediment. These data imply that many areas of deposition during the Paleozoic where calcium carbonate bioclastic materials formed the dominant sediment type had unique organismal associations.

In oxygen-deficient areas of the Gulf floor the prominent foraminiferans are species of *Uvigerina*. In this setting, the volume of decaying organic matter in the bottom sediment is great enough to consume most of the available oxygen in the bottom water so that only very tolerant foraminiferan species can survive.

Modern Ostracods, Northwest Gulf of Mexico

Van Morkhoven (1972) provides an example of biogeographic variation in the ostracod fauna of the Gulf. He examined 171 species of ostracoda in environments ranging from fresh-water lakes to marine depths of 4000 m in the offshore area of

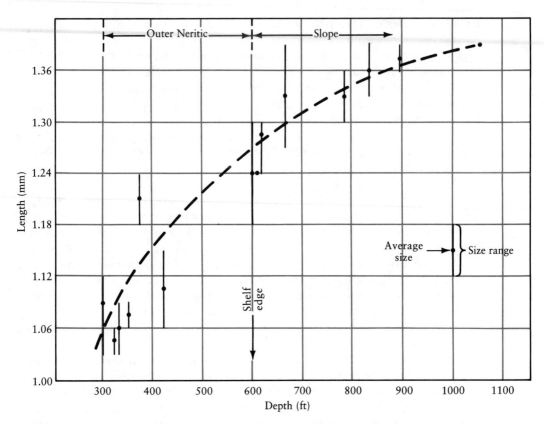

FIG. 5-11 • Relationship between depth and size of *Echinocythereis* species in the northwestern Gulf of Mexico. The two variables are fairly well correlated. (Reprinted by permission of the publisher from van Morkhoven, *Gulf Coast Association of Geological Societies*, v. 22, 1972.)

Texas and Louisiana. He, like Poag (1981), found that many species have rather restricted depth ranges and that ostracod faunal assemblages varied at the generic level with bathymetry. Also, the size of some genera varied with depth in a regular manner (Fig. 5-11). Many of the ostracod genera studied occur regularly throughout the entire Tertiary and Upper Cretaceous sediments of the Gulf coast and are of great help in reconstructions of paleodepth changes during the numerous transgressions and regressions that characterize this depocenter.

Cambrian Trilobite Biofacies, Laurentia, Canada

The studies by Dörjes (1972), Poag (1981), and van Morkhoven (1972) deal either with modern organisms or with organisms that have living close relatives. But paleo-bathymetric data can be obtained as well from extinct groups with no modern close relatives. For example, Ludvigsen et al. (1986) recognized a depth-dependent shelf trilobite fauna in Upper Cambrian rocks in Laurentia, the depth dependence being also related to lithology (Fig. 5-12). Unfortunately, the dependence of biofacies on

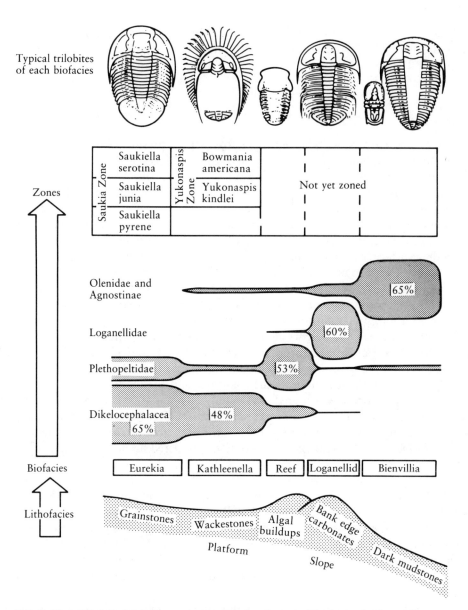

FIG. 5-12 • Trilobite biofacies and associated carbonate lithofacies in Upper Sunwaptan strata (Upper Cambrian), Laurentia, Canada. (Reprinted by permission of the Geological Association of Canada from Ludvigsen et al., *Geoscience Canada*, v. 13, 1986.)

lithofacies resulted in not enough taxa from the shelf facies extending to platform edge and slope facies, and other biozones will need to be developed for these areas of the ancient sea floor to complete temporal correlations of strata in the region.

Communities

A community is a group of organisms living together, all presumably adapted to the same habitat. Communities tend to be recurring assemblages at any particular time and so can be distinguished from random assemblages of fossils. As was noted earlier, because of irreducible stratigraphic condensation, it is not possible to establish that all organisms in an apparent community actually coexisted at a particular location at the same instant. But if the assemblage recurs with some frequency, we can assume that the taxa existed for a longer period of time than that represented by the numerous bedding surfaces in the rock and thus are useful in biogeographic interpretations.

It should be kept in mind, however, that not all organisms are fossilizable. Only those with hard parts, which may be only a small fraction of the community, have appreciable preservation potential, and many of them will be destroyed before burial by either abrasion, predation, or dissolution on the sea floor. The environmental interpretation of presumed paleocommunities is fraught with difficulties. Nevertheless, in some cases, biostratigraphers have produced very convincing evidence for accurate interpretations of ancient environments (McKerrow, 1978). It should be kept in mind that reconstructions of the soft parts and the mode of life of extinct forms are interpretations, which are based mainly on comparisons with the modern relatives of the organisms. For some extinct groups the mode of life has to be deduced from a study of the function of those hard parts that are preserved, the ichnofossils formed by the organism, or a study of unrelated animals that appear to resemble the fossils.

Among his descriptions of ancient paleocommunities, McKerrow (1978) gave relatively complete summaries of Silurian marine communities, describing those that lived close to shore and those that lived well offshore. He pointed out that these marine communities passed laterally one to the next from the shoreline seaward across the shelf in bands parallel to the shoreline (Table 5-2). These Silurian depth-reflective communities have been used to develop basin histories, tectonic histories, and shoreline changes through time in regional studies of Silurian rocks and communities.

Because McKerrow (1978) studied communities through all subdivisions of the Silurian Period, he was able to recognize changes in the communities that lived in one habitat or depth through time. He called a sequence of communities that lived in a similar habitat over a long time interval an **ecogroup**. An *Eocoelia* Community, for example, lived on sandy bottoms in shallow seas during the early part of the Silurian. A *Salopina* Community replaced the *Eocoelia* Community in similar habitats late in the Silurian. The finely ribbed brachiopod *Salopina* is rare in the *Eocoelia* Community. Through time, *Eocoelia*, a coarsely ribbed brachiopod, disappeared,

TABLE 5-2 ● The Silurian ecogroups and chief animal communities and their evolution with time

Time ↓ / Depth →	Ecogroup						
	Lingula (increases)	Eocoelia-Salopina (→)	Pentamerus-Sphaerirhynchia	Stricklandia-Isorthis	Clorinda-Dicoelosia	Visbyella	Graptolite
Pridoli and Ludlow	Lingula	Salopina	Sphaerirhynchia	Isorthis	Dicoelosia	Visbyella	Graptolite
Wenlock	Lingula	Salopina / Eocoelia	Sphaerirhynchia (sometimes Homoeospira in clastics)	Isorthis	Dicoelosia	Visbyella	Graptolite
Late Llandovery	Lingula	Eocoelia	Pentameroides / Pentamerus	Costistricklandia / Stricklandia	Clorinda	Marginal Clorinda	Graptolite
Early Llandovery	Lingula	Cryptothyrella	Pentamerus	Stricklandia	Clorinda		Graptolite

(Reprinted by permission of the publisher from McKerrow (ed.), *The Ecology of Fossils*, Gerald Duckworth and Co., Ltd., 1978.)

and *Salopina* became more and more prominent. Changes among ecogroup taxa permit histories of basins, platforms, and shorelines to be determined. Changes in species from one community to the next result in the new taxa used to establish zones. Ecogroup communities give information about relative ocean depth and other environmental conditions through time. Communities and ecogroups are thus useful tools in showing transgressions and regressions as well as changes in basin size, shape, and rate of basin filling or subsidence.

Summary

An understanding of modern biogeography is critical for adequate biostratigraphic studies, which are essentially four-dimensional biogeography. There is a clear need to supplement geologic training with collateral studies in oceanography, meteorology, and biology and perhaps astronomy as well. Within geology it appears that many more details of plate outlines and movements will be required before ancient biogeographies can be satisfactorily unraveled for organisms with living close relatives, and the difficulties of interpretation and inference for extinct taxa are even greater. It will require very talented and well-trained scientists to succeed as biostratigraphers in the future.

Today, construction of paleogeographic maps must consider not only remanent magnetism but also the positions of major rock suites such as reefs and platform carbonates. Biogeographic patterns such as those described here assist in interpreting the positions of lands and seas on the plates that form the Earth's crust at any one interval of time. By understanding the complexities of biogeography, biostratigraphers can make significant contributions toward revealing the tectonic history of the Earth's crust as well as the history of life on the crust.

References

Berry, W. B. N. and Boucot, A. J. (eds.), 1970. Correlation of the North American Silurian Rocks. Boulder, Geological Society of America, Spec. Paper 102, 289 pp.

Berry, W. B. N. and Boucot, A. J., 1973. Correlation of the African Silurian Rocks. Boulder, Geological Society of America, Spec. Paper 147, 83 pp.

Briggs, J. C., 1987. Biogeography and Plate Tectonics. New York, Elsevier Science Publishers, 204 pp.

Cox, C. B. and Moore, P. D., 1985. Biogeography: An Ecological and Evolutionary Approach, 4th ed. Boston, Blackwell Scientific Publications, 272 pp.

Critchfield, H. J., 1983. General Climatology, 4th ed. Englewood Cliffs, N.J., Prentice-Hall, 453 pp.

Dietz, R. S. and Holden, J. C., 1970. The breakup of Pangaea. Scientific Amer., v. 223, no. 4, pp. 30–41.

Dörjes, J., 1972. Georgia coastal region, Sapelo Island, U.S.A.: Sedimentology and biology. Senkenbergiana Maritima, v. 4, pp. 183–216.

Hallam, A. (ed.), 1973. Atlas of Paleobiogeography. New York, Elsevier Science Publishers, 531 pp.

Hay, W. W., 1988. Paleoceanography: A review for the GSA Centennial. Geol. Soc. Amer. Bull., v. 100, pp. 2011–2034.

Hays, J. D., 1971. Faunal extinctions and reversals of the Earth's magnetic field. Geol. Soc. Amer. Bull., v. 82, pp. 2433–2447.

Hsü, K. J. and Weissert, H. J. (eds.), 1985. South Atlantic Paleoceanography. New York, Cambridge University Press, 350 pp.

Ludvigsen, R., Westrop, S. R., Pratt, B. R., Tuffnell, P. A. and Young, G. A., 1986. Paleoscene #3: Dual biostratigraphy: Zones and biofacies. Geosci. Canada, v. 13, pp. 139–154.

McKerrow, W. S. (ed.), 1978. The Ecology of Fossils. London, Gerald Duckworth, 384 pp.

Murray, J. W., 1973. Distribution and Ecology of Living Benthic Foraminiferids. London, Heinemann Educational Books, 288 pp.

Owen, H. G., 1983. Atlas of Continental Displacement, 200 Million Years to the Present. New York, Cambridge University Press, 159 pp.

Pielou, E. C., 1979. Biogeography. New York, John Wiley & Sons, 351 pp.

Pierrot-Bults, A. C., van der Spoel, S., Zahuranec, B. J., and Johnson, R. K. (eds.), 1986. Pelagic Biogeography. Paris, UNESCO, Tech. Papers in Marine Sci. No. 49, 295 pp.

Poag, C. W., 1981. Ecologic Atlas of Benthic Foraminifera of the Gulf of Mexico. New York, Van Nostrand Reinhold, 174 pp.

Schmidt, K. P., 1954. Faunal realms, regions, and provinces. Quart. Rev. Biology, v. 29, pp. 322–331.

Schopf, T. J. M., 1980. Paleoceanography. Cambridge, Mass., Harvard University Press, 341 pp.

Scotese, C. R., Bambach, R. K., Barton, C., Van der Voo, R., and Zieglar, A., 1979. Paleozoic base maps. J. Geol., v. 87, pp. 217–277.

Smith, A. G., Hurley, A. M., and Briden, J. C., 1981. Phanerozoic Paleocontinental World Maps. New York, Cambridge University Press, 102 pp.

Smith, P. L., 1988. Paleoscene #11: Paleobiogeography and plate tectonics. Geosci. Canada, v. 15, pp. 261–279.

Stehli, F. G. and Webb, S. D. (eds.), 1985. The Great American Biotic Interchange. New York, Plenum Press, 532 pp.

van Morkhoven, F. P. C. M., 1972. Bathymetry of recent marine ostracoda in the northwest Gulf of Mexico. Gulf Coast Assoc. Geol. Soc., v. 22, pp. 241–252.

Ziegler, A. M., Bambach, R. K., Parrish, J. T., Barrett, S. F., Bierlowski, E. H., Parker, W. C., Raymond, A., and Sepkoski, J. J., Jr., 1981. Paleozoic biogeography and climatology. In K. J. Niklas (ed.), Paleobotany, Paleoecology, and Evolution. New York, Praeger Publishers, pp. 231–266.

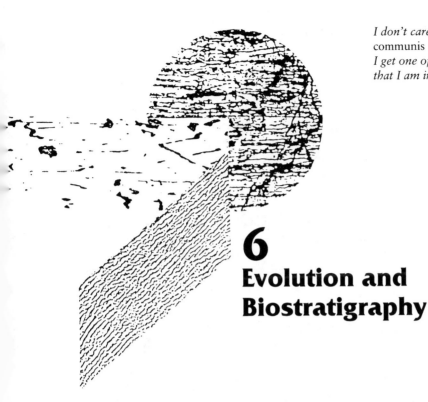

6
Evolution and Biostratigraphy

Evolving species, the changing players in the Earth's evolutionary theater, form the basis for **biostratigraphy**, the subdivision of strata based on fossil content. Biostratigraphers collect and use information about the morphological evolution of individual species as determined from their paleobiogeographic and stratigraphic distributions. A major goal of biostratigraphy is the arrangement of strata in chronological order of formation for use in studies of correlation. The great success of early 19th century workers in formulating the periods of the geologic time scale was due in no small part to easily described gross differences of fossil content in the various systems of rock. Even the development of sophisticated geochemical, isotopic, and paleomagnetic systems for dating strata has not displaced fossils as the most important tool for stratigraphic research, because fossils are consistently the most usable and reliable dating tool in Phanerozoic sedimentary strata.

But most species are extinct, and the sedimentary record, which was formed by sedimentation at variable rates, is fragmented by unconformities. How are we to choose species for study that may prove most useful for correlation? What are the limits to biostratigraphic resolution? Should we modify the degree of sampling based on rock type and facies? Should we work with single species or species assemblages? Answers to these and other questions are still being developed, but we will examine some approaches to the uses of species in the study of biostratigraphy.

Biogeography and Evolution

An **index fossil** is a species that is useful for delimiting a relatively restricted interval of geologic time. Features that make index fossils useful in biostratigraphic studies include abundant, easily recognizable remains, widespread geographic distribution, independence from facies changes, and short temporal duration (Shimer and Shrock, 1944). All these characteristics, except the last, are reflections of basic ecologic properties of species, which develop through evolutionary adaptations to physical and biologic environments.

On ecologic grounds, unfortunately, we shouldn't expect all of these properties to occur in the same species. That pessimistic tone, however, should not be grounds for despair because the evolutionary record of species is rich with many types of organisms. From them the biostratigrapher may more efficiently target the species that hold the most promise for in-depth study and stratigraphic use.

Environmental Fluctuations and Organism Tolerances

Species that are useful for long-distance correlation exhibit geographically dispersed distributions. How do these distributions develop? In a study of Caribbean venerid pelecypods, Jackson (1974, *in* Kauffman and Hazel, 1977) found that many species in this family were abundant in shallow, subtidal environments and were geographically dispersed about the oceans to varying degrees. Jackson hypothesized that species inhabiting very shallow waters (which he arbitrarily classified as less than or equal to 1 meter) were subject to more extreme environmental fluctuations, such as temperature, than were populations of species living in deeper waters (greater than 1 meter, but not at much larger depths).

A possible evolutionary consequence for shallow-water species is the development of adaptations for greater environmental tolerances. If such tolerances preferentially developed in very shallow-water species, Jackson hypothesized, then these species should exhibit wider geographic distributions than offshore species. Jackson found just this pattern for the Caribbean venerids. Jablonski and Valentine (*in* Scudder and Reveal, 1981) found similar correspondences for many of the bivalves and gastropods on the northeastern Pacific shelf.

Modes of Larval Development and Dispersal Potential

If some species possess adaptations for tolerating wide environmental fluctuations, we would expect these species to occur throughout widespread geographic ranges. How do marine invertebrates disperse over hundreds or thousands of kilometers? Species may achieve **eurytopic** (geographically widespread) distributions by either **larval transport** or **rafting** of either larvae or adults (Fig. 6-1). Of the two means of dispersal, the former is probably more important. Although larval development in many marine invertebrate species may last from several hours to a few days, the

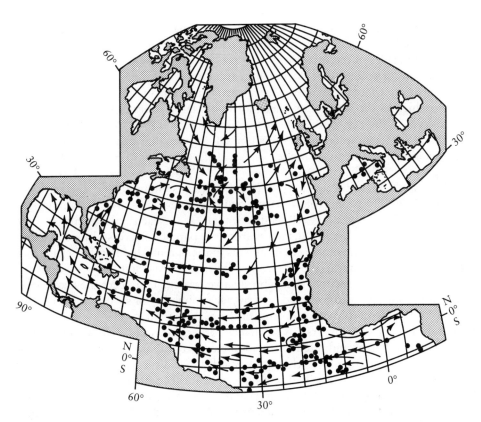

FIG. 6-1 • Filled circles indicate locations where bivalve larvae were found in the North and Tropical Atlantic Ocean. Larvae were trapped at a depth about 100 m below the surface of the sea. Arrows show the direction of surface currents. (Adapted from Scheltema, *in* Kauffman and Hazel, 1977.)

development of temperate zone organisms may take as long as a year (Scheltema, *in* Kauffman and Hazel, 1977). Extended larval development in the water column (the **pelagic** stage) promotes the wide dispersal during geologic time of many marine invertebrates, including biostratigraphically important groups such as gastropods, bivalves, and echinoderms.

The egg sizes of pelagic larvae are small in comparison to those of nonpelagic larvae. Pelagic larvae possess structures for capturing and feeding upon other microscopic organisms and organic detritus found floating in plankton. This mode of feeding is termed **planktotrophic**. Planktotrophic larvae may be widely dispersed by ocean currents. Scheltema (*in* Kauffman and Hazel, 1977) discovered that many species of bivalves on the shallow continental shelves of the Atlantic Ocean produce planktotrophic larvae that are carried across the entire ocean basin (**teleplanic** larvae). He estimated that the larvae of many species may spend nearly a year in the plankton prior to metamorphosis and settlement, resulting in distributions of the adults on both sides of the Atlantic.

For larvae of species that do not feed in or on plankton, nutrients for larval development are contained within a large egg. This mode of larval feeding is called **lecithotrophic**. As the feeding of a lecithotrophic larva is restricted to the limited food reserves in the egg, development time in the plankton is brief. Therefore the opportunity for geographic dispersal is limited. Species that exhibit lecithotrophic and other nonplanktotrophic modes of larval development (Levinton, 1982) are **stenotopic** (narrowly distributed).

Why should there be two very different modes (stenotopic and eurytopic) of larval dispersal? From an evolutionary viewpoint we would have to assume that there must be advantages (as well as disadvantages) for each type. Lecithotrophic larvae are dispersed only short distances from the parents. Successful reproduction means the presence of favorable environments in the parents' habitat. If the environment doesn't change rapidly, the offspring will probably find suitable conditions for their growth and reproduction. Any reproductive strategy that keeps the larvae near the parents would seem to be based on a higher probability of success there relative to the risks of unfavorable environments far from the parent.

But if long-distance dispersal is so chancy for lecithotrophs, why should planktotrophic development and attendant widespread geographic dispersal be so common among many marine invertebrates? Long-distance dispersal may guard against local environmental disasters. If the frequency of local catastrophes is substantial, then eurytopy guarantees that some larvae will by chance avoid lethal locations. But even eurytopy would fail to protect against a truly worldwide catastrophe, and a mass extinction would likely result.

Correlation of Shell Characters with Ecology

Although the correspondence between larval ecology and capabilities for geographic dispersal may be observed in living organisms, how is the paleontologist to identify these characteristics in fossil species? For species that exhibit accretionary skeletal growth (nonarthropodan invertebrates), some shell characteristics related to larval ecology are occasionally preserved in the skeleton of the adult organism (Jablonski and Lutz, *in* Scudder and Reveal, 1981). For example, a well-preserved fossil gastropod or bivalve mollusc shell may terminate in the larval shell (called a **protoconch** for gastropods and a **prodissoconch** for bivalves) secreted by the larval organism before metamorphosis into young adulthood. Because planktotrophic species possess a relatively small egg, the first formed larval shell is also relatively small (Fig. 6-2a). Any subsequently formed shell is larger. For lecithotrophic species the first formed larval shell is relatively larger, owing to the large nutrient-rich egg (Fig. 6-2b). Because these species spend little or no time in the plankton, any additional larval shell subsequently formed before metamorphosis is relatively smaller.

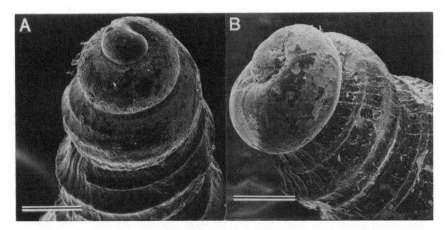

FIG. 6-2 • Larval shell morphology for two gastropods that differ in developmental type. (Left) *Haustator trilira*, (planktotrophic). (Right) An undescribed cerithopsid (nonplanktotrophic). Each scale bar is 100 μm in length. (Photographs courtesy of David Jablonski, from Jablonski, 1986).

Pattern Confirmation in the Fossil Record

If this correspondence between larval ecology and larval shell morphology holds for fossil species, then their paleobiogeographic potential may be determined. Several studies (Jackson, *in* Kauffman and Hazel, 1977; Jablonski, *in* Elliott, 1986) put the determination of paleobiogeographic potential to the test. For example, Jablonski (*in* Elliott, 1986) studied the rich Late Cretaceous bivalve and gastropod fauna of the Gulf and Atlantic coastal plains. He separated from the sediment well-preserved young molluscs that had shells only several millimeters long. For each specimen, Jablonski was able to determine the relative sizes of the larval shells. From the several hundred species collected, Jablonski selected a random sample of 50 species, noted for each the developmental type as determined by the larval shells, and determined the paleogeographic range over which the species occurred. For species identified as planktotrophic the paleogeographic range is 1860 km; for the non-planktotrophic species the range is 380 km. Perhaps the same adaptations that control the biogeography of living species also constrained the paleobiogeography of these Late Cretaceous molluscs.

Species Longevity and Biogeographic Distribution

Jackson (1974) originally hypothesized that species living in shallower waters would be subject to greater environmental fluctuation and extremes for such characteristics as temperature and food supply than species living in deeper offshore waters. If

Jackson is correct, these shallow-water species, dominantly eurytopic in geographic distribution, should be more resistant to extinction than stenotopic species for the following reasons.

1. Eurytopic species are geographically widespread and therefore less likely to be entirely extinguished during the environmental changes that cause extinctions.
2. Furthermore, even if most species populations are drastically reduced during extinction events, eurytopic species will more likely repopulate vacant environments from the refuges in which they survive because of their long-distance dispersal capabilities.
3. Finally, because eurytopic species evolved in shallow-water environments, which are subject to a greater degree of continuing environmental change, they will more likely already possess genetic adaptations that reduce the impact of environmental fluctuations that may cause extinction events.

If a species possesses adaptations that reduce the impact of extinction events, we would expect it to exhibit a longer geologic lifetime. Eurytopic species should therefore exhibit larger stratigraphic ranges than stenotopic ones. For example, Jablonski (*in* Elliott, 1986) discovered that the Late Cretaceous eurytopic planktotrophic species of the Atlantic and Gulf Coastal Plains existed on the average nearly three times as long as the stenotopic nonplanktotrophic species (an average of 6 million years versus an average of 2 million years, respectively).

Evolution—Speciation Through Time

In the analysis of fossiliferous strata, biostratigraphers look for levels at which evolutionary changes occur, within either single species lineages or fossil assemblages. Because not all changes of fossils represent time-significant events, an understanding of the nature of a varying fossil record is important in the study of biostratigraphy. In this section we will discuss the biologic basis of species, environmental, and genetic causes of morphologic variation, the processes of speciation and extinction, and paleontologic models of evolution.

Concepts of Species

Typological Species Concept. To the biostratigraphers of the 19th century, every taxon (one of the taxonomic groups in the Linnean hierarchy, from species to kingdom; plural: taxa), is characterized by an ideal form or "type," envisioned earlier by Linnaeus as God's perfect creation. All members of a species shared characteristics of an ideal plan. Because each species might consist of nearly uncountable numbers of individual specimens, each of which varies somewhat from the others, assignment of an unknown specimen by reference to a species description was not an easy task. A single specimen, called simply the **type**, served as the name-bearer and final standard. The philosophy of classification in which the type is the basis of a taxon's definition is known as **typology**.

In modern taxonomic practice, according to the rules of zoological nomenclature, a single representative specimen is designated the **holotype** of a species to stabilize species name use. But with the advent of evolutionary theory (mid-19th century), morphologic variation is no longer seen as a nuisance to identification; instead, variation among specimens is critical for evaluating evolutionary relationships. Sample variation has replaced the individual specimen as a basis for modern species definition.

Biologic Species Concept. The typologic concept of species, rooted in the concept of an ideal form, was largely abandoned by 20th-century taxonomists because they recognized the importance of genetic relationships among interbreeding individual organisms. **Biologic species** came to be defined as "groups of interbreeding organisms which are reproductively isolated from other such groups" (Mayr, 1970). Because of practical difficulties in assessing the occurrence of interbreeding among the individuals of various populations, biologists recognized populations as members of different species if they found substantial differences in morphologic, behavioral, ecologic, and/or other characteristics.

The development of genetic differentiation among populations does not always proceed at the same rate as morphologic differentiation. There are some species pairs, known as **sibling** or **cryptic species** (Futuyma, 1979), that are morphologically nearly identical but lack the ability to interbreed (Wake, *in* Scudder and Reveal, 1981). In contrast, some species exhibit large amounts of geographic variation among their various populations (Fig. 6-3). To the untrained eye, the various forms may appear so different as to imply separate species or even genera. Yet even these populations may belong to the same web of interbreeding populations. The implication for biostratigraphy from these studies is that the determination of fossil species is a difficult task best accomplished by the taxonomic specialist.

Although type specimens are still designated under the rules of nomenclature, taxonomists today identify an unknown fossil by reference to a wide variety of population samples, if available. Requirements for detailed taxonomic studies now include collection of numerous stratigraphically ordered samples of well-preserved fossils from many localities (Fig. 6-4). Taxonomic studies are becoming increasingly statistical in nature and commonly use electronic devices such as digitizers for quantitative measurements and computers for analysis and interpretation. Under the biologic species concept the unknown specimen is compared not only to the range of morphologic variation for the species, but also to its growth characteristics, paleoenvironmental habitats, and associated fauna. If the characteristics of the specimen are consistent with known material, it may be classified as a member of that species.

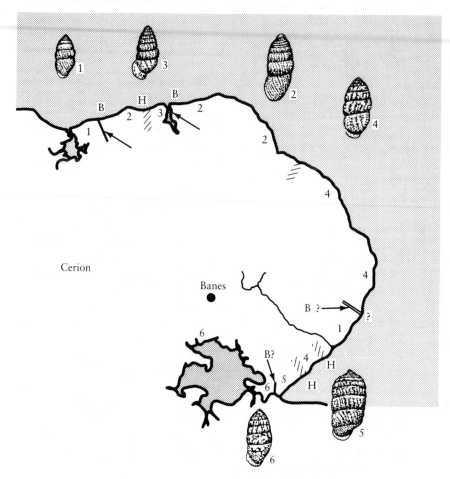

FIG. 6-3 • Morphologic variation exhibited by the land snail *Cerion* across the Banes Peninsula, eastern Cuba. Numbers refer to distinctive races, nearly all of which hybridize (zones indicated by H) when in contact (contact zones indicated by slashes), regardless of degree of difference. Some barriers (B) prevent interbreeding and gene flow between adjacent populations. (From Mayr, 1970.)

Importance of Morphologic Differentiation.   In the philosophy of typology, morphologic variation reflected the degree to which perfection of form was not achieved. In the biologic species concept, morphologic variation may be controlled in part by adaptation of populations of organisms to local environments. When such morphologic trends can be related to differences of environment, it is called **ecophenotypic** or **geographic variation**. For example, a snail living in the intertidal

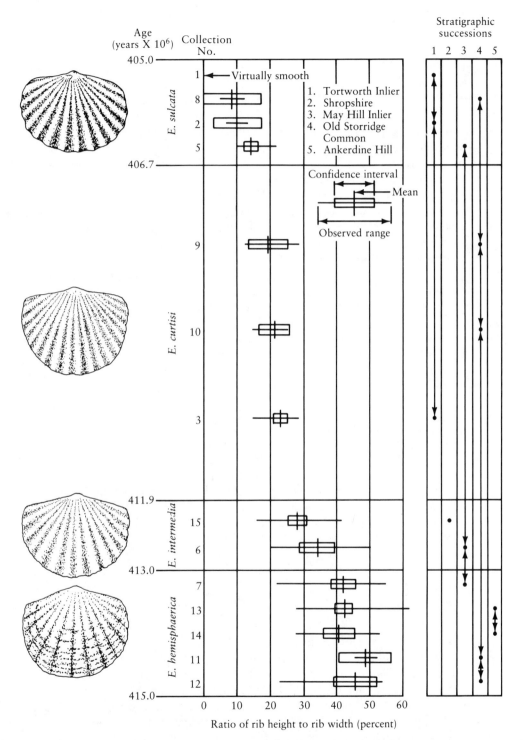

FIG. 6-4 ● Stratigraphically ordered populations of the Silurian brachiopod *Eocoelia*. Evolution of ribs provides for differentiation of species during 10-m.y. interval. Note statistical reporting of mean, confidence interval, and range for each measured population. (After Ziegler *in* Dobzhansky et al., 1977.)

zone of the Pacific Northwest, *Thais lamellosa*, develops smooth sculptured forms in the turbulent exposed coasts of Washington and more delicate forms within the quieter, protected San Juan Islands of Puget Sound (Spight, 1973). The predominance of the smooth form along the outer coast may result from the selective death of the delicate type, the shell of which is more susceptible to breakage by turbulence. Genetic differences built up in isolation during adaptation to local environments may eventually prevent one local population from interbreeding with populations elsewhere when contact is reestablished.

A Darwinian explanation for the development of geographic differentiation of populations of *Thais lamellosa* is that long-term local selection eventually eliminates some genetic systems of development. The buildup of geographic differentiation may ultimately result in the formation of geographic trends in morphology, reproductive barriers and species formation.

But not all morphologic differentiation is caused by a buildup of genetic differences among populations. Nongenetic causes may account for morphologic differentiation in some species (Etter, 1988; Hecht, 1974; Alexander, 1975). Furthermore, for most living species, detailed studies of environmental variation of morphology are lacking. Many "species" may be nothing more than populations interpreted as morphologically distinct because the intergrading morphologic forms that would otherwise connect the end members have not yet been found or recognized.

Species and Evolution

Because biozones are based on the stratigraphic distribution of fossil species, an understanding of the nature of the species and mechanisms of species formation may help to sharpen our insight into their limits of applicability and usefulness. A discussion of the biologic and evolutionary details of speciation may be found in Bush (*in* Milkman, 1982), Futuyma (1979), Levinton (1988), Mayr (*in* Barigozzi, 1982), and Stanley (1979). The two evolution models that we discuss—gradualism and punctuated equilibrium—have received much attention in recent years (Gould and Eldredge, 1977; Gingerich, 1985; Levinton, 1988), and reviews of their relative merits may be found in Gould and Eldredge (1977, 1986), Gould (*in* Milkman, 1982), Gingerich (1985), Levinton and Simon (1980), and Stanley (1985).

Species are the units of evolution.   The species is the fundamental unit of biologic classification and occupies a central position in evolutionary theory as well. What ties the individual organisms together into a species is the potential for the circulation of genetic material. The limits to this circulation circumscribe the gene pool (Fig. 6-5). For example, a favorable mutation may spread rapidly among the progeny of the interbreeding parents within a given species but not normally to the gene pool of another, even closely related, species (Fig. 6-5). Through time, the gene

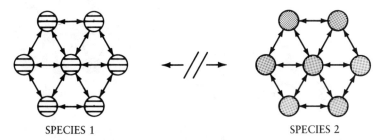

SPECIES 1   SPECIES 2

FIG. 6-5 • Hypothetical geographic distributions of local populations within two species. Arrows indicate the potential for interbreeding between individuals that migrate from one population to another. For biologic species, reproductive isolation prevents individuals from interbreeding successfully.

pool of a species retains its individual identity from other species, leading to its individual and unique evolution. These concepts lead Simpson (1961) to a description of the species for the paleontologist: "An evolutionary species is a lineage (an ancestral-descendant sequence of populations) evolving separately from others and with its own unitary evolutionary role and tendencies."

Speciation.   The great diversity of living organisms on the Earth developed from some one or few original living forms. The term **speciation** implies the increase in species abundance by the fragmentation of preexisting species populations into separate, reproductively isolated groups (Fig. 6-6a). But a new species may also arise by the accumulation of evolutionary changes without fragmentation of populations. Long-term climatic change could modify selection intensity, resulting in the modification of characteristics for species populations. Eventually, such accumulated changes of the descendant populations would be sufficient to imply hypothetical reproductive isolation from the ancestral species populations (Fig. 6-6b). Such reproductive isolation is hypothetical because, since ancestor and descendant populations are separated in time, no test of reproductive isolation is possible. Under this concept, no change in species abundance occurs, because only a single species exists at any one time.

Although species, the end product of speciation, are easy to observe, many mechanisms may be associated with the process of speciation. By whatever mechanism speciation proceeds, it results from the genetic differentiation of populations (Fig. 6-6). Futuyma (1988) perceptively comments, "Speciation [is] a breakdown of communication." We would therefore expect, on average, that as populations differentiate, their genetic resemblances would decrease. From the results of a variety of studies we see that such a prediction has been verified. For example, in Table 6-1 we see the extent to which genetic dissimilarity (genetic distance) increases with stage

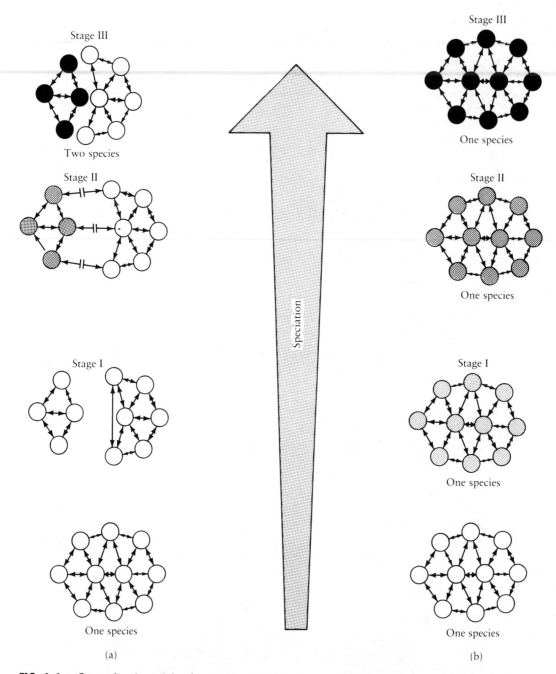

FIG. 6-6 ● Generalized models of speciation. Circles represent local populations, and arrows represent the potential for interbreeding. (a) Speciation by population fragmentation. During the speciation process, gene flow is restricted or prevented between some local populations (Stage I), resulting in genetic differentiation (Stage II), and fragmentation of the ancestral species into new descendant species (Stage III). (b) Speciation by phyletic transformation. Selection progressively transforms ancestral species populations (Stages I, II, III). Resulting species would be unable to interbreed successfully with ancestral type in hypothetical cross. (Adapted from Ayala, *in* Barigozzi, 1982.)

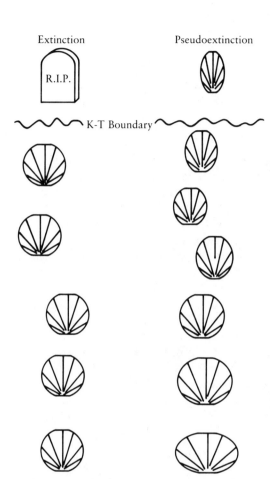

FIG. 6-7 • Extinction and pseudoextinction of hypothetical species at the Cretaceous-Tertiary boundary. Biologic extinction is due to death of all members of a lineage. Pseudoextinction is caused by the transformation of ancestral species into descendant species.

TABLE 6-1 • Measure of genetic divergence with increasing stage of evolutionary development in some groups of organisms. Note that as evolutionary divergence increases (from local population to species), genetic differences become larger.

	Genetic Distance			
Organisms	Local Populations	Subspecies	Semispecies	Species and Closely Related Genera
Drosophila	0.013	0.163	0.239	1.066
Other invertebrates	0.016	—	—	0.878
Fish	0.020	0.163	—	0.760
Salamanders	0.017	0.181	—	0.742
Reptiles	0.053	0.306	—	0.988
Mammals	0.058	0.232	0.263	0.559
Plants	0.035	—	—	0.808

(From Ayala, *in* Barigozzi, 1982.)

of evolutionary divergence for a variety of organisms. Notice that within each group of organisms the index increases from the local population to the taxonomic levels of species and genera.

Relationship between morphologic change and speciation. Genetic differentiation among species populations is a necessary factor in the development of a new species from an ancestral one. But should a new species diverge *morphologically* in concert with genetic divergence? On this subject the evidence is less consistent. For some species pairs or complexes, little morphologic differentiation accompanies reproductive isolation (Mayr, 1970; Wake, *in* Scudder and Reveal, 1981). In others, substantial morphologic differences develop among populations without noticeable reproductive isolation (Coyne and Barton, 1988). One view is that "Although speciation is not required for morphological change, it can preserve it, because reproductive isolation can perpetuate local adaptations that would normally disappear with ecological change and gene flow" (Coyne and Barton, 1988). Therefore the number of fossil species will be underestimated, owing to the existence of **cryptic** species (closely related species that are nearly morphologically indistinguishable yet differentiated genetically because they lack the ability to interbreed). The number of fossil species will be overestimated by geographically variable populations for which intermediate morphologies are missing. The extent of this bias, currently unknown, will be a function of the relative proportions of these two factors, given, of course, the absence of all other potential biases.

Extinction

In the long run, extinction is an inevitable consequence of evolution, since the Malthusian multiplication of individuals (Malthus, [1798] 1970) would rapidly exhaust resources and the Malthusianlike multiplication of species (Erwin et al., 1987) would rapidly deplete habitats. Although there are varying estimates of the number of species alive today, because of species turnover the cumulative number of extinct species is certainly much, much greater.

In stratigraphy, extinctions are important because they have been used to define biostratigraphic boundaries over a wide range of scales. For example, marine micropaleontologists place the Pliocene-Pleistocene boundary in the interval between the extinction of *Globigerinoides obliquus* and the location of the first appearance of *Globorotalia truncatulinoides*. Eras of the geological time scale and their boundaries were proposed in 1841 by John Phillips, who believed in a unique history of life characterized by a "great law of succession" (Wilmarth, 1925). Phillips suggested that an analysis of fossil species would reveal the era to which systems of strata belong (Wilmarth, 1925).

Just as speciation refers to the appearance of reproductively isolated populations of organisms, **biologic extinction** refers to the disappearance of every living member of a taxon (a species or higher taxonomic group). Two types of biologic extinction are recognized: extinction and pseudoextinction (Fig. 6-7). **Extinction** refers to the disappearance by death of all individual members of a species. The once-living lineage dies out. However, a species may evolve into a different species by phyletic transformation. **Pseudoextinction**, also called **phyletic extinction**, refers to the disappearance of a species because it evolves into something else. By whatever mechanism a species becomes extinct, it is most unlikely that some future species will ever evolve the exact same set of characteristics. Once extinct, always extinct.

Under one special circumstance, however, species or higher taxa may sometimes appear to become extinct, only to reappear later in a stratigraphically higher interval. For example, the coelacanthine fishes were first recognized with Agassiz's 1839 naming of the Permian fossil *Coelacanthus*. The geologically youngest species of that group was then known only from Upper Cretaceous rocks. In 1938 the first living specimen of *Latimeria chalumnae* (Fig. 6-8), a living coelancanth, was captured in 80 meters of water off South Africa. The geologic range of the coelacanthine fishes was therefore extended to the present and now includes an interval some 60 million years long that is barren of known fossils. The lineage must have survived, perhaps in marine environments that are not commonly preserved or possibly in areas that are currently inaccessible or unexplored.

In contrast to such species that are absent from the geologic record for millions of years, some may seem only temporarily extinct. For example, in a study of the brachiopods that were seemingly extinct at the Cretaceous-Tertiary (K-T) boundary, Surlyk and Johansen (1984) found that six of 26 Upper Maastrichtian species reappeared within 10 meters stratigraphically above the K-T boundary (Fig. 6-9). This

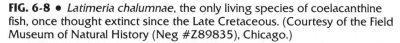

FIG. 6-8 ● *Latimeria chalumnae*, the only living species of coelacanthine fish, once thought extinct since the Late Cretaceous. (Courtesy of the Field Museum of Natural History (Neg #Z89835), Chicago.)

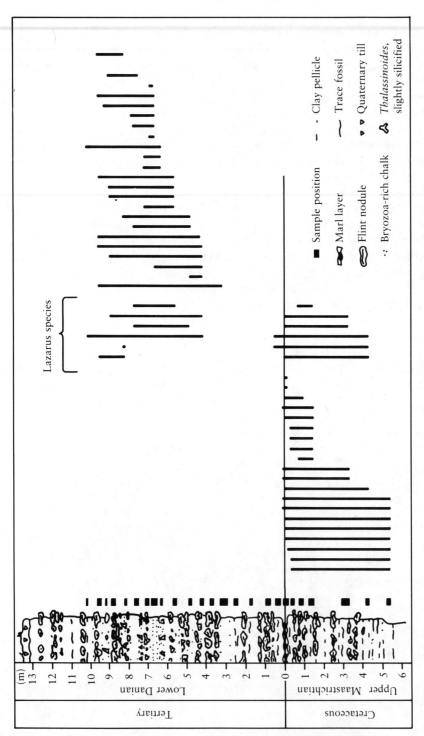

FIG. 6-9 • Biostratigraphic range chart of brachiopods found near the Cretaceous-Tertiary (K-T) boundary sequence in Denmark. Note the **Lazarus** species, which reappear higher in the section after vanishing near the K-T boundary. (Adapted from Surlyk and Johansen, 1984.)

phenomenon, appropriately termed the **Lazarus effect** (John 11:43–44; Jablonski, 1984; Jablonski, *in* Elliott, 1986), is not unexpected. After all, many species avoid permanent extinction through widespread geographic dispersal into a variety of environments. Local disasters may extinguish many populations and millions of individual organisms but leave others untouched in habitat refuges. Migration may then replenish local populations and reextend the species' geographic range.

Finally, patterns of biologic extinction may be modified by purely sedimentologic processes. For example, marine regressions frequently expose large areas of continents to subaerial weathering and erosion. The removal of previously deposited strata creates a stratigraphic **unconformity** (if the interval of missing time is large) or **hiatus** (if the interval of missing time is small). For example, on the continental shelf or in the deep-sea environment, sedimentation may be discontinuous, or currents may erode existing portions of the sea floor (Pinet and Popenoe, 1985), causing the juxtaposition of different sedimentary facies. Therefore whatever the pattern of species stratigraphic ranges throughout the original stratigraphic column, the resulting unconformity creates a common surface of truncation, the extent of which may greatly accentuate the appearance of a common surface of extinction or facies change (Fig. 6-10). This result, termed the **truncation effect**, has been detected in a series of Danish sections encompassing the Cretaceous-Tertiary boundary (Birkelund and Hakansson, *in* Silver and Schultz, 1982).

FIG. 6-10 ● Truncation effect on hypothetical biostratigraphic ranges near the K-T boundary. (Top) Larger truncation (unconformity) may cause species loss and range shortening from stratigraphic record. (Bottom) Smaller truncation (hiatus) may only shorten species ranges. Truncation always causes range ends to coincide. (Reprinted by permission of the author from Birkelund and Hakansson, *in* Silver and Schultz, *Geological Implications of Impacts of Large Asteroids and Comets on the Earth*, 1982.)

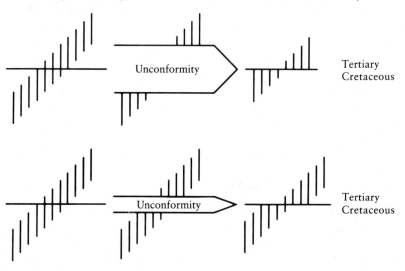

Models of Evolution

Are the basic principles and practices of biostratigraphers changed by a knowledge of the various models of speciation and evolutionary theory? To the stratigraphers of the earlier 19th century (prior to 1859, the year in which Darwin's theory of evolution was published), fossil species were either present or not present in the geologic record, and the discontinuities in their stratigraphic distributions were used to demarcate time-stratigraphic units of the geologic column. Because the utility of fossils for stratigraphic subdivision depended only on the presence or absence of species and their assemblages, rather than on evolutionary sequences, the practice of 19th century biostratigraphy may be interpreted as theory-free.

Although Darwin lamented the fossil record for its fragmentary nature of preservation (Darwin, [1859] 1964), some paleontologists have suggested that it might not be so poor after all (Eldredge and Gould, *in* Schopf, 1972; Gould and Eldredge, 1977). Eldredge and Gould reinterpreted the spotty fossil record as a fair representation of the waxing and waning of species populations caused more by evolutionary processes than by geologic effects and the preservation potential of fossils. Although Eldredge and Gould's hypothesis stimulated research into the nature of paleontologic speciation, most of it was not, in fact, directed toward direct application to biostratigraphy. However, even Gould and Eldredge (1977) suggest that an acquaintance with the various speciation models should help to enhance an appreciation for the nature of biozones. We will first briefly review the most common models of paleontologic speciation—gradualism and punctuated equilibrium—and then explore implications for the practice of biostratigraphy.

Phyletic Gradualism.   Phyletic gradualism (or simply gradualism) means that evolution proceeds by both phyletic transformation (at relatively even and slow rates of morphologic change) and by speciation (the fragmentation of populations). Its various elements were articulated by Eldredge and Gould (*in* Schopf, 1972), who proposed as its keystone concept the Darwinian prediction of "interminable varieties, connecting together all the extinct and existing forms of life by the finest graduated steps" (Darwin, [1859] 1964). The geometry of evolution by gradualism (Fig. 6-11a) is typically portrayed as a branching tree, with upwardly acute angles at branch intersections. According to this model, most morphologic change is generated by phyletic transformation, rather than concentrated in speciation events. Species that arise by phyletic transformation are considered to be arbitrary segments on a continuous scale of morphologic change. Continuous morphologic changes are due to long-term, directional selection by the change of environments—physical, chemical, or biological. As a consequence, long-term morphologic trends are due to phyletic gradualism. In the context of gradualism, sudden morphologic shifts result from nonpreservation, erosion, or migration of the species away from the local section.

Punctuated Equilibrium.   It is Eldredge and Gould's opinion (Eldredge and Gould, *in* Schopf, 1972; Gould and Eldredge, 1977) that the fossil record reveals few, if any, examples of such gradual evolutionary transformations, not because of the vagaries of sedimentation and preservation, but because speciation proceeds by

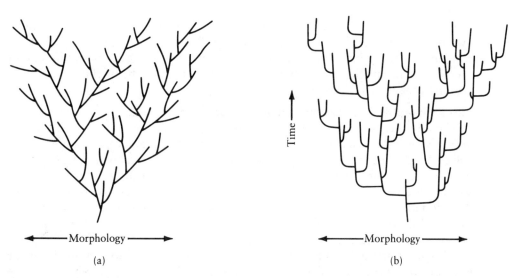

Time →

Morphology ← →

(a)

Morphology ← →

(b)

FIG. 6-11 • Geometry of evolution under two models, (a) gradualism and (b) punctuated equilibrium. Under gradualism, most morphologic change is generated by phyletic transformation; under punctuated equilibrium, most morphologic change takes place at speciation events. (From Stanley, 1979.)

an inherently different mode, which they termed **punctuated equilibrium**. Speciation by punctuated equilibrium is the evolution of one species from another by means of population isolation and rapid evolutionary rates during only a small portion of the species's existence (Fig. 6-11b). The geometry of evolution by punctuated equilibrium is typically depicted as a branching tree in which branch intersections are perpendicular; most morphologic change occurs at the speciation event during a brief interval of geologic time. It is interpreted as beginning in small, ecologically isolated populations (termed **peripheral isolates**). These populations, living on the ecologic fringe of conditions of survival, adapt to surroundings different from those experienced by the main species populations. Whether by the reorganization of the genome (the genetic revolution of Mayr, 1970) and/or the breaking of genetic continuity (gene flow) with the rest of the species populations, a new species is eventually formed. Because the populations that become the daughter species usually live in different places than the main species populations, their distribution is termed **allopatric**. The outcome of evolution in these marginal environments is speciation by the splitting of the ancestral species populations into two or more daughter species. The characteristics of the new daughter species are determined by chance events, unrelated to those of the parental species, and so, unlike speciation by gradualism, do not arise out of within-species variability. Long-term morphologic trends result from the differential success of populations, rather than directional selection for individual characteristics.

Gould and Eldredge (1977) emphasize that the greatest proportion of evolutionary changes is associated with the brief process of speciation (Fig. 6-11b). Gould (*in* Milkman, 1982) estimates the minimum period of species stability, known as **stasis**, to be 99% and specifies that changes associated with speciation are concentrated "within 1% or less of later existence in stasis." Thus according to the model of punctuated equilibrium, the evolutionary history of a lineage is primarily one of genetic (and morphologic) stability, interrupted occasionally by brief periods of speciation or the final alternative—extinction.

Biostratigraphic implications of speciation models. These two views of the speciation process result in rather different interpretations of vertical sequences of an evolving lineage, correctly emphasized by Sylvester-Bradley (*in* Kauffman and Hazel, 1977): The theoretical consequence of speciation by gradualism (exclusive of branch intersections) is Darwin's graded chain of intermediates (Fig. 6-11a), along which it is difficult to pick a point at which species boundaries may be recognized. But speciation by punctuated equilibrium theoretically (and hence practically) results in discrete points in time of species origination (Fig. 6-11b), since most morphologic change occurs at branch intersections.

The biostratigraphic implications for these two models of speciation also differ in a fundamental way. Under phyletic gradualism, boundaries between successive

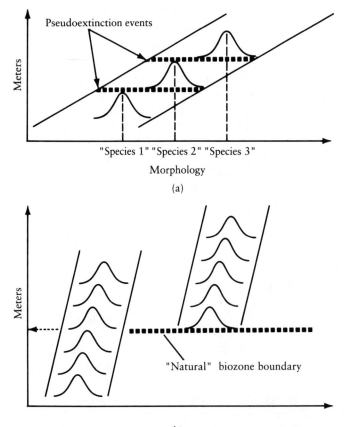

FIG. 6-12 • Biostratigraphic subdivision of geologic column under two models of evolution. (a) Lineage and strata subdivided by pseudoextinction events, which are arbitrarily chosen. (b) Strata subdivided at the level of appearance of new species arising by punctuated equilibrium. The location of the first observed occurrence is not subject to arbitrary decision by the taxonomist, as the descendant taxon is distinct from the ancestral one at time of first occurrence.

TABLE 6-2 • Estimates of mean species durations for well-studied extinct invertebrate taxa. Values computed from known stratigraphic ranges.

Biologic Group	Estimated Mean Species Duration (m.y.)
Ammonites	~5 (but with a mode in the 1–2 m.y. range)
Trilobites	>1
Graptolites	2–3

(From Stanley, 1985.)

species are arbitrary (Fig. 6-12a), and any biostratigraphy derived from it must be equally arbitrary (Bown and Rose, 1987). The origination points for species produced by punctuated equilibrium, however, represent objective boundaries (Fig. 6-12b) that are not subject to the arbitrary whims of the stratigrapher.

One might quickly conclude that biozonation based upon the first occurrences of new taxa that arise by punctuated equilibrium would be both easier to detect and less subject to error in comparison to biozones erected throughout a continuously varying lineage. However, as emphasized by Gould and Eldredge (1977), although the origin of a new daughter species in a single peripheral isolate may be nearly instantaneous in geologic time, its migration could span large intervals of time. Therefore the first occurrences of such daughter species in a series of local sections may well be significantly diachronous.

An example of diachronous first occurrences is seen for the radiolarian *Theocorythium trachelium* in cores from the Indian and Pacific Oceans (Baker, 1983). By using the record of magnetic reversals to calibrate in each core the passage of absolute time, Baker found that the migration spanned approximately 1 million years. This interval is nearly as long as the entire average species duration for such rapidly evolving organisms as trilobites, graptolites, and ammonites (Table 6-2).

Such problems of stratigraphic diachroneity are theoretically inapplicable to the establishment of species by gradualism, according to Eldredge and Gould (*in* Schopf, 1972), because the entire set of populations for the species undergoes transformation simultaneously throughout the species, geographic range. If this is true for any given lineage, then its subdivision into arbitrary species may result in a potentially higher resolution, provided that vertically adjacent populations can be distinguished. Biostratigraphic subdivision based on such stratigraphic patterns is called **stage-of-evolution**. Unfortunately, this method of biostratigraphy is probably inapplicable to temporally long sequences in the fossil record of species that undergo evolution by phyletic transition, because long-term morphologic trends are unlikely to maintain precise constant direction and rate. As population-based studies become more sophisticated (Sheldon, 1987), frequent fits and starts and even reversals of short-term morphologic trends may become the normal pattern, rather than the exception.

In contrast, speciation by punctuated equilibrium results in temporally invariant morphologies, according to Eldredge and Gould (*in* Schopf, 1972) and Gould and

Eldredge (1977). Therefore biozone boundaries may be chosen only at points at which species either disappear or appear. The resolution of such biozonations will generally be limited by the rate of speciation. A greater rate of speciation would permit (all other factors being equal) the construction of smaller biozones based on single species.

Ecophenotypy and Evolution. The biostratigrapher must proceed with caution in the subdivision of strata based on evolution within a continuously varying lineage. If a series of populations within a single species varies in some regular fashion (such as the common observation, known as Bergmann's rule, that mammals' body sizes generally increase toward more polar latitudes), then a simple pattern of geographic migration may lead to vertical sequences of fossils exhibiting gradual changes in morphology (Fig. 6-13). Successive replacements of ecophenotypic variants may account for incorrect interpretations of evolutionary change in biostratigraphic studies.

Few field studies meet the stringent requirements (Gould and Eldredge, 1977) for testing the evolutionary significance of stratigraphic trends in morphology: (1) long sequences of stratigraphically closely spaced samples, (2) statistically defensible sample sizes, (3) unambiguous definitions of taxa, and (4) widespread collections providing for assessment of geographic variation. In two exemplars of such studies, Johnson (1979) and Baarli (1986) found temporal morphologic trends in a lineage of *Stricklandia lens*, a Silurian brachiopod first described from the Llandovery Series of Wales and subsequently discovered in Iowa (Johnson, 1979), Estonia, and Norway (Baarli, 1986). Relative sizes of internal shell structures, such as the outer plates and brachial process, and measurements of shell shape are used to distinguish various subspecies within the lineage (Fig. 6-14). Wherever these subspecies have been found, their stratigraphic order of appearance has been invariant. Although morpho-

FIG. 6-13 • Hypothetical morphologic trends may be caused by migration rather than evolution, as shown by a temporal sequence of hypothetical populations at five (A through E) geographic locations. At time t_1, geographic variation is clinal; notice the increase in morphologic mean from A to E. The subsequent migration of all populations in direction E to A results in stratigraphic trends in morphology at any one locality. (Adapted from Bookstein et al., 1978.)

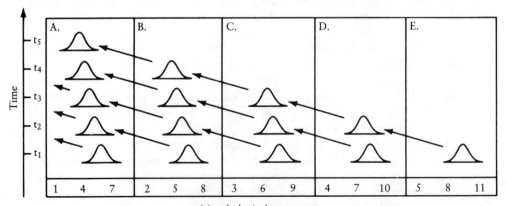

Morphological measurement

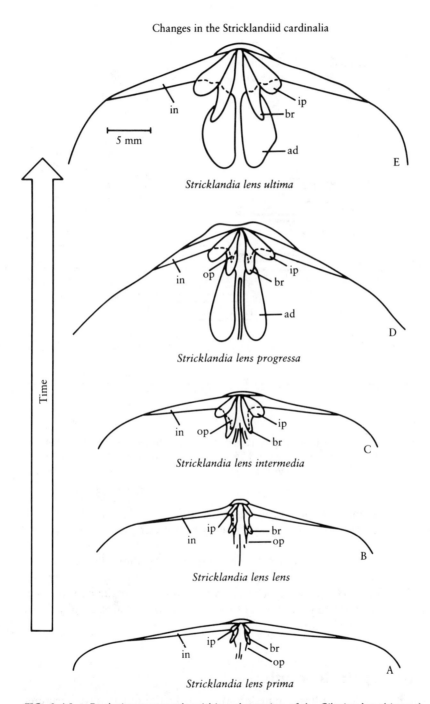

Changes in the Stricklandiid cardinalia

Stricklandia lens ultima

Stricklandia lens progressa

Stricklandia lens intermedia

Stricklandia lens lens

Stricklandia lens prima

FIG. 6-14 • Evolutionary trends within subspecies of the Silurian brachiopod *Stricklandia lens*. Note the relative enlargement of the interarea (in), brachial process (bp) and inner plate (ip). (Adapted from Johnson, 1979.)

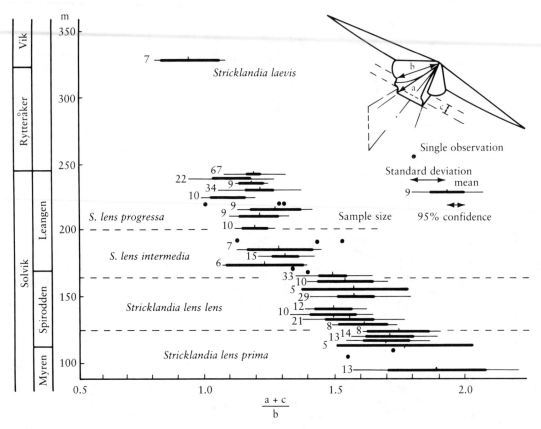

FIG. 6-15 • Stratigraphic trend in populations of *Stricklandia lens* collected from the Solvik Formation of Norway. The trend of morphology, measured by the ratio $(a + c)/b$ is toward relative enlargement of the inner process. (After Baarli, 1986.)

logic variation within any one population may be relatively large (Fig. 6-15), a modest statistical sample (of about 10–15 specimens) is sufficient to differentiate the various subspecies.

The hypothesis of true evolutionary change (phyletic transition) among the subspecies of *Stricklandia lens* may well be falsified by the future discovery of temporal patterns reversed from those presently known. But until then, the *Stricklandia lens* lineage stands as one of the few examples of high-resolution biostratigraphy in the shelly fauna of the Lower Silurian.

Limits to Biostratigraphic Resolution

With increasingly detailed studies in paleontology, biostratigraphers have been able to subdivide the geologic column more finely, from the hundred-million-year eras of the Phanerozoic Eon defined by John Phillips in terms of the "ancientness" of life to

Johnson's (1979) brief zones of the Silurian based on evolution within a single brachiopod lineage. Is there any ultimate minimum limit to the size of a useful biozone? If so, what factors affect maximum biostratigraphic resolution? What, if any, are the implications of the various models of evolution for biostratigraphy? Is zone resolution affected by sedimentologic processes of burial or preservation of fossils? In this section we will examine biologic and geologic factors that affect the resolution of the biostratigraphic record.

Rates of Evolution

To the biostratigrapher, heaven on earth might be a perfect geologic record of constant and continuous sedimentation with complete preservation of the entire fossil fauna and flora. If the goal is biozonation, however, such a perfect world would not necessarily permit the construction of infinitesimally small biozones. Should species evolve at exceedingly slow rates, there would be little differentiation of them with time, and the lack of unique species assemblages would prevent the establishment of high-resolution biostratigraphies. But different groups of species evolve at different rates (e.g., Stanley, 1985). Average species durations differ by about an order of magnitude, from about 1–2 million years to 20–30 million years (Table 6-3).

We have already noted that average species durations based upon the direct measurement of stratigraphic ranges (Table 6-2) for some well-known invertebrate groups are characteristically short. For example, from studies of Late Cambrian trilobites of the Great Basin of the western United States, Stanley (1979) constructed a histogram of the durations of individual species. Although there is significant vari-

TABLE 6-3 • Estimates of mean species durations for a variety of taxonomic groups

Biologic Group	Estimated Mean Species Duration (m.y.)
Benthonic foraminifers	20–30
Marine diatoms	25
Bryophytes	>20
Planktonic foraminifers	>20
Marine bivalves	11–14
Marine gastropods	10
Higher plants	>8
Fresh-water fishes	3
Beetles	>2
Snakes	>2
Mammals	>1

(After Stanley, 1985, Table 1; for discussion of method and references, see Stanley, 1985.)

ability in individual species durations, the average span is somewhat less than 1 million years. This biostratigraphic temporal resolution is far greater than that of dating methods for Cambrian rocks using radiometric clocks, whose errors are estimated at ± 12 m.y. (Harland et al., 1982).

This biostratigraphic base for Late Cambrian trilobites permits quantitative estimates of the rate of appearance of new trilobite species. Stanley estimated the net rate of increase for the Late Cambrian trilobite species within the Pterocephaliid Biomere to be between about 0.6 and 1.2 species per million years, a figure that he notes (Stanley, 1979) is much higher than that for comparable expansions of bivalves, gastropods, planktonic forams, or echinoids. Thus within intervals during which rates of speciation are high (Vendian through the Lower Cambrian, Ordovician, and Mesozoic), various fossil species are excellent stratigraphic tools (Lewin, 1988).

Sampling Resolution

Coarse versus Fine Sampling.   Even if the accumulation of strata were continuous and the preservation of fossils unbiased, the resolution of a biostratigraphic framework would depend in part upon the care and resolution of stratigraphic sampling. For example, in a study of thousands of meters of strata, perhaps only a coarse sampling scheme is feasible, owing to limitations of labor and funds. The geologist may therefore decide to sample only some small percentage of the strata (Fig. 6-16a, b). The probable consequences of such a sampling scheme are (1) the detection of fewer species due to smaller sample sizes and (2) shortened species ranges because the true lower and upper limits to the occurrences of species' ranges will tend to remain undetected.

In many cases a better sampling strategy would be to collect more frequent but smaller samples (Fig. 6-16c, d), resulting in the same total sampled volume and effort. Such a sampling scheme would generally yield a more accurate estimate of the lower and upper biostratigraphic limits. This method, unfortunately, is sensitive to small sample size; species that are uncommon or rare will generally remain undetected (Koch, 1987).

Paleontologists have recently recognized the importance of high-resolution sampling in tracing the patterns of evolutionary change between species (Bookstein et al., 1978; Johnson, 1979; Sheldon, 1987; Schindel, 1980; Williamson, 1981). When such detailed studies are made, what were formerly simple stratigraphies may be complicated by complex evolutionary patterns. For example, several successional species of Silurian trilobites within individual genera (such as *Ogygiocarella*) were distinguished by differing number of pygidial ribs (Fig. 6-17a). However, with careful, fine-scale sampling (average stratigraphic sample thickness of 23 cm), and very large sample sizes (collection of nearly 15,000 trilobites and measurements on nearly 3500 individual specimens), Sheldon (1987) found intergradations in rib morphology for the various "species" (Fig. 6-17b). Rib count was not even con-

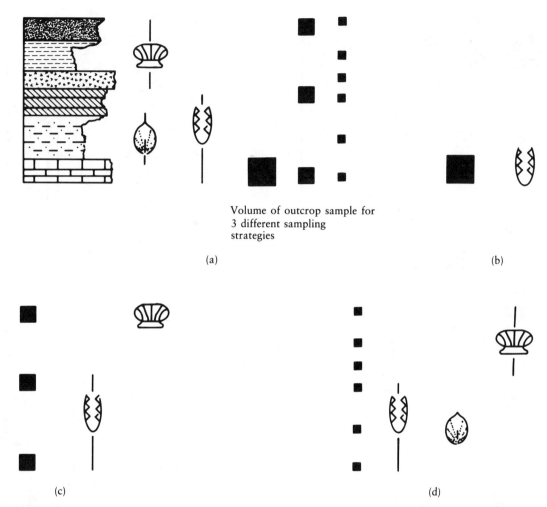

FIG. 6-16 ● Biostratigraphic ranges may be controlled in part by sampling design. (a) Hypothetical ranges and three different sampling strategies; each strategy produces an equal volume of rock. (b) Single bulk sample from the bottom of the section detects only a single species. (c) and (d) Smaller but more frequent samples detect other species and a greater proportion of biostratigraphic range.

stant in some stratigraphically successional population samples of *Cnemidopyge* (Fig. 6-17c), increasing from a mode of seven to a mode of eight in less than a meter of section.

Aggregation and Lumping of Samples.   Biostratigraphic resolution may also be affected by aggregation of samples from various stratigraphic intervals. For example,

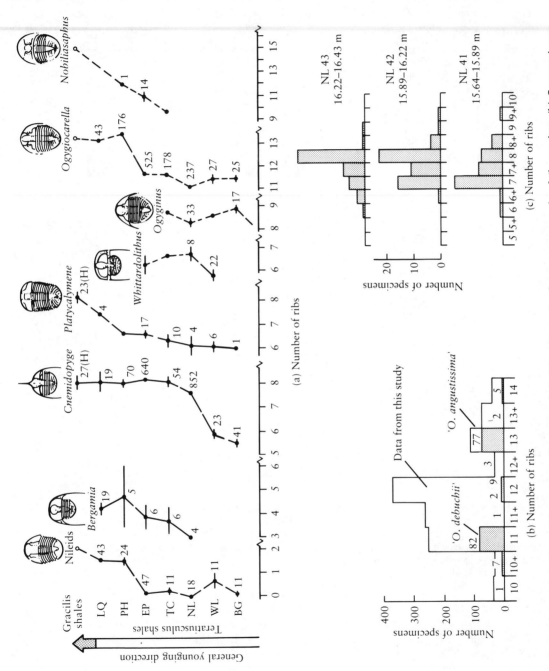

FIG. 6-17 • (a) Stratigraphic trends in lineages of trilobites. Note the frequent reversals in rib number. (b) Comparison of data from study of small number of measurements (black histogram) and larger study (white histogram) of *Ogygiocarella*. Taxonomic conclusions of species separated by morphologic gaps in rib number from small scale study are not carella. (c) Stratigraphic increase in rib number of *Cnemidopyge*. (All adapted from Sheldon, 1987.)

172

Bookstein et al. (1978) illustrate several very different interpretations of the phyletic history of a single species based upon aggregation of specimens from within single beds (see Fig. 6-18). Notice that aggregation may hide more highly resolved patterns of morphologic change that would be revealed by finer sampling.

From this example you might conclude that samples should be drawn from infinitesimally thin stratigraphic intervals (Fig. 6-19a). But the nature of sedimentation units may also affect the resolution of evolutionary sequences. For example, thick-bedded muddy units may be extensively bioturbated, resulting in a naturally mixed

FIG. 6-18 ● Potential effect of aggregation of specimens collected within single beds of a stratigraphic section. (a) Morphologic trend for specimens collected by high-resolution (closely spaced) sampling. (b) Morphologic trend of specimens aggregated by bed. (Adapted from Bookstein et al., 1978.)

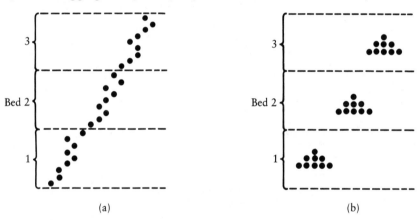

FIG. 6-19 ● Potential effects on resolution of biostratigraphic range caused by bioturbation or mixing of sedimentary unit (hypothetical data). (a) True biostratigraphic ranges from laminated strata may be highly resolved. (b) The same ranges from a mixed unit.

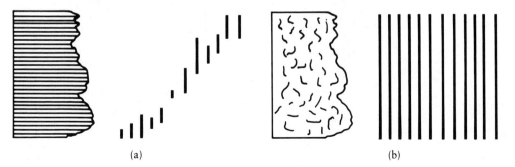

and aggregated sample (Fig. 6-19b). Any original morphologic sequences will then be obscured, and as a consequence, biostratigraphic resolution will be decreased.

Taphonomy

Biostratigraphic resolution may also be affected by the processes of burial and preservation of fossils. The study of what happens to fossils after the organism's death and before discovery, is part of the paleontologic subfield termed **taphonomy**. Kidwell (1986) has recently reviewed the literature on the stratigraphic accumulation of fossils and suggests that their concentrations, and consequently patterns of morphologic change, may sometimes be controlled by changes in sedimentation rate. In Figure 6-20, notice that a decreasing rate of sedimentation results in an increasing concentration of fossils. As a consequence, samples of constant volume (for example, collected from 10 cm of section) include larger numbers of specimens accumulated over longer intervals of time, compared to samples from smaller stratigraphic inter-

FIG. 6-20 ● Stratigraphic morphologic trends may be controlled by taphonomic or ecologic processes, rather than evolution. For each example, assume an upwardly decreasing rate of sedimentation. (a) Shell populations are increasingly time-averaged and therefore exhibit larger morphologic variance. (b) Upward decreasing sedimentation causes increased exposure time of fossils, leading to differential destruction or transportation of smaller or more fragile specimens. (c) Upwardly increasing shell/sediment ratio changes character of substrate; species undergoes nonevolutionary adaptation by ecophenotypic response. (Adapted from Kidwell, 1986.)

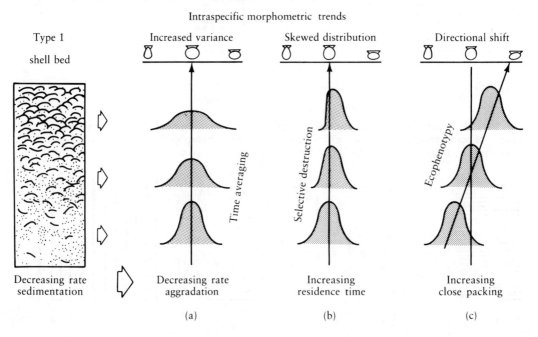

Intraspecific morphometric trends

vals. Morphologic variation per sample would therefore increase (Fig. 6-20a). Possible bias might also increase (Fig. 6-20b), because some specimens, such as small ones, may be more susceptible to dissolution or destruction during the longer periods of exposure caused by smaller sedimentation rates. Finally, the increasing concentration of shells relative to matrix may control the local occurrence of species or morphologic types. For example, we would expect more highly inflated forms with greater surface areas to resist better the force of gravity tending to drag the organisms into muddy bottoms (Fig. 6-20c).

Stratigraphic Completeness

Stratigraphically incomplete sections may seriously affect the quality and usefulness of the fossil record for biostratigraphic studies. Stratigraphic incompleteness is caused by variation in the constancy and rate of sedimentation (Chapter 3). The implications of an incomplete stratigraphic record for biostratigraphy are relatively straightforward.

1. Punctuations in the stratigraphic record such as those caused by intervals of erosion, and halts in sedimentation, produce punctuated-looking lineages (Kellogg, 1983).
2. Sedimentary punctuations decrease the probability of finding the true maximum stratigraphic limits to the occurrences of species.

Constancy and Rate of Sedimentation.   The rates at which sediments accumulate vary over many orders of magnitude (Schindel, 1980; Sadler, 1981; Chapter 3). The highest rates of sedimentation occur in environments in which sedimentation is more intermittent than constant, such as during intermittent flooding in deltaic systems. The lowest rates of sedimentation are associated with environments in which sedimentation is more constant than intermittent, such as in the deep sea, where microscopic debris constantly rains onto the abyssal plains. For biostratigraphic studies in which the goal is to derive a highly resolved view of morphologic evolution, speciation, or biogeographic range shifts, different microstratigraphic sampling schemes may be employed. For example, Schindel (1980) suggests that sampling on the scale of 1–10 cm is required for observation of potential speciation events in the deep sea and inland lakes, which exhibit lower yet more constant rates of sedimentation than fluvial systems. For the latter environments the same short time intervals of speciation may be spread out over 10–100 m of section because of larger sedimentation rates.

Punctuated Sedimentation.   Although environments that exhibit high rates of sedimentation, such as fluvial systems, might potentially record microevolutionary changes and speciation events with greater fidelity, it is an unfortunate empirical fact that these systems are notoriously intermittent; sedimentation may alternate with erosion events in a series of irregular fits and starts (Schindel, 1980, Fig. 2). When the rocks of these paleoenvironments are sampled, any patterns of continuous morphologic change will be interrupted by an intermittent sedimentary record.

We have already discussed the truncation effect (see Fig. 6-10), which artificially enhances the appearance of abrupt extinction. A similar effect may artificially enhance the apparent abruptness of speciation. For example, Williamson (1981) studied many lineages of Lake Turkana molluscs and interpreted the simultaneous punctuational changes in 13 of them as caused by isolation and environmental stress. The simultaneity of these changes, however, may have been exaggerated because "the speciation events occur at times of lacustrine regressions" (Williamson, 1982), in which the sedimentary column is compressed and the details of evolutionary changes are "lost in the unconformities" (Lindsay, 1982).

Biozones

Though the geologic record is replete with unconformities and the history of life is biased by the differential preservation of species, biostratigraphers require practical guidelines for subdividing strata on the basis of fossil content. A framework for the practice of stratigraphy in North America was published by the North American Commission on Stratigraphic Nomenclature (NACSN) as the North American Stratigraphic Code (NACSN, 1983). Biostratigraphic terms and concepts are defined in the code and form the modern basis for biostratigraphy.

The **biozone**, according to the NACSN, is the fundamental biostratigraphic unit and is "a body of rock defined or characterized by its fossil content" (NACSN, 1983, Articles 48, 53). The boundaries (lateral and vertical) of a biozone are defined by the limiting occurrences of the defining fossil events (Fig. 6-21). The lateral limits of a biozone are determined by the biogeographic distributions of the defining species; the vertical limits result from the geologic persistence of the defining species. Because species persisted for different intervals of time and were geographically distributed to different extents, the limits to a biozone may vary widely. No quantitative restriction as to minimum or maximum limits have been set by the NACSN.

The NACSN defines three types of biozones: interval, assemblage, and abundance biozones. An **interval zone** is a body of strata found between the limiting occurrences of any two taxa. The NACSN recognizes three major types of interval zones (Fig. 6-22). The **taxon range zone** is defined as the body of strata between the lowest and highest occurrences of a single taxon (Fig. 6-22a). The **concurrent range zone** is the body of strata found between the lowest occurrence of one taxon and the highest occurrence of a second taxon (Fig. 6-22b). The limits of a **lineage zone** are the lowest and highest occurrences of a named taxon that forms part of an evolutionary series (Fig. 6-22c). For example, we previously discussed *Stricklandia lens*, a Lower Silurian brachiopod. A key to its biostratigraphic utility is the progressive change in morphology of the cardinal processes (Fig. 6-14). Johnson (1979) used such changes to correlate beds of the Hopkinton Dolomite (Iowa, USA) with Llan-

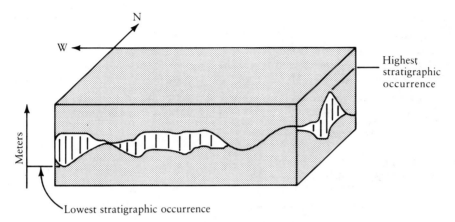

FIG. 6-21 • Hypothetical biozone of a taxon indicated by geographic and vertical limits of fossil-bearing section. Lateral limits are established by the biogeographic range of the taxon. Vertical limits are controlled in part by the geologic persistence of the taxon.

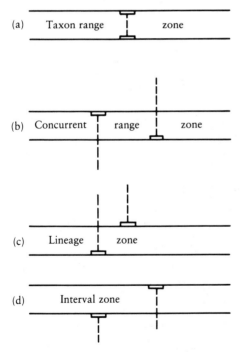

(a) Taxon range zone

(b) Concurrent range zone

(c) Lineage zone

(d) Interval zone

FIG. 6-22 • Examples of various types of interval biozones defined by NACSN. (Adapted from North American Commission on Stratigraphic Nomenclature, 1983.)

dovery Series strata in Wales (Fig. 6-23). Finally, the limits between successive lowest or highest occurrences of any unrelated taxa define unqualified interval zones (Fig. 6-22d).

The NACSN defines the **assemblage zone** as a "biozone characterized by the association of three or more taxa" (NACSN, 1983, Article 51). Two types of assemblage biozones are recognized: the unqualified term assemblage zone and the Oppel zone (also called a concurrent range zone). The **assemblage zone** is simply the inter-

FIG. 6-23 ● Biostratigraphic ranges of various brachiopods within Llandovery Series strata in Wales. (Reprinted by permission of the publisher and author from Johnson, *Palaeontology*, v. 22, 1979.)

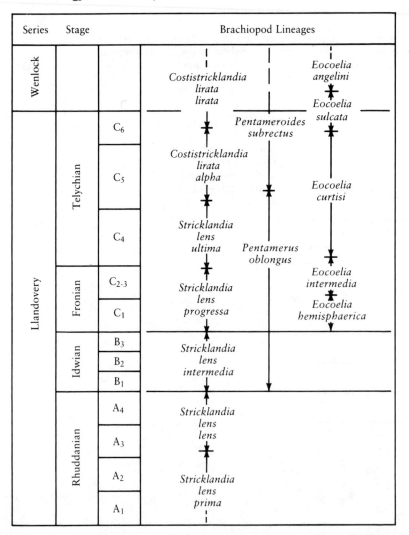

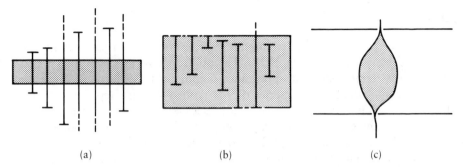

(a)   (b)   (c)

FIG. 6-24 • Assemblage biozones defined by NACSN. (a) Assemblage biozone; note the overlap of species occurrences. (b) Oppel biozone; note the coincident limits of species occurrences. (c) Abundance biozone; the width of the bar represents the abundance of fossils in the strata. (Adapted from North American Commission on Stratigraphic Nomenclature, 1983.)

val throughout which three or more taxa are all found (Fig. 6-24a); the lower and upper limits of the assemblage zone are placed without regard to the limits of the defining taxa (except that these limits must extend to, or beyond, the zone boundaries). In contrast to the assemblage zone, the **Oppel zone** boundaries (Fig. 6-24b) are located where limits of two or more of the characterizing taxa nearly coincide. Furthermore, the defining taxa may be, but are not necessarily, present throughout the entire interval of the Oppel zone.

An **abundance zone** (Fig. 6-24c) "is a biozone characterized by quantitatively distinctive maxima of relative abundance of one or more taxa" (NACSN, 1983, Article 52). Such a pulse in population abundance of a species is usually infrequent and may occur because of the conjunction of favorable or optimum environmental variables, such as temperature, abundance of food, or lack of predators. But beds of concentrated remains may also be due to differential preservation, varying rates of sedimentation, or other taphonomic processes. Because these environmentally sensitive conditions may recur from time to time, their use in time correlation may be subject to significant error.

Chronostratigraphic Significance of Biozones

Since the time of William Smith and the making of his geologic map (1815), stratigraphers have depended on the unique history of species throughout geologic time to help them chronologically order strata of the geologic record. But you cannot assume that a boundary of a biozone is everywhere the same age; for example, some time must pass for species populations to disperse throughout their habitat. The study of the time significance of biozones is **biochronology**.

Single Species

Because no species yet known has endured forever, every species is characterized by its unique occurrence in time (Fig. 6-25). The point in time at which it first came into existence is termed the **first appearance datum** (**FAD**). The point in time at which the last individual of the species dies is termed the **last appearance datum** (**LAD**). The interval in time during which the species lived is called its **range**.

We discussed the species as an array of populations distributed throughout the environment (this chapter) and whose biogeographic distribution varies with changing physical and biotic factors (Chapter 5). The dynamic nature of a species biogeographic distribution (Fig. 6-25a) makes improbable the time equivalence of all **first observed occurrences** (**FOO**) and **last observed occurrences** (**LOO**) throughout various local sections.

Because the FOO and LOO may vary in age from place to place, the biozone of the species (a taxon range zone, for example) will not necessarily be as large as the **biochronozone**, the total body of rock formed during the range of the defining taxon. In Figure 6-25b the largest biozone is located at section D but, because of a migrational delay relative to the FAD, does not entirely coincide with the biochronozone.

Species Assemblages

For a single species the biozone may represent different intervals of geologic time from one local section to another (Fig. 6-25b), introducing chronostratigraphic uncertainty into correlation studies. Biostratigraphers therefore utilize species assemblages in hopes of decreasing such uncertainty, because the greater the number of species with restricted stratigraphic ranges, the larger the probability that time-significant boundaries or intervals may be recognized.

In a study of Lower Mississippian rocks of northern Alabama, Ruppel (1979) discovered several conodont taxa within the Fort Payne Chert and Tuscumbia Limestone formations upon which biostratigraphically important biozones were previously established (Fig. 6-26). In the Upper Mississippi Valley the base of the *Gnathodus–Taphrognathus* biozone is recognized by the first abundant occurrence of the species *Gnathodus texanus*, whereas in southwestern Missouri the base is marked by the transition of *G. bulbosus* to *G. texanus*. The *Gnathodus-Taphrognathus* biozone is overlain by the *Taphrognathus varians–Apatognathus* biozone, the base of which is marked by the lowest occurrence of *Apatognathus* and its co-occurrence with *T. varians*. Overlying this zone is the *Apatognatus porcatus–Cavusgnathus* biozone, marked at its base by the lowest occurrence of *A. porcatus* and *Spathognatodus scitulus* and their first common occurrence with various species of *Cavusgnathus*.

At Ruppel's locality 1, a 15-m section of Fort Payne Chert yielded *Gnathodus texanus* and, near its base, some *G. bulbosus* (Fig. 6-27). Although *G. texanus* ranges throughout all three biozones, the occurrence of *G. bulbosus* restricts the

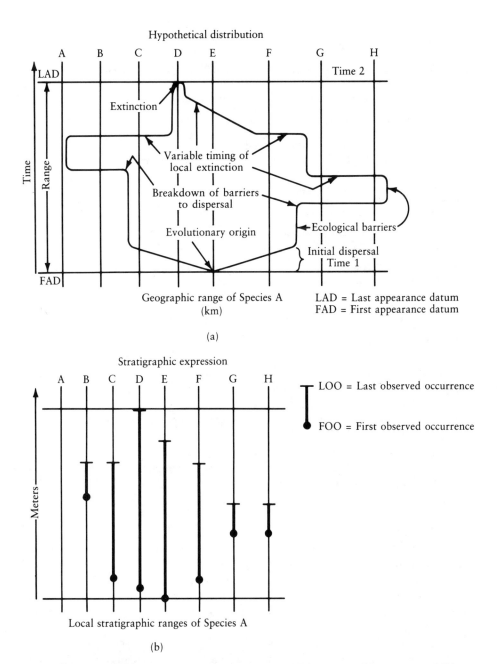

FIG. 6-25 • Hypothetical species lateral and vertical (a) stratigraphic ranges and (b) stratigraphic expression. Note that first and last observed occurrences (FOO and LOO, respectively), and biostratigraphic range vary with locality, and depend on history of migration and time of local extinction. (Adapted from Taylor, *in* Boardman, Cheetham, and Rowell, 1987.)

North American Series	Mississippi Valley Formations	Mississippi Valley conodont zones	Northwest Alabama conodont zones (Ruppel, 1979)	Northwest Formations
Valmeyeran	St. Louis	*Apatognathus scalenus* *Cavusgnathus*	*Apatognathus porcatus* *Cavusgnathus*	Tuscumbia
	Salem	*Taphrognathus varians*	*Taphrognathus varians*	
	Warsaw	*Apatognathus*	*Apatognathus*	
	Keokuk	*Gnathodus texanus* *Taphrognathus* / *Gnathodus bulbosus*	*Gnathodus texanus* *Taphrognathus*	Fort Payne

FIG. 6-26 ● Correlation of conodont biozones of northwestern Alabama with Upper Mississippi Valley formations. (Adapted from Ruppel, 1979.)

correlation of the lower part of this section to the *Gnathodus-Taphrognathus* biozone of southwestern Missouri. At the Vulcan Quarry, Ruppel sampled a 49-m section of the uppermost Tuscumbia Limestone and collected a diverse fauna that included *Apatognatus porcatus, Spathognatodus scitulus,* and three species of *Cavusgnathus.* These species indicate that this section correlates with the *Apatognatus porcatus–Cavusgnathus* biozone of the Upper Mississippi Valley. At two intermediate localities, Ruppel sampled sections that ranged through the Fort Payne Chert– Tuscumbia Limestone boundary. The most common conodonts found were *G. texanus* and *T. varians.* Unfortunately, these taxa range through all three biozones and are therefore unable to provide evidence for precise correlation to the Upper Mississippi Valley–Missouri sections. Furthermore, a key indicator genus, *Apatognatus,* is not present in these sections, and so precise placement of them on the biozonation chart (Fig. 6-27) is not possible.

Because these conodont faunas lack some key taxa necessary for more precise correlations to the Upper Mississippi Valley and southwestern Missouri regions, an alternative working biostratigraphy could be proposed. Instead of futilely trying to recognize separately the lower two biozones, a conodont worker could lump the taxa present in this region and define a broader biozone based on just those common species.

Abundant numbers of species do not always guarantee the presence of biostratigraphically significant ones. In Maryland and adjacent states an abundant invertebrate fauna of over 200 species is found throughout the Keyser Limestone. Because the Keyser Limestone is similar both paleontologically and lithologically to the overlying Helderberg limestone (Devonian) and the underlying Tonolaway limestone (Silurian), the age of this formation was controversial. From a study of the biostratigraphic distribution of brachiopods in the Keyser Limestone, Bowen (1967) located the Silurian-Devonian boundary between two of its biozones (Fig. 6-28) by the following analysis. Of the 26 genera of brachiopods, two (*Meristella* and *Nanothyris*) are found only in the upper zone, and these are known only from Lower Devonian or younger rocks elsewhere in North America. One coral (*Cystihalysites*) is found only in the lower biozone, and corals of its type are known only from pre-Devonian rocks in other areas. Bowen therefore concluded that the Silurian-Devonian boundary must be located between the two biozones.

Just because a biostratigraphic boundary is located between two biozones defined by species assemblages, there is no guarantee that such a boundary is everywhere the same age. After all, environmental conditions that change enough to cause a turnover in the local fauna may spread slowly from one region to another. Scott (1985) reviews the problem of diachronous biostratigraphic horizons and sug-

FIG. 6-27 • Inferred conodont biozonation of northwestern Alabama Fort Payne Chert and Tuscumbia Limestone formations. There is no horizontal scale, but localities 1 and 4 are about 60 km apart. (Adapted from Ruppel, 1979.)

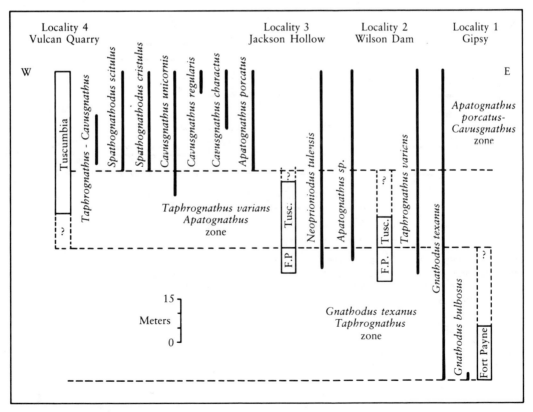

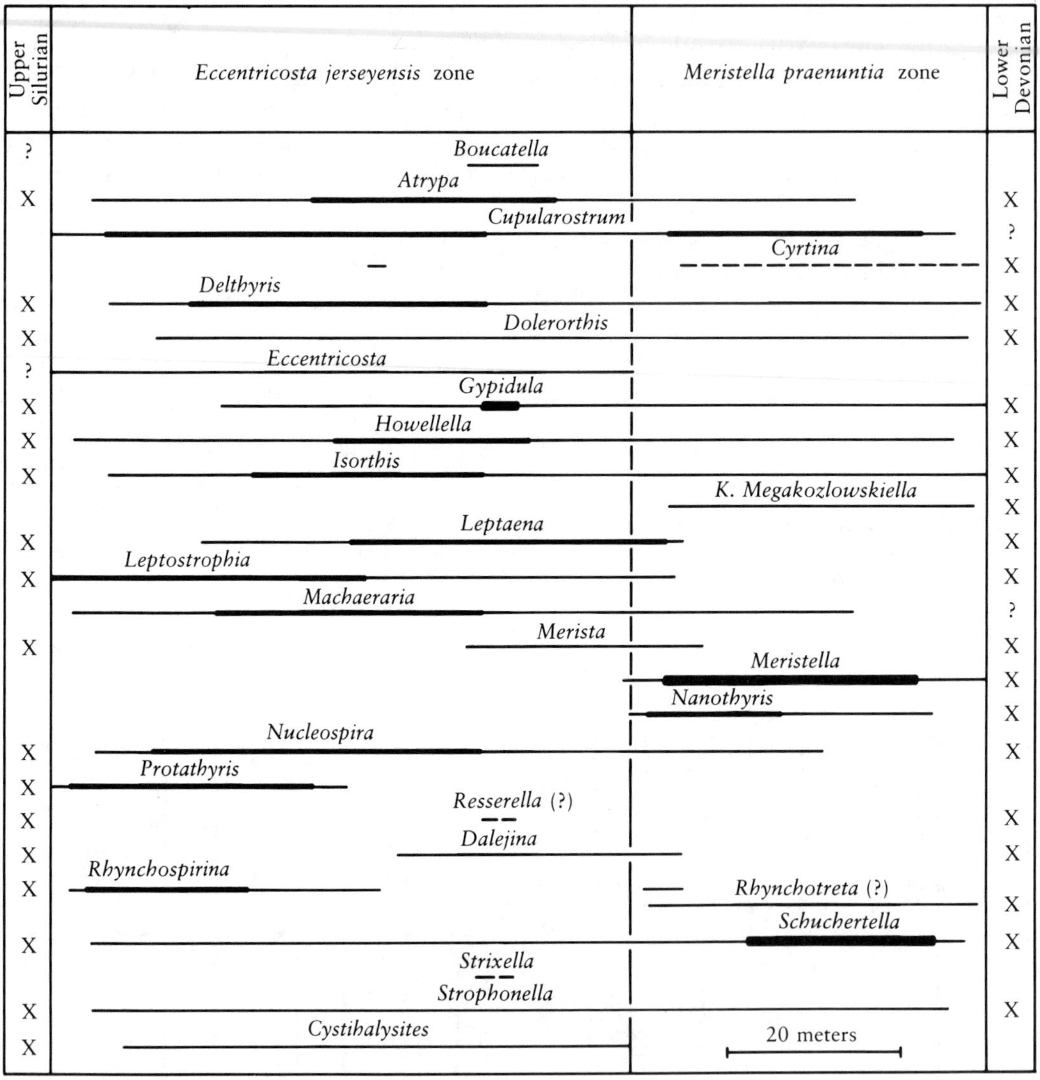

FIG. 6-28 ● Biozonation of composite stratigraphic section of Keyser Limestone in Maryland and adjacent areas. All taxa except *Cystihalysites* are brachiopods. The Silurian-Devonian boundary must lie between the two zones based on known geologic ranges of same taxa from other areas. (Adapted from Bowen, 1967.)

gests a test for their recognition: diachronous boundaries will be revealed by reversals of order in FOOs of two or more taxa from two or more local sections. For example, if the true migrational history of two species, A and D, is that illustrated in Figure 6-29a, then their stratigraphic ordering will appear the same in a series of three local sections. If the order of first appearances is the same, the FOOs of the two species are said to be **homotaxial**.

FIG. 6-29 • Patterns of first observed occurrences (FOOs) are controlled in part by migrational history of taxa. True migrational histories and pattern of stratigraphic FOOs for (a) two, (b) three, and (c) four hypothetical species. Note that as FOOs become closer together (in time and in stratigraphic proximity), detection of reversal of FOOs (c) becomes more probable.

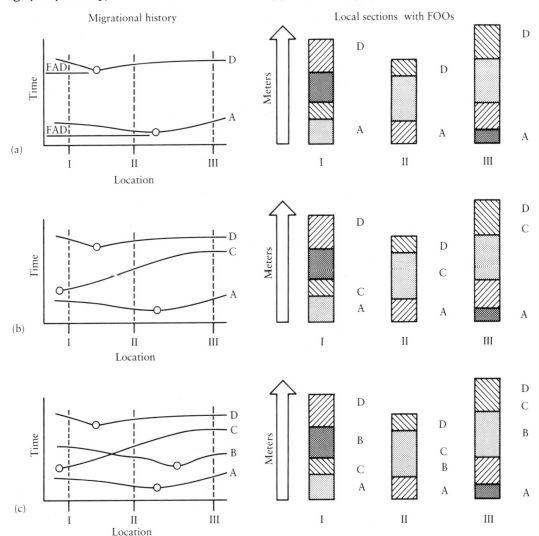

If an additional species (C) is available for analysis, and its true temporal history is that shown in Figure 6-29b, then the pattern of stratigraphic ordering may appear as shown in the local sections. The order of all observed occurrences (FOOs and LOOs) will again be homotaxial. If a fourth species (D) is available for analysis and the true temporal history is that of Figure 6-29c, we should detect a reversal of the stratigraphic ordering of species B and C at local section I, compared to sections II and III.

Scott argues that as FOOs are found closer together and the number of species increases, the hypothesis of homotaxy becomes easier to falsify. Therefore homotaxial sequences comprised of many species are more likely indicators of a smaller, unique interval of time.

Assemblage biozones are likely to be smaller on average than the biozones of individual taxa because the probability is small that several or more species evolve or become extinct at nearly the same time (Fig. 6-30a). But some workers (e.g., Murphy, 1977) criticize the use of assemblage zones because either gaps or overlaps are likely to occur between them (Fig. 6-30b). Murphy (1977) suggests the use of interval zones defined by only the first observed occurrence of a single species to permit the development of an unbroken time-stratigraphic scale (Fig. 6-31). The North American Commission on Stratigraphic Nomenclature (1983, Article 67 remark) also emphasizes the use of only lower boundaries for all chronostratigraphic units.

For example, in studies of the Middle Cambrian trilobite, *Ptychagnostus*, Robison (1982) and Rowell et al. (1982) found several species present in Nevada and Utah (Fig. 6-32). Because two species, *P. gibbus* and *P. atavus*, are found nearly worldwide (Great Basin, Scandinavia, Australia, Siberia, Greenland, China), they may be used to define biozones of potentially great importance in long-distance correlation. In the Great Basin deposits, Robison (1982) defined the base of the *P. gibbus* interval biozone as the first observed occurrence of *P. gibbus* and the base of the overlying *P. atavus* biozone as the first observed occurrence of *P. atavus*.

FIG. 6-30 • Distinctive assemblages of hypothetical species (a) may occur sporadically throughout a stratigraphic section (b), thereby leading to gaps in the biostratigraphic zonation scheme.

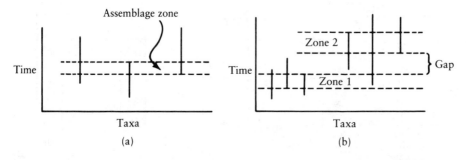

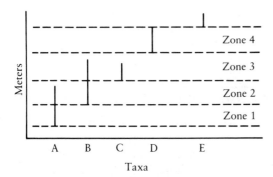

FIG. 6-31 ● Biozonation of hypothetical species based only on first observed occurrences.

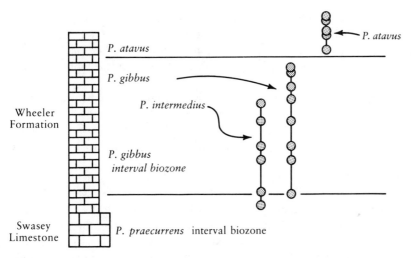

FIG. 6-32 ● Biozonation by first observed occurrences of Middle Cambrian trilobites *Ptychagnostus gibbus* and *P. atavus* in strata of the Great Basin. (After Rowell et al., 1982.)

With what confidence can we believe that the location of the top of the *P. gibbus* interval zone (bottom of the *P. atavus* interval zone) will indicate about the same point in time, whether this zone is studied in rocks of China, Greenland, or Nevada? Robison (1982) and Rowell et al. (1982) have found another species in the *P. gibbus* zone, *P. intermedius*, which they hypothesize to be the ancestor of *P. atavus* because (1) *P. intermedius* is always homotaxial with *P. atavus* (*P. intermedius* is always found stratigraphically below *P. atavus*), and (2) nowhere in the world have *P. intermedius* and *P. atavus* been found together in the same rocks. If this hypothesis of ancestor-descendant relationship is true, then the evolutionary event that led to the extinction of *P. intermedius* and the evolution of *P. atavus* is exactly the same event in time. Therefore we may use the FOO of *P. atavus* to approximately locate that point.

Tests of Chronostratigraphic Significance

It may be difficult to discern in stratigraphic sequences the true evolutionary changes within and among species because such processes as migration, preservation, compaction, and erosion may bias the fossil record. How is the biostratigrapher to sort out environmental and preservational effects from those of evolutionary change? From our previous discussions of the properties of species and the processes of preservation we make the following hypotheses and suggest programs for further study.

1. If in the vertical distribution of a species, morphologic trends reverse direction, migration of ecophenotypic varieties into the local section may be the cause. Collect additional stratigraphic sequences from a variety of other local sections. If you cannot confirm the same pattern of morphologic trends, ecophenotypy is the likely explanation.
2. If a species reappears higher in the stratigraphic column, it may be only temporarily extinct in one or more local sections owing to the Lazarus effect. Its renewed presence is caused by migration into the local section from a refuge elsewhere. Collect other widely distributed stratigraphic sections to hunt for the refuges.
3. If the limits of stratigraphic occurrence (FOO or LOO) of one or more species coincides with a change in lithology, the distribution of the species may be controlled by environmental factors. The species may be ecologically restricted to substrates of specific grain size or regions of particular temperature or salinity or be affected by the distributions of predators. Collect other stratigraphic sections of varying lithologies. If the species is found only in one facies, ecologic control is likely.
4. If the limits of occurrence (FOO or LOO) of one or more species coincides with a change in lithology or a confirmed unconformity, the stratigraphic distribution of the species may have been truncated by nondeposition, nonpreservation, or erosion (the truncation effect). Some breaks in the stratigraphic history of a local section are apparently undetectable in the lithologic record. Collection of stratigraphic sections in other areas or regions may reveal the missing interval and extend the stratigraphic range.

Quantitative Methods

One of the ultimate uses of biozones is for correlating one local section to another on the basis of geologic time. The U.S. biostratigrapher is free to subdivide the stratigraphic column with any type and combination of zones from those recognized by the North American Stratigraphic Code. For example, Figure 6-33 illustrates two en-

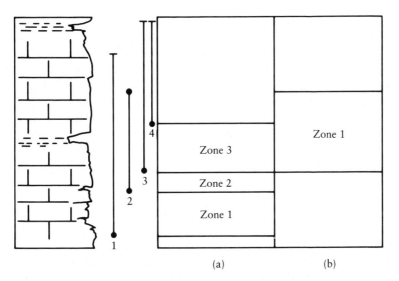

FIG. 6-33 • Two different biostratigraphic zonations based upon the same taxa. (a) Interval biozones based on first occurrences of species 1, 2, 3, and 4. (b) Assemblage biozone based on co-occurrence of species 1, 2, and 3.

tirely different zonal schemes for a hypothetical section. It is likely that few biostratigraphers would zone the same section of rock in exactly the same way. Because a zonal scheme may depend upon the tastes and views of the individual biostratigrapher, the correspondence of "zones" from composite sections to the reality of fossil distributions in the rocks has been questioned by Hazel (*in* Kauffman and Hazel, 1977), who facetiously defines "biochartigraphy" as the study of chart biozones.

As stratigraphers have become more quantitative and receptive to computer-based techniques, a variety of mathematical methods have been applied to the analysis of biostratigraphic data. Applications range from the construction of zonal subdivisions by Q-mode factor analysis (Hazel, *in* Kauffman and Hazel, 1977) to the probabilistic determination of the true chronologic ordering of species ranges (Edwards, *in* Cubitt and Reyment, 1982). The field of quantitative biostratigraphy is rapidly progressing (for a recent review, see Agterberg and Gradstein, 1988), but the lack of consensus at the present time on the best methods makes it difficult to predict which techniques will provide useful results in any given study.

Conclusions

Our understanding today of the biological species is vastly different from that of the geologists who, during the pre-Darwinian decades, constructed the major subdivisions of the geologic time scale on the basis of gross faunal differences (for example, dominance of reptiles in the Mesozoic versus mammals in the Cenozoic). That a detailed understanding of the evolution of species was not necessary for the successful establishment of the coarse geologic time scale is supported by the fact that no modifications to the serial order of its periods have yet been necessary. But if we look to more regional areas and shorter intervals of time, the construction of high-resolution biostratigraphic zones becomes dependent upon an increasingly detailed knowledge of a species' responses to ecologic conditions and its course of evolution. The biostratigrapher today may function more effectively as a member of a team, which might include the taxonomic specialist, ecologist, and biometrician. The potential payoff is a resolution in the Earth's history undreamed of by our intellectual forebears.

Summary

Biozones are based on evolutionary changes in a single lineage, species turnover, or assemblages of species, the composition of which may vary owing to extinction and evolution. Many species of geographically widespread distribution also tend to possess adaptations that confer a high degree of extinction resistance and geologic longevity in comparison to narrowly distributed species. Although many eurytopic species are more suitable for long-distance correlation than stenotopic ones, their lengthened geologic longevity makes them individually less suitable for the construction of high-resolution biozones.

High-resolution biozonation requires extensive studies of species' morphologic variation, ecology, and geographic distribution. Assemblages of species for which such information is available may be used to define relatively small intervals of geologic time of great use in long-distance correlation.

Processes of sedimentation may greatly alter the quality of the fossil record and hence biozone resolution. The number and sizes of samples collected for biostratigraphic analysis should be adjusted according to the sampled lithofacies.

References

Agterberg, F. P. and Gradstein, F. M., 1988. Recent developments in quantitative stratigraphy. Earth-Sci. Rev., v. 25, pp. 1–73.

Alexander, R. R., 1975. Phenotypic lability of the brachiopod *Rafinesquina alternata* (Ordovician) and its correlation with the sedimentologic regime. J. Paleontol., v. 49, pp. 607–618.

Baarli, B. G., 1986. A biometric re-evaluation of the Silurian brachiopod lineage *Stricklandia lens/S. laevis*. Palaeontology, v. 29, pp. 187–205.

Baker, C., 1983. Evolution and hybridization in the radiolarian genera *Theocorythium* and *Lamprocyclas*. Paleobiology, v. 9, pp. 341–354.

Barigozzi, C. (ed.), 1982. Mechanisms of Speciation. New York, Alan R. Liss, 546 pp.

Boardman, R. S., Cheetham, A. H., and Rowell, A. J. (eds.), 1987. Fossil Invertebrates. Boston, Blackwell Scientific Publications, 713 pp.

Bookstein, F. L., Gingerich, P. D., and Kluge, A. G., 1978. Hierarchical linear modeling of the tempo and mode of evolution. Paleobiology, v. 4, pp. 120–134.

Bowen, Z. P., 1967. Brachiopoda of the Keyser Limestone (Silurian-Devonian) of Maryland and Adjacent Areas. New York, Geological Society of America, Mem. 102, 103 pp.

Bown, T. M. and Rose, K. D., 1987. Patterns of dental evolution in Early Eocene Anaptomorphine primates (Omomyidae) from the Bighorn Basin, Wyoming. J. Paleontol., v. 61, p. 162.

Coyne, J. A. and Barton, N. H., 1988. What do we know about speciation? Nature, v. 331, pp. 485–486.

Cubitt, J. M. and Reyment, R. A. (eds.), 1982. Quantitative Stratigraphic Correlation. New York, John Wiley & Sons, 301 pp.

Darwin, C., [1859] 1964. The Origin of Species. Reprint. Cambridge, Mass., Harvard University Press, 513 pp.

Dobzhansky, T., Ayala, F. J., Stebbins, G. L., and Valentine, J. W., 1977. Evolution. San Francisco, W. H. Freeman & Co., 572 pp.

Elliott, D. K. (ed.), 1986. Dynamics of Extinction. New York, John Wiley & Sons, 294 pp.

Erwin, D. H., Valentine, J. W., and Sepkoski, J. J., 1987. A comparative study of diversification events: The early Paleozoic versus the Mesozoic. Evolution, v. 41, pp. 1177–1186.

Etter, R. J., 1988. Asymmetrical developmental plasticity in an intertidal snail. Evolution, v. 42, pp. 322–334.

Futuyma, D. J., 1979. Evolutionary Biology. Sunderland, Mass., Sinauer Associates, 565 pp.

Futuyma, D. J., 1988. *Sturm und drang* and the evolutionary synthesis. Evolution, v. 42, pp. 217–226.

Gingerich, P. D. 1985. Species in the fossil record: Concepts, trends, and transitions. Paleobiology, v. 11, pp. 27–41.

Gould, S. J. and Eldredge, N., 1977. Punctuated equilibria: The tempo and mode of evolution reconsidered. Paleobiology, v. 3, pp. 115–151.

Gould, S. J. and Eldredge, N., 1986. Punctuated equilibrium at the third stage. Syst. Zool., v. 35, pp. 143–148.

Harland, W. B., Cox, A. V., Llewellyn, P. G., Pickton, C. A. G., Smith, A. G., and Walters, R., 1982. A Geologic Time Scale. New York, Cambridge University Press, 131 pp.

Hecht, A. D., 1974. Intraspecific variation in recent populations of *Globigerinoides ruber* and *Globigerinoides trilobus* and their application to paleoenvironmental analysis. J. Paleontol., v. 48, pp. 1217–1234.

Jablonski, D., 1980. Apparent versus real biotic effects of transgressions and regressions. Paleobiology, v. 6, pp. 397–407.

Jablonski, D., 1984. Keeping time with mass extinctions. Paleobiology, v. 10, pp. 139–145.

Jackson, J. B. C., 1974. Biogeographic consequences of eurytopy and stenotopy among marine bivalves and their evolutionary significance. Amer. Naturalist, v. 108, pp. 541–560.

Johnson, M. E., 1979. Evolutionary brachiopod lineages from the Llandovery Series of eastern Iowa. Palaeontology, v. 22, pp. 549–568.

Kauffman, E. G., and J. E. Hazel (eds.), 1977. Concepts and Methods of Biostratigraphy. Stroudsburg, Pa., Dowden, Hutchinson & Ross, 658 pp.

Kellogg, D. E., 1983. Phenology of morphologic change in radiolarian lineages from deep-sea cores: Implications for macroevolution. Paleobiology, v. 9, pp. 355–362.

Kidwell, S. M., 1986. Models for fossil concentrations: Paleobiologic implications. Paleobiology, v. 12, pp. 6–24.

Koch, C. F., 1987. Prediction of sample size effects on the measured temporal and geographic distribution patterns of species. Paleobiology, v. 13, pp. 100–107.

Lane, N. G., 1986. Life of the Past. Columbus, Charles E. Merrill Publishing Company, 326 pp.

Levinton, J. S., 1982. Marine Ecology. Englewood Cliffs, N.J., Prentice-Hall, 526 pp.

Levinton, J., 1988. Genetics, Paleontology, and Macroevolution. Cambridge, Mass., Cambridge University Press, 637 pp.

Levinton, J. S. and Simon, C. M., 1980. A critique of the punctuated equilibria model and implications for the detection of speciation in the fossil record. Syst. Zool., v. 29, pp. 130–142.

Lewin, R., 1988. A lopsided look at evolution. Science, v. 241, pp. 291–293.

Lindsay, D. W., 1982. Punctuated equilibria and punctuated environments. Nature, v. 296, pp. 611–612.

Malthus, T. R., [1798] 1970. An Essay on the Principle of Population. Reprint. Baltimore, Pelican Books, 291 pp.

Mayr, E., 1970. Populations, Species, and Evolution. Cambridge, Mass., Harvard University Press, 453 pp.

Milkman, R. (ed.), 1982. Perspectives on Evolution. Sunderland, Mass., Sinauer Associates, 241 pp.

Murphy, M. A., 1977. On time-stratigraphic units. J. Paleontol., v. 51, pp. 213–219.

North American Commission on Stratigraphic Nomenclature, 1983. North American stratigraphic code. Amer. Assoc. Petroleum Geol. Bull., v. 67, pp. 841–875.

Pinet, P. R. and Popenoe, P., 1985. Shallow seismic stratigraphy and post-Albian geologic history of the northern and central Blake Plateau. Geol. Soc. Amer. Bull., v. 96, pp. 627–638.

Rhoads, D. C. and Lutz, R. A. (eds.), 1980. Skeletal Growth of Aquatic Organisms. New York, Plenum Press, 750 pp.

Robison, R. A., 1982. Some middle Cambrian agnostoid trilobites from western North America. J. Paleontol., v. 56, pp. 132–160.

Rowell, A. J., Robison, R. A., and Strickland, D. K., 1982. Aspects of Cambrian agnostoid phylogeny and chronocorrelation. J. Paleontol., v. 56, pp. 161–182.

Rudwick, M. J. S., 1976. The Meaning of Fossils. New York, Science History Publications, 287 pp.

Ruppel, S. C., 1979. Conodonts from the Lower Mississippian Fort Payne and Tuscumbia Formations of Northern Alabama. J. Paleontol., v. 53, pp. 55–70.

Sadler, P. M., 1981. Sediment accumulation rates and the completeness of stratigraphic sections. J. Geol., v. 89, pp. 569–584.

Schindel, D. E., 1980. Microstratigraphic sampling and the limits of paleontologic resolution. Paleobiology, v. 6, pp. 408–426.

Schopf, T. J. M. (ed.), 1972. Models in Paleobiology. San Francisco, Freeman, Cooper & Company, 250 pp.

Scott, G. H., 1985. Homotaxy and biostratigraphical theory. Palaeontology, v. 28, pp. 777–782.

Scudder, G. G. E. and Reveal, J. L. (eds.), 1981. Evolution Today. Pittsburgh, Hunt Institute for Botanical Documentation, Carnegie Mellon University, 486 pp.

Sheldon, P. R., 1987. Parallel gradualistic evolution of Ordovician trilobites. Nature, v. 330, pp. 561–563.

Shimer, H. W. and Shrock, R. R., 1944. Index Fossils of North America. Cambridge, Mass., The M.I.T. Press, 837 pp.

Silver, L. T. and Schultz, P. H. (eds.), Geological Implications of Impacts of Large Asteroids and Comets on the Earth. Denver, Geological Society of America, Spec. Paper 190, 528 pp.

Simpson, G. G., 1961. Principles of Animal Taxonomy. New York, Columbia University Press, 247 pp.

Spight, T. M., 1973. Ontogeny, environment, and shape of a marine snail *Thais lamellosa Gmelin*. J. Exp. Mar. Biol. Ecol., v. 13, pp. 215–228.

Stanley, S. M., 1979. Macroevolution: Pattern and Process. San Francisco, W.H. Freeman and Company, 332 pp.

Stanley, S. M., 1985. Rates of evolution. Paleobiology, v. 11, pp. 13–26.

Surlyk, F. and Johansen, M. B., 1984. End-Cretaceous brachiopod extinctions in the chalk of Denmark. Science, v. 223, pp. 1174–1177.

Williamson, P. G., 1981. Paleontological documentation of speciation in Cenozoic molluscs from Turkana basin. Nature, v. 293, pp. 437–443.

Williamson, P. G., 1982. Williamson replies: Punctuated equilibria and punctuated environments. Nature, v. 296, pp. 611–612.

Wilmarth, M. G., 1925. The Geologic Time Classification of the United States Geological Survey Compared with Other Classifications. Washington, D.C., U.S. Geological Survey, Bulletin 769, 137 pp.

7
Depositional Environments and Facies
I: Continental Environments

The term **facies** has been variously defined and used since the original definition by Amand Gressly in 1838. Gressly's concept was that a facies of sedimentary rocks is defined by an aggregate of lithologic and organic characteristics that are not separable and that are distinguishable from adjacent facies. Thus we might recognize a facies characterized by texturally supermature quartz sand containing molluscan carbonate shells and an adjacent facies of illitic mud-shale containing only ichnofossils. Gressly interpreted his facies in environmental terms, so the sand facies might be a beach deposit and the shale facies a shallow marine deposit. A key point is that the facies itself is defined on objective grounds and includes all features of the rocks. An environmental interpretation that is later shown to be incorrect will not invalidate the facies designation.

During the past 150 years, however, other definitions of the term facies have appeared and have been accepted by stratigraphers and sedimentologists. For example, many workers accept a subdivision of Gressly's concept into *lithofacies* and *biofacies*. Thus there exist red-bed facies, black shale facies, and evaporite facies based on lithologic characteristics; *Calamites* facies, *Skolithus* facies, and radiolarian facies based on biologic features. Sometimes, chemical facies or mineralogic facies are distinguished on the basis of either field or laboratory differentiation; hence there are aragonitic facies, magnesian facies, or phosphatic facies. Finally, a large proportion of geologists use the term facies as genetic grouping, giving rise to categories such as basinal facies, fluvial facies, or deep-water facies. It seems clear that such usages cannot be legislated out of existence, regardless of the desires of the purists in our profession. The varied usages are too entrenched in the literature. We see no harm in categories such as evaporite facies, zeolite facies, gastropod facies, or cross-bedded sandstone facies, as long as it is recognized that such descriptions refer only to an aspect of a unit that is particularly distinctive and meaningful in the region being considered. The guiding principle should be functionality. If a usage serves a legitimate function in a scientific description, it is permissible.

The importance of facies in stratigraphy is that the analysis of facies is the basis for an environmental interpretation of stratigraphic units. Furthermore, the interrelationships among sedimentary environments, and therefore of facies, are not chaotic or random; they are subject to controls imposed by geological setting, tectonics, and climate. Thus the distribution of facies is subject to a number of regularities. The key to the interpretation of facies is accurate identification of lithologic and biologic attributes of individual rock units and their integration into geologically meaningful patterns. Even when this is done, however, a large proportion of facies as seen in the field, on seismic records, or on electric logs do not have unique interpretations. It is necessary to relate descriptions of individual facies to adjacent, broadly conformable facies that appear to be genetically related. This concept of **facies associations** is fundamental to environmental interpretation. Sometimes, patterns of facies are repeated in a stratigraphic sequence, giving rise to the concept of **cyclic sedimentation** (Chapter 10), one of the more fruitful insights of recent research. Cycles vary greatly in clarity, so statistical analysis of presumed regularities is commonly required. One psychological problem that typically arises when cycles are being analyzed is that noncyclic features (irregularly occurring or random events) tend to be overlooked or ignored, and such features may provide invaluable clues to the meaning of the cyclicity. Do not become so locked into a single method of analysis that you fail to exercise your powers of observation.

Walther's Law

Much of the progress in stratigraphic analysis during the past 30 years has resulted from its integration with sedimentology and the emphasis on the study of modern depositional environments ("actualism"). This approach was originally suggested by Johannes Walther in 1894. Walther's contribution to the principles of stratigraphic facies analysis was recognized in the English-speaking world relatively recently (Middleton, 1973) but is fundamental. **Walther's Law** states that facies that overlie each other in a conformable vertical sequence must have been laterally adjacent to one another at the time of deposition. Thus to find out what is possible in a vertical sequence of sedimentary rocks, a stratigrapher must study modern depositional environments. Without information from modern environments, stratigraphy quickly degenerates into a barren cataloging of rock and fossil sequences.

Facies in Modern Environments

Facies, as the term is variously used, can refer to a geographic unit as large as a continental glacier or oceanic abyssal plain ($\sim 10^7$ km²) or as small as lakes or beaches, which may be restricted to perhaps 1 km². Many books have been written discussing in great detail the various depositional environments and the facies they contain, and a few of the more recent ones are given in the references at the end of this chapter. In a stratigraphy text we must condense in the extreme our consideration of things such as fluid mechanics and sediment transport, water chemistry, and many aspects of modern environments that are rarely preserved in the stratigraphic record. For the interested reader we recommend Blatt et al. (1980) and Middleton and Southard (1978) for fluid mechanics and sediment transport, Drever (1988) and Stumm and Morgan (1981) for geochemistry, and Pettijohn (1975) for a description of many sedimentologic features rarely preserved in ancient rocks.

The classification of depositional environments commonly represented in ancient rocks that we use in this book is a standard one based on the threefold division of continental, transitional, and fully marine environments (Table 7-1). Within each of these three groups, environments are distinguished on the basis of physical distinctiveness, for example, fluvial versus eolian. Each of the dozen or so environments that we consider can, of course, be subdivided almost indefinitely: fluvial facies into braided and meandering streams; meandering streams into channel facies, overbank facies, and floodplain facies; and from there to various sand bar facies, crevasse-splay facies, and floodplain lake facies. In an introductory book, generalizations and omissions are inevitable. Nevertheless, the topic is of great importance to stratigraphic interpretation and deserves extensive coverage. This chapter will consider continental environments, Chapter 8 will cover transitional environments, and Chapter 9 will cover marine environments. The basic reference books for the discussions are Reading (1986), Walker (1984), and Scholle and Spearing (1982).

TABLE 7-1 • Classification of depositional environments and deposits

Continental	Transitional	Marine
Alluvial fans	Deltaic environments	Neritic siliciclastic
Braided stream deposits	Nondeltaic coastlines	Neritic carbonate
Meandering stream deposits		Evaporite
Eolian deposits		Deep water
Lacustrine deposits		
Swamp deposits		
Glacial deposits		

Alluvial Fans

An **alluvial fan** is a wedge-shaped accumulation of sediment formed at a place where a stream issues from a narrow mountain valley onto a surface that is relatively unconfined laterally. The fan is thus a result of a sharp decrease in stream gradient (from perhaps 1° to 0.1°) and a loss of horizontal confinement; it is independent of climate. Alluvial fans are most prominent in semiarid regions but are known from most other continental environments as well (Kochel and Johnson, *in* Koster and Steel, 1984). Most of the alluvial fans described from ancient rocks are from probable semiarid climates, as interpreted from the presence of vertebrate fossils, plant fragments, and freshness of chemically unstable minerals such as feldspars and ferromagnesians. Well-studied fans of this type in the United States include those in Triassic rocks of New England (Newark Group) and in Pennsylvanian rocks of central Colorado (Fountain and Minturn Formations). Thick alluvial fans such as these are orogenic deposits and can attain great thicknesses, commonly 10^3-10^4 m. Optimum conditions for thick accumulations occur where the rate of uplift exceeds the rate of downcutting of the trunk stream channel at the mountain front. Modern examples include the Himalayas and the plain of India to the south and the Sierra Nevada and California lowlands to the west.

In nonglacial settings, the deposition of an alluvial fan is characterized by grossly intermittent sedimentation on a human time scale. At any time, much of the fan surface is inactive, being the site of weathering, pedogenesis, and erosion. Nondepositional horizons are much more significant temporally than depositional periods. For example, the growth of the modern White Mountain fan in California can be explained by a single typical debris flow every 350 years for the past 700,000 years. In more humid climates, debris avalanches initiate transport of the sediment from high-relief areas to lowlands, where it is distributed by braided streams. The frequency of the debris avalanches varies with annual rainfall, bedrock type and coherency, and vegetation, intervals of 10^1-10^3 years being most common. In glacial regions, fans are formed by constantly shifting braided streams on an aggrading outwash plain, sediment deposition being continuous and regulated only by the rate of melting of the glacial ice.

As a result of the sharp change in slope at the foot of mountains, alluvial fan deposits have distinctive textural and sedimentary structural features in outcrop. The features are those characteristic of debris flows, sheetfloods, and sieve deposits. A **debris flow** is a moving mass of rock fragments, soil, and mud in which more than half the particles are of gravel size. As grain size decreases, debris flows grade into mudflows. Beds formed by debris flows are extremely poorly sorted, are unlaminated, and have a matrix-supported fabric, and the coarse particles are randomly oriented with respect to the major bedding surfaces (Fig. 7-1). Because the beds were deposited as lobes in a viscous medium flowing downslope, they tend to be tongue-shaped, each bed (individual flow) having a uniform thickness of up to several meters.

FIG. 7-1 • Bedding of Late Tertiary alluvial fan deposits, south side of the Santa Catalina Mountains, Arizona. (Reprinted by permission of the publisher from Bull, *Society of Economic Paleontologists and Mineralogists Spec. Pub. 16,* 1972.)

Debris flow

Sheetflood
Debris flow

Interbedded with the debris flow beds are the deposits of sheetfloods, layers formed by deposition of separate particles from water rather than by flow of a viscous mass. A **sheetflood** is a broad expanse of storm-generated water that spreads as a continuous film over a large area in an arid region; it is unchanneled. The bedding thickness of sheetflood deposits is variable, as is particle sorting. As sheetfloods give way to channelized flow, sorting improves to that of nonfan fluvial deposits.

Near the apex of an alluvial fan the sediments are very coarse grained and commonly are permeable because only small amounts of clay are produced in dry climates. Consequently, it is not unusual for some of the water rushing out of canyon mouths in semiarid areas to filter through the porous zone near the fan apex, perhaps carrying fine-grained sand and silt with it. The sediment thus stranded on the fan surface can be either lobate or linear in plan and is termed a **sieve deposit**. When linear, such deposits are oriented normal to the mountain front, parallel to depositional strike.

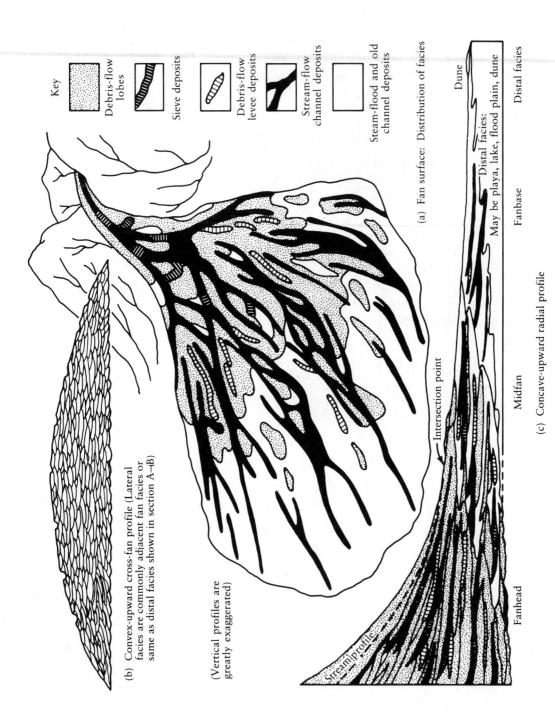

Key

Debris-flow lobes

Sieve deposits

Debris-flow levee deposits

Stream-flow channel deposits

Steam-flood and old channel deposits

(a) Fan surface: Distribution of facies

Intersection point

Stream profile

Fanhead   Midfan   Fanbase   Distal facies

(c) Concave-upward radial profile

Distal facies: May be playa, lake, flood plain, dune

Dune

Distal facies

(b) Convex-upward cross-fan profile (Lateral facies are commonly adjacent fan facies or same as distal facies shown in section A–IB)

(Vertical profiles are greatly exaggerated)

FIG. 7-2 ● Distribution of sediment facies and morphological profiles of an ideal alluvial fan. (Reprinted by permission of the publisher from Klein, *Sandstone Depositional Models for Exploration for Fossil Fuels*, 3rd ed., International Human Resources Development Corporation, 1985.)

The interbedding of debris flow with sheetflood deposits is a characteristic feature of alluvial fan deposits in outcrop. The massive thick bed of uniform thickness above the hat in Figure 7-1 consists of clayey gravel and was deposited as a viscous debris flow. Beneath the debris flow are beds of well-sorted water-laid sand. A 1-cm bed of water-laid clay immediately above the sands may be deposited during the waning phase of ephemeral water flooding on a fan, when the competence is sufficient to transport only silt and clay. Beds of poorly sorted silty gravel occur above the debris flow bed and beneath the water-laid beds. These beds may be interpreted as low-viscosity debris flow deposits or as poorly sorted water-laid deposits.

Considerable variety is present in radial and cross-fan stratigraphic relations (Fig. 7-2). Along radial sections of a fan, individual beds may be traced for long distances, and channel fill deposits are rare. In contrast, cross-fan sections reveal overlapping beds of limited extent that are interrupted by cut-and-fill structures. These structures are most common near the fan apex or midfan area and are rare near the toe of the fan. Bed thickness and maximum grain size decrease downfan, and pebble roundness may increase, and their trends may be mapped to determine distance to the fan apex (mountain front). Variations in rock fragment and mineral composition and paleocurrent indicators such as pebble imbrication and cross-bedding may reveal the boundaries of individual fans. In modern alluvial fans the ratio of fan length to fan width is about 1.5/1.

The downslope edge of an alluvial fan may interfinger with a variety of other depositional environments. If rainfall is adequate, fan deposits may grade into floodplain deposits of a perennial stream flowing normal to the fan radius (parallel to the mountain front), as exemplified today in central California. If the fan is located near a coastline, fan deposits can grade into marine sands or shales, as in Pennsylvanian deposits of southern Oklahoma. In arid regions, the toe of an alluvial fan may interfinger with playa lake muds, possibly containing evaporites as in desert areas of southern California.

The facies model for alluvial fans is straightforward and is well described as a pie-sliced accumulation of detrital sediment with the pie apex located at the mountain front. The thickest fan sections are located there, and thicknesses may exceed 10^3 m. The fan surface is subdivided into fanhead, midfan, and distal fan areas (Fig. 7-3), each area being dominated by different types of beds reflecting different depositional processes. Because of their distinctive three-dimensional geometry, abundance of coarse gravel, poor sorting, and association with orogenesis, alluvial fan deposits are generally easily recognized in outcrop. Commonly occurring coarsening-upward and fining-upward sequences with scales on the order of 10^2 m can result from either recurrent tectonic activity or climatic variation. Rapid increase in relief between the mountain area and the basin floor causes the coarse proximal fan deposits to prograde over more distal facies (Steel et al., 1977). The same result is obtained by increased sporadic and torrential rainfall within the mountainous region. Fining-upward sequences result from source area retreat during atectonic periods as fan regression superimposes more distal facies over more proximal facies (Elmore, 1984).

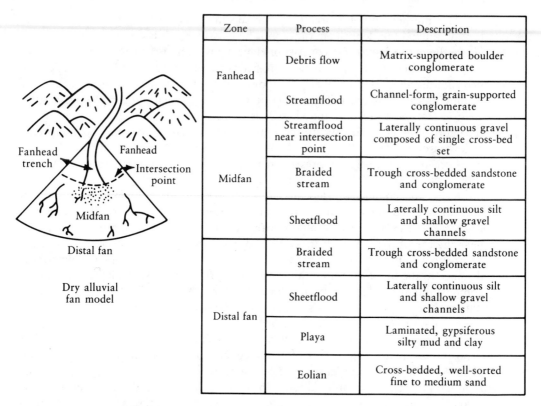

Zone	Process	Description
Fanhead	Debris flow	Matrix-supported boulder conglomerate
	Streamflood	Channel-form, grain-supported conglomerate
Midfan	Streamflood near intersection point	Laterally continuous gravel composed of single cross-bed set
	Braided stream	Trough cross-bedded sandstone and conglomerate
	Sheetflood	Laterally continuous silt and shallow gravel channels
Distal fan	Braided stream	Trough cross-bedded sandstone and conglomerate
	Sheetflood	Laterally continuous silt and shallow gravel channels
	Playa	Laminated, gypsiferous silty mud and clay
	Eolian	Cross-bedded, well-sorted fine to medium sand

FIG. 7-3 • Depositional model for arid and semiarid alluvial fans. (Reprinted by permission of the author from Mack and Rasmussen, *Geological Society of America Bulletin*, v. 95, 1984.)

Braided Stream Deposits

Braided streams commonly form on the surfaces of alluvial fans and can extend outward for hundreds of kilometers. The braided pattern results from a dominance of bedload sediment (gravel and sand) over suspended sediment (silt and clay) and from a widely fluctuating rate of water discharge. It has been suggested by Cotter (*in* Miall, 1978) that almost all streams were braided during the early Paleozoic because of the absence of vegetative cover, consequent low abundance of clay minerals, and lack of soil cover on the land surface. A region of high relief is required for braided stream development (in nonglacial areas), but the braided pattern may extend for many hundreds of kilometers outward from the orogenic belt. Modern examples include tropical rivers such as the Ganges, Brahmaputra, and Mekong, all of which have their headwaters in the Himalayan Mountains.

Because the development of braiding depends so heavily on the availability of coarse detritus and fluctuations in stream velocity, the sediment typically is deposited within a wide but shallow fluvial system as a series of longitudinal gravel or sand bars (Fig. 7-4). Gravel is more common closer to the source terrane; sand is

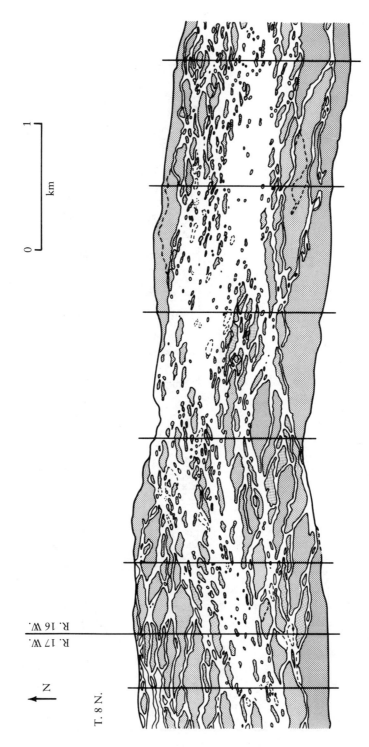

FIG. 7-4 ● Braided channel of the Platte River near Kearney, Nebraska, about 450 km east of the mountainous source area in Wyoming. Shaded areas are vegetated and underlain by older sand bars. As a result of damming and irrigation, the width of the river here has decreased by 50–90% since 1860. (Adapted from Eschner et al., 1983.)

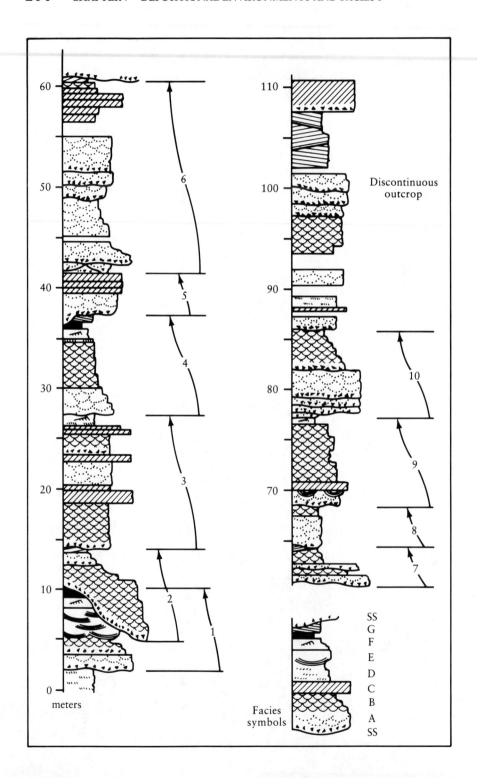

more common in the distal region. The sedimentary structures within braided stream deposits are very variable because of the great fluctuations in water velocity. Fining-upward sequences dominate. Bars may show parallel lamination or nearly unidirectional cross-bedding, and cut-and-fill structures are common. Sediment sorting is moderate to good in sand-dominated bars because of the absence of clay. Figures 7-5 and 7-6 illustrate braided stream subfacies of the Battery Point Sandstone (Devonian), Quebec, part of the clastic wedge shed westward from the rising Acadian (Caledonian) Mountains.

Braided streams are gradational into anastomosing streams, which have smaller width-depth ratios and, as a result, more stable channels. Cut-and-fill structures are less abundant in anastomosing channels. In post-Silurian humid climate terranes, channel stability can be enhanced by vegetation in interchannel areas, and extensive

FIG. 7-6 • Scour surface cutting down about 5 m stratigraphically from the notebook at top left to the man's hand at lower right. The light-colored slab in foreground is a regional bedding surface. The photo is the equivalent of 5 to 10 m on Figure 7-5. Trough cross-bedding in the lower left foreground is overlain by a prominent isolated scour (facies E) overlain by rippled sandstone and siltstone, with interbedded mudstone (facies F). (Reprinted by permission of the publisher from Cant and Walker, *Canadian Journal of Earth Science*, v. 13, National Research Council of Canada, 1976.)

◄ **FIG. 7-5** • Measured section of the Devonian Battery Point Sandstone, Quebec. Numbers indicate individual fluvial sequences, and letters define facies: SS = scoured surface; A = poorly defined trough cross-bedding; B = well-defined trough cross-bedding; C = large planar tabular cross-beds; D = small planar tabular cross-beds; E = isolated scour fills; F = ripple cross-laminated silts and muds; and G = low-angle inclined stratification. (Reprinted by permission of the publisher from Cant and Walker, *Canadian Journal of Earth Science*, v. 13, National Research Council of Canada, 1976.)

levees, marshes, and lakes may develop. As the proportion of suspended load in the stream increases and the variability of water discharge decreases, anastomosing streams are transformed into meandering streams.

Meandering Stream Deposits

The meandering alluvial valley system has probably received more study by sedimentologists than any other type of fluvial environment. Studies of modern examples began in the mid-1940s on the lower Mississippi River and have grown enormously since that time in both number and sophistication. Meandering streams occur in low-gradient areas of perennial streamflow where mud is the major component in transit. Subfacies of meandering stream deposits include the main channel, which is asymmetric in cross-section, abandoned channels, point bars, crevasse splays, and floodplains (Fig. 7-7). The abandoned channel deposits may be overlain by lacustrine sediments (oxbow lakes), and the floodplains may contain swamp sediments and economically valuable coal deposits. Many of the bituminous and anthracite coals of the Appalachian region were formed in this type of environment (Chapter 12).

FIG. 7-7 ● Block diagram showing morphological elements of a meandering river system. Erosion on the outside bend of a meander loop leads to lateral accretion on the opposite point bar. The dunes and ripples in the channel give rise to trough cross-bedding and ripple cross-lamination, respectively (inset, lower right), which are preserved in a fining-upward sequence. (Reproduced with the permission of the Geological Association of Canada from Walker and Cant, *in* Walker, *Facies Models*, 1984.)

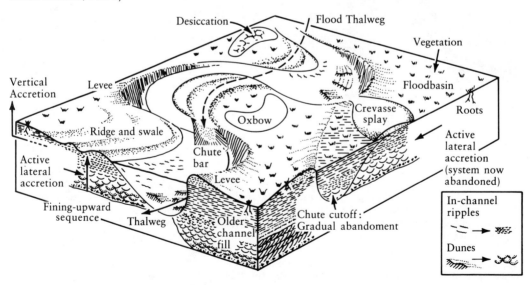

It commonly is difficult to distinguish between ancient braided and meandering stream systems in outcrop (Jackson, *in* Miall, 1978). The most generally applicable criteria for meandering stream deposits are (1) substantial mud content in sandy units, (2) muddy units present and thicker than sandy members, (3) asymmetrical channel-fill deposits that contain mud, and (4) absence of coarse gravel-size particles and wide variance in imbrication direction in any gravel present. These criteria all imply a highly sinuous channel with lateral migration, contrasted to the straight and very sandy channels that are characteristic of braided and anastomosing streams. As is true of most natural processes and environments, the boundaries between braided streams and anastomosing streams and between anastomosing streams and meandering streams are gradational, so although it is not difficult to write criteria for distinction in a textbook, applying them in the field setting is not always as easy.

The vertical section of meandering stream deposits is generally more variable in aspect than the corresponding section for braided streams, a difference resulting from the difference in mud content, predominance of lateral migration of the stream channel in a meandering system, and greater variety of subfacies in the meandering stream system (oxbows, swamps, etc.). A typical example from the Lower Old Red Sandstone (Lower Devonian) of the Anglo-Welsh Basin is shown in Figure 7-8. The deposits are the normal fining-upward sequence with a basal scoured surface recording the erosive wandering of the river channel, sandstones representing the deposits formed in the channel, and muddy clastics formed by floods overflowing their suspended load through and over levees bordering the channels (crevasse-splay deposits). The section at Mitcheldean contains, in addition to the extensive muds and soil traces characteristic of meandering stream deposits, evidence of braided stream reaches as well. The primary current lineation in zones 1–3 suggests high-velocity currents that exist in shallow, sandy river channels during floods but are absent in meandering parts of a river system.

Within the past few years, several workers have studied the soils developed in ancient meandering stream sequences as sources of stratigraphic information (Kraus and Bown, *in* Wright, 1986; Retallack, 1986b; Bown and Kraus, 1987; Kraus, 1987). The concept of a vertical soil profile in which lithologic properties differ has long been recognized; less commonly recognized is the fact that soil properties vary laterally as well, in response to differing environmental conditions. *Pedofacies* are as valid and can be as useful to an alluvial stratigrapher as lithofacies and biofacies. For example, Kraus and Bown (*in* Wright, 1986) reported that the Willwood Formation (Eocene) of Wyoming contains 500–1200 superposed paleosols and is a very complete alluvial sequence and that probably the only time not represented by sediment is in the scours in local sections. They were thus able to discern individual vertebrate fossil accumulations at a 2000- to 14,000-year level of discrimination. Their worst levels of faunal resolution were approximately 56,000 years, far better than is possible using other stratigraphic techniques and sufficient to distinguish punctuated and gradual evolution in the Willwood mammal record.

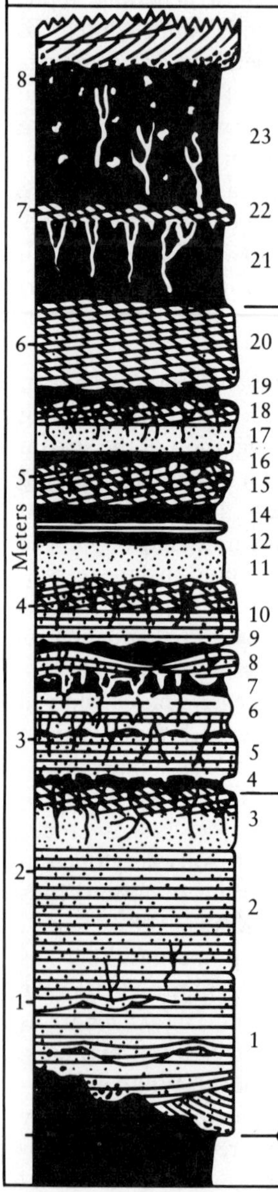

	Main facts	Interpretation
	Red sandy coarse siltstones with traces of ripple-bedding, abundant carbonate concretions, and invertebrate burrows. Thin ripple-bedded, very fine sandstone overlying sun-cracked surface.	Vertical accretion deposit from overbank floods. Backswamp deposit with intercalated levee tongue. Fluctuating groundwater table and periodic exposure.
	Alternation of thin sandstones and siltstones. Red sandy coarse siltstones with invertebrate burrows and rare carbonate concretions. Very fine to fine poorly sorted sandstones, flat- or ripple-bedded or massive. Commonly rest on sun-cracked or eroded surfaces. Tops gradational or sharp with ripples. Invertebrate burrows.	Vertical accretion deposit from overbank floods. Deposition of suspended load via bed load on levees, crevasse splays, and in backswamps. Repeated scour aggradation, and exposure of floodplain top stratum. Flow at times in direction away from earlier channel.
	Local lenses of siltstone clasts. Well-sorted fine sandstones with cross-stratification, flat-bedding, primary current lineation and fluted scoured surfaces. Top sandstone ripple-bedded with sharp top. Rare invertebrate burrows.	Probably mixed channel-fill and lateral accretion deposit. Deposition of bed load in channels, shallow and probably shifting and braided, with some wave action on exposed banks and bars. Local channel lag deposits.
	Scoured surface with 40 cm relief cut on siltstone. Small channels, flute casts, and current crescents.	Erosion at floor of wandering river channel.

FIG. 7-8 ● Generalized stratigraphic section of Lower Old Red Sandstone at Mitcheldean, England. (Reprinted by permission of the publisher from Allen, *Sedimentology*, v. 3, Blackwell Scientific Publications, 1964.)

FIG. 7-9 • Red fine-grained fluvial deposits, Moenkopi Formation (Triassic), Arizona. The upper flat-bottomed sand beds at the top of the outcrop are crevasse splays (no basal scour). The four thin white bands below the splay deposits are caliche beds within the red floodplain mud sequence. The lower half of the photo shows a mixture of splay sands and point bars (concave bases) with a large swale or scour fill at the base with large cross-beds. The thickness of the section is about 6 m. (Reprinted by permission of the American Association of Petroleum Geologists from Cant, *in* Scholle and Spearing, *American Association of Petroleum Geologists Memoir*, v. 31, 1982.)

Eolian Deposits .

Sediments deposited by wind include the great sand seas such as those in North Africa and elsewhere (Fig. 7-10), the silts termed **loess** that occur in North China and adjacent to the lower Mississippi River in the United States, coastal dune deposits found worldwide, and the silts that are deposited in the ocean basins by the larger-scale atmospheric circulation. In ancient rocks the sand seas (ergs) are volumetrically significant, ancient coastal dunes are less common, pre-Pleistocene loess is rare, and wind-deposited oceanic silt grains are widely scattered over the ocean floor and do not form discrete stratigraphic units.

The characteristic desert sedimentary deposit is the sand dune, and most geologic studies of modern deserts have concentrated on these structures. The focus of the studies has been either the mechanism of formation of sedimentary structures in individual dunes or the geomorphology of different dune shapes, and these investigations have greatly increased our understanding of atmospheric fluid mechanics. Many different types of dunes have been defined, but unfortunately, dune types are not commonly identified in ancient deposits.

Stratigraphically significant desert sand accumulations occur in paleotradewind belts in cratonic basins (Porter, 1987; Kocurek, 1988). One well-studied example is

FIG. 7-10 • A sand sea showing a complex linear ridge formed of several dune types. Aerial photo taken southwest of the Kuiseb River, Namib Desert, Namibia. (From McKee, 1979.)

the Navajo Sandstone in the western United States (McKee, 1979), which at present covers an area of 300,000 km², has an average thickness of about 300 m, and has a maximum thickness exceeding 800 m. The original extent of the Navajo dune field is unknown because of Cenozoic erosion. The formation represents about 5 million years of Jurassic time and thus an average rate of accumulation of about 60 μm/yr, illustrating once again the dominance of erosional over depositional events in the stratigraphic record.

Interpretation of the depositional environment is relatively straightforward for the bulk of the Navajo Formation. It is recognized as eolian by the stereotypical field criterion for desert dune deposits, the dominance of large-scale, steeply dipping, wedge-planar and tabular-planar cross-bedding (Fig. 7-11). The scale of the cross-strata, up to 15 m in thickness, results from the high relief of eolian dunes and is much larger than those formed in any other environment, with the possible exception of some modern subaqueous shallow marine dunes. Eolian cross-bedding in dipmeter plots is usually easy to recognize because of the great thicknesses of the cross-bed sets, the relatively high-angle dips of the foresets, and the usual decrease in angle of dip at the base of the set (Fig. 7-12). Other diagnostic field criteria for desert dunes include sand ripples with wavelength-to-amplitude ratios (ripple index) exceeding 20, excellent grain sorting and medium-to-fine grain size, well-rounded quartz grains, and frosted surfaces on many of the grains. In some outcrops of the Navajo Formation, sediments interpreted as interdune deposits have been recognized. They are much thinner than the dune sands and contain discontinuous and irregular ripples, rare dinosaur tracks, ventifacts, and occasional thin-bedded and mud-cracked microcrystalline limestones containing ostracods. The carbonate beds are interpreted as deposits of ephemeral lakes.

Another type of eolian sand deposit that is common in modern dune areas and identified in the Navajo Sandstone is sand sheets (Fig. 7-13), which occur marginal to dune fields. They are characterized by excellent parallel lamination and some low-angle ripples. Scour-and-fill structures are common, and bioturbation structures may occur as well.

At their margins, eolian deposits grade into other arid climate environments such as alluvial fans and playa lakes. If the dune field is adjacent to a coastline, as in the modern Persian Gulf area, adjacent deposits may be those of supratidal flats (sabkhas), evaporites and dolomites (Chapter 9).

FIG. 7-11 • Outcrop of Navajo Sandstone in Zion National Park, Utah, showing large-scale wedge-planar cross-strata (middle) between tabular-planar cross-strata (above and below). (From McKee, 1979.)

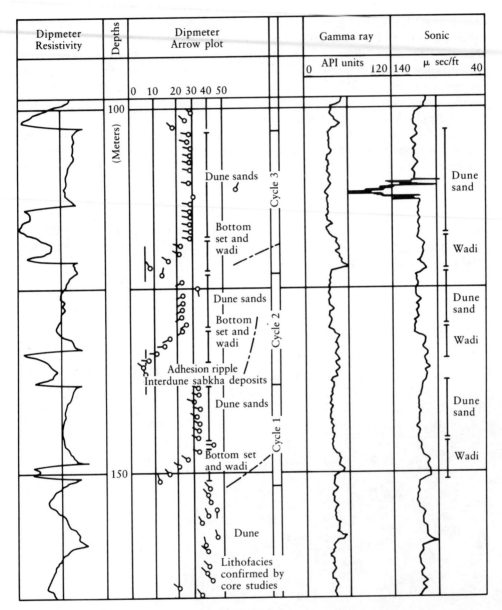

FIG. 7-12 • Dipmeter, gamma-ray, and sonic log data for a section through eolian cross-bedding and associated facies in the Rotliegendes Sandstone (Permian), southern North Sea. (Reprinted by permission of the publisher from Nurmi, *Society of Economic Paleontologists and Mineralogists Short Course Notes No. 4*, 1978.)

FIG. 7-13 • Sand sheet, Navajo Formation, Dinosaur National Monument, Colorado. The beds are inversely graded with occasional ripple foreset laminae. (Reprinted by permission of the American Association of Petroleum Geologists from Ahlbrandt and Fryberger, *in* Scholle and Spearing, *American Association of Petroleum Geologists Memoir*, v. 31, 1982.)

Lacustrine Deposits

Lakes are among the most varied of all depositional environments, ranging in depth from ephemeral films on playa surfaces to a modern maximum of 1.7 km (Lake Baikal) in more humid climates. The grain size of detrital lacustrine deposits ranges from gravel to fine clay, depending on topography surrounding the lake and distance from shore. Smaller lakes are primarily formed in nontectonic topographic depressions and are short-lived; larger lakes appear to form as a byproduct of crustal rifting, may exist for periods of 10^7-10^8 years, and can accumulate sediments up to perhaps 5 km thick (Baikal rift, USSR).

Picard and High (1972, 1981) have discussed the general stratigraphic relationships of lacustrine deposits and the most useful criteria for recognizing lake sediments (Fig. 7-14). Although lake sediments are by definition continental deposits, they resemble shallow seas in their physical processes of deposition, and the resulting facies patterns of detrital sediments are similar. An important difference is that shoreline deposits tend to be narrow and poorly developed in lacustrine sequences because lake levels normally are fairly unchanging; that is, transgressions and regressions are minimal. If the lake is shallow, however, so that climatic variations are able to cause major changes in shoreline position, areally significant transgressive and regressive facies can occur. Another important distinction between lacustrine and marine facies patterns is that lakes have a much shallower wave base than marine shoreline areas because of very limited wind fetch and resulting low wave amplitudes. This means that undisturbed, apparently deep-water sediments can occur within a stone's throw of the shoreline.

Varves are a type of repetitive and cyclic deposit found largely but not exclu-

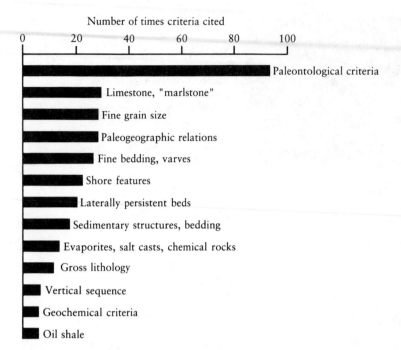

Number of times criteria cited

Paleontological criteria
Limestone, "marlstone"
Fine grain size
Paleogeographic relations
Fine bedding, varves
Shore features
Laterally persistent beds
Sedimentary structures, bedding
Evaporites, salt casts, chemical rocks
Gross lithology
Vertical sequence
Geochemical criteria
Oil shale

FIG. 7-14 ● Cited criteria for the recognition of lacustrine deposits, based on a review of 148 papers and books. All criteria cited for a lacustrine origin in each reference were noted. (Reprinted by permission of the publisher from Picard and High, *Society of Economic Paleontologists and Mineralogists Spec. Pub. 16*, 1972.)

sively in lakes. The most common type of varve consists of a couplet (Fig. 7-15) formed by seasonal variation in sedimentation rate, a "summer" layer composed of light-colored fine sand and silt produced by rapid settling of sediment carried to the lake during the rainy season and an overlying "winter" layer composed of clay and dark organic matter deposited slowly from suspension during the dry season. The discontinuity between the two units occurs because sand and silt are relatively subequant in shape but clay is platey in shape and has a slightly lower specific gravity than quartz and feldspar. If clay were also subequant in shape and had the same specific gravity as quartz, a graded bed would be produced rather than two laminae (or thin beds).

Varves may also consist of a carbonate lamina alternating with a clastic layer. For example, in Lake Superior a highly calcareous layer is formed during summer, when the rate of precipitation is high, and the carbonate is rapidly buried. During winter, when the rate of sedimentation is low, the carbonate is partially dissolved, and a low-carbonate winter layer is formed.

Carbonate-evaporite varves can also occur in stratigraphic sequences. Anderson and Kirkland (1960) examined the Todilto Formation (Jurassic) in New Mexico, a laminated sequence of limestone and gypsum of probable lacustrine origin. By plotting varve thicknesses (in millimeters) against time (in centimeters above a reference datum) they were able to detect the effects of the 10- to 13-year sunspot cycle and 60-, 85-, 170-, and 180-year cycles (Fig. 7-16) that presumably reflect other climatic changes. The presence of the sunspot cycle indicates that the laminations were deposited annually.

FIG. 7-15 • Photomicrograph of varves, Lake Kassjön, Sweden. Each varve is about 0.6 mm thick. (Photo courtesy of I. Renberg.)

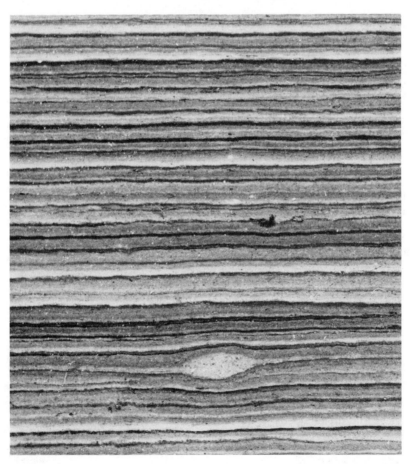

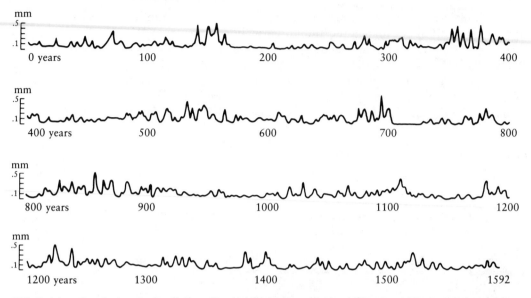

FIG. 7-16 ● Graph showing variation of varve thickness with time. Many prominent peaks are the result of the 10- to 13-year sunspot cycle. The 180-year cycle is most prominent in the older part of the graph with a tendency to break down into a 60-year pattern near the younger end. Curve was constructed by use of smoothing formula: thickness = (*A* + 2*B* + *C*)/4. (Reprinted by permission of the American Association of Petroleum Geologists from Anderson and Kirkland, *American Association of Petroleum Geologists Bulletin*, v. 44, 1960.)

Preservation of an annual cycle in sediments requires that four criteria be satisfied (Kelts and Hsu, *in* Lerman, 1978):

1. Processes in the upper part of the lake must produce a flux of seasonally differentiated particulate material that settles through a stagnant, stratified water mass.
2. No bottom dwellers can be present to homogenize the sediment and destroy the seasonal laminae. This condition is usually satisfied by anoxic conditions and toxicity of sulfides.
3. Bottom currents must be minimal or absent, and detrital inflow must not overprint or dilute the seasonal cycle.
4. Degradation of organic matter must be unable to produce excessive gas bubbles.

Inasmuch as all lakes are ultimately filled, the gross aspect of an ideal lacustrine sequence grades upward from fine-grained rocks deposited below the wave base into coarser near-shore and fluvial deposits. In thick sequences, however, the coarsening-upward pattern might not be apparent because of the limited thickness of the outcrop.

Salinities of lake waters range from those of tap water to values ten times as salty as ocean water (the Dead Sea). Because water compositions in most lakes are locally controlled, nondetrital sediments can be calcitic, dolomitic, evaporitic, zeolitic, or cherty. Some accumulations are rich in organic matter because of a seasonal or perennial deficiency of oxygen during the life of the lake, and economically valuable oil shales may result (Chapter 13). As with detrital lacustrine sediments, laminations and/or varves are common in chemical lacustrine facies, and couplets of almost any mineralogy are possible. Carbonate laminae are microcrystalline, evaporitic layers commonly contain minerals not seen in marine deposits (such as trona, shortite, and borates), analcime layers can be abundant, and nonmarine cherts occur in lakes where large changes in pH have occurred. All of these facies are discussed in some detail by Reading (1986), but even from the brief consideration possible here it is clear that no single facies model can apply to all or even most lacustrine deposits. Lake environments are simply too variable (Fig. 7-17). Their recognition and interpretation can pose some of the most challenging problems in stratigraphic analysis.

Swamps

Swamps are areas of poor drainage where the groundwater table intersects the ground surface. Thus most extensive swamps form in lowland coastal areas, and the waters in them can vary from meteoric to saline, depending on nearness to a coastline. Low- to mid-latitude swampy areas are characterized by abundant organic growth, and because of the chemically reducing conditions that result from poor drainage, decay of organic matter is very slow. As a result, the abundant vegetation does not oxidize and is preserved as dispersed or laminated organic matter, peat, lignite, or coal. In fact, richness in organic matter is the main distinguishing feature of ancient swamp deposits. Interbedded nonorganic detrital muds are typically structureless because of intense bioturbation, but occasional parallel laminations of silt and clay may be present. Syngenetic minerals characteristic of reducing conditions, such as pyrite and siderite, commonly are sufficiently abundant to be visible in outcrop.

Swamp sediments form practically at sea level, so accumulation of a thick sequence (many coal beds) means that the basin they fringe must sink progressively, standing still for a time while vegetable material accumulates, then subsiding so that the incipient lignite can be buried. Sporadic uplifts or transgressions are recognized by interfingering of the swamp sequence with coarser clastics or marine beds. It is not clear how accumulations in pre-Devonian swamps can be recognized. They probably would be identified as lacustrine rocks, because the chief difference between swamps and lakes is the growth and accumulation of woody plant debris in the shallower waters of the swamp.

(a)

(b)

FIG. 7-17 • Lacustrine facies, Late Triassic (Norian), south Wales. (a) Well-sorted breccias interpreted as a beach deposit from the coarse end of the clastic wedge. (b) Dolomitic fine sandstone and siltstone from the fine end of the clastic wedge, interpreted as shallow sublittoral in origin, located 50 m from (a). (c) Calcilutite with persistent calcisiltite laminae, the whole partly affected by desiccation (acetate peel). (d) Nodular dolomite formed by replacement of anhydrite. (Reprinted by permission of the publisher from Tucker, *in* Matter and Tucker, *International Association of Sedimentologists Spec. Pub. No. 2*, Blackwell Scientific Publications, 1978.)

Glacial Deposits

The formation of glacial deposits in ancient rocks (Hambrey and Harland, 1981) was controlled by the paleolatitude and relief of the continents during each geological period (continental drift). Continental glaciation requires that the bulk of the land mass be at latitudes greater than about 60° so that great thicknesses of ice can accumulate, be highly stressed in their lower parts, and flow outward to lower latitudes. In the United States, for example, Pleistocene glacial deposits occur as far

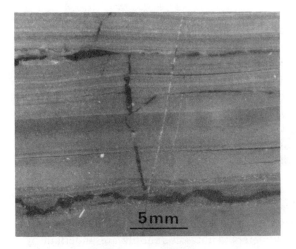

(c)

(d)

south as 40° latitude. The more localized mountain glaciers can, because of elevation, form at relatively low latitudes, but the preservation potential of their deposits is low. Recent evidence (Frakes and Francis, 1988) indicates that large masses of ice centered in polar regions have existed throughout the geological past and that only their latitudinal extent has varied.

The distinctive sediment of ancient glaciers is till, the unstratified and unsorted deposit of gravel, sand, and mud dropped by the glacier as it melts. Ancient tills (tillites) are easily mistaken for avalanche or landslide deposits or alluvial fan debris, and many of the unstratified and unsorted deposits (diamictites as shown in Fig. 7-18) described as tillites by early workers have since been reinterpreted. What are the defining characteristics of tillite in outcrop? The key features are (1) lack of stratification; (2) extremely poor sorting with a volumetric dominance of mud but a prominent content of coarse gravel (cobbles and boulders); (3) faceted, striated, polished, or fractured gravel clasts, the percentage that show striations varying greatly from one tillite to another, ranging from 1–2% to more than 30% (Edwards, *in* Reading, 1986); (4) a wide variety of clast types because of the enormous "drainage basin" of continental glaciers. Tillites also may contain lenses or discontinuous beds of stratified and better-sorted sandstone and conglomerate formed by flow of subglacial meltwater. Outward from the terminus of the glacier, the amount of such sediment increases, and proglacial meltwater deposits resemble those of humid climate alluvial fans dominated by braided streams.

The thickness of tillites is measured in meters or tens of meters, and the units can be traced for at least several kilometers. Tillites commonly are underlain by a glaciated bedrock surface containing striations, grooves, or other erosional gouges,

FIG. 7-18 • Typical massive matrix-supported diamictite. (Reprinted by permission of the author from Mustard and Donaldson, *Geological Society of America Bulletin*, v. 98, 1987.)

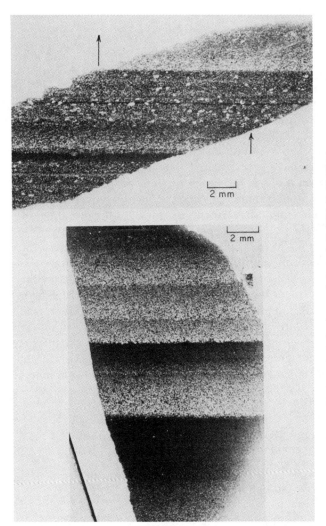

2 mm

2 mm

FIG. 7-19 • Glacial laminites from the late Proterozoic Smalfjord Formation, in northern Norway, which display distinct graded rhythmic laminae (arrowed). (a) Laminated mudstone shows both well-graded silty layers and ungraded laminated units with scattered sand grains. The graded layers may have been rapidly deposited from surface plumes, while the ungraded layers with scattered sand grains may represent slower deposition and ice rafting. (b) Rhythmite consists solely of graded units, some of which show multiple grading. These rhythmites may have formed by settling from surface meltwater plumes. (Reprinted by permission of the publisher from Edwards, *in* Reading, *Sedimentary Environments and Facies*, Blackwell Scientific Publications, 1986.)

but these features usually cannot be seen because outcrops tend to be covered by modern soil.

Continental glaciers commonly terminate at the continental margin but, because of their buoyancy, can fragment and drift into low latitudes before melting completely. As the icebergs drift and melt, they release sediment that falls to the sea floor, where the gravel particles deposited in the fine-grained sea floor sediment are preserved as "dropstones." If the sea floor debris is dominated by sediment released from the melting glacier, the sediment is termed glaciomarine. Such deposits may be graded because of the different settling rates of the variety of grain sizes dropped into the sea (Fig. 7-19).

Paleosols

Ancient soil horizons, termed **paleosols**, are stratigraphically very important lithofacies. They record an exposure surface, an unconformity, and should therefore be sought during field work. For example, Retallack (1983) found 87 fossil soils in 143 m of stratigraphic section in Eocene and Oligocene nonmarine units in South Dakota (Fig. 7-20). The buried Eocene forests of the northeastern part of Yellowstone National Park, Wyoming, consist of about 27 successive buried surfaces containing petrified tree trunks in about 400 m of stream gravel, mudflows, and tuffs. As many as 500 growth rings have been counted in some of these buried trees, indicating at least that many years of soil formation on some of the surfaces (Retallack, 1981). Paleosols may be more common in thick continental sequences than as the topmost unit in a transition from nonmarine to marine beds because the transgressing sea typically reworks the soil before deposition of the marine units.

The abundance of fossil soils in the geologic record is uncertain because their gross characteristics may differ only subtly from those of normal sedimentary units (Table 7-2, Figs. 7-21, 7-22, 7-23, 7-24). Most fossil soils are characterized by their poor sorting and absence of textures and structures that characterize detrital deposits, such as cross-bedding, grain rounding, and other current features. The regolith may grade downward into unaltered bedrock and may also grade upward into the overlying sedimentary rocks. Such graded unconformities commonly occur where arkose overlies its parent granite (granite wash). Petrographic and geochemical studies may be needed to identify a paleosol, and even then the identification may be

FIG. 7-20 • Zisa Series paleosols (dark bands, which are red in the outcrop) in swales (especially evident at the arrow) of nearstream sandstones (light colored), Scenic Member of Brule Formation, beside South Dakota highway 240, 1.8 km southeast of Pinnacles Lookout, Badlands National Park. Thickness of the Scenic Member (shown here completely) is 38 m. (Reprinted by permission of the author from Retallack, *Geological Society of America Spec. Paper 193*, 1983.)

TABLE 7-2 • Some criteria for identification of paleosols

1. Stratiform
2. Relatively thin (usually <20 m)
3. Transitional lower boundary, sharp upper boundary
4. Color variations
5. Destruction of primary rock textures often accompanied by formation of soil textures including clay coatings on grains, calcite or silica nodules or iron and manganese crusts, etc.
6. Mineralogical variations, destruction of primary minerals and formation of clay minerals or their metamorphic equivalents
7. Major and trace element distributions; depletion of most cations; enrichment of Al, Ti, Zr, and other elements forming insoluble compounds
8. Dikes of material from overlying sediments washed down into desiccation cracks in the soil
9. Rip-up clasts in overlying sediments

(Reprinted by permission of the publisher from Grandstaff et. al., *in* Retallack, *Precambrian Research*, v. 32, Elsevier Science Publishers B.V., 1986.)

FIG. 7-21 • Granular ped structure outlined by clay skins (argillans, especially prominent at arrow) from type Conata clay paleosol of mid-Oligocene (Orellan or about 32 m.y.), Scenic Member of Brule Formation, in Pinnacles area, Badlands National Park, South Dakota. Scale is in both centimeters (figures) and millimeters (fine gradations). (Reprinted by permission of the author from Retallack, *in* Reinhardt and Sigleo, *Geological Society of America Spec. Paper 216*, 1988.)

FIG. 7-22 • Sharp upper contact and diffuse lower horizons and drab haloed root traces (in upper part of red B horizon), in forested paleosol, type Long Reef clay paleosol. Black and white scale is 1 ft, graduated in inches. (Reprinted by permission of the author from Retallack, *in* Reinhardt and Sigleo, *Geological Society of America Spec. Paper 216,* 1988.)

only tentative because of soil stratigraphy. The preservation potential of the C-horizon is greater than that of the B-horizon, and that of the B-horizon is greater than that of the A-horizon. Thus the part of a paleosol that is preserved is likely to be composed of fragmental material that differs only slightly from the underlying bedrock. And if the soil was in an early stage of development, it will be particularly difficult to recognize in the field (Fig. 7-21). It is clear from these considerations that investigators with some training in soil science will find more paleosols than will those without much training. Given the importance of soil recognition in stratigraphic work, a basic course in soil science should be part of the education of all stratigraphers.

Paleosols are one of the best indicators of paleoclimate preserved in ancient rocks. They can provide evidence of the nature of the vegetation, depth of the water table and character of the groundwater, rates of subsidence and sedimentation, and

FIG. 7-23 • Downward and laterally branching root trace fossil (light colored areas). Purple (5RP 4/2) mottle of root trace fossil contrasts with red (2.5YR 4/6) color of very fine sandstone to siltstone host rock. Jacob staff is numbered in centimeters. (Reprinted by permission of the author from Blodgett, *in* Reinhardt and Sigleo, *Geological Society of America Spec. Paper 216*, 1988.)

paleotopography. Of course, paleosols can be altered during diagenesis, as can other lithofacies, but frequently, enough of their initial features remains to permit interpretation of original conditions. The most easily recognized paleosols are those that are most unusual in field appearance, the duricrusts. **Duricrusts** are soils markedly enriched in and well lithified by a specific mineral. Thus there are ferricretes (ferruginous soil conglomerates and laterites), alumina soils (bauxite or aluminous laterite), silcretes (silicified soil conglomerates and sandstones), calcretes (calcite-cemented soil conglomerates and caliche), and, more rarely, phoscretes (phosphatized soils).

Ferricrete and Bauxite

Paleosols composed largely or entirely of gibbsite, hematite, and ferric hydroxides are known from rocks as old as Proterozoic but are uncommon in pre-Devonian

FIG. 7-24 • Rootlet trace fossils (light-colored areas) associated with larger, purple, root trace fossils. Branching frequency and angles suggest that these are trace fossils of rootlets rather than bleached haloes around fungal rhizomorphs. The coin is 24 mm in diameter. (Reprinted by permission of the author from Blodgett, *in* Reinhardt and Sigleo, *Geological Society of America Spec. Paper 216,* 1988.)

sequences. Apparently, vegetation is essential for their widespread development. The work of soil scientists has documented that such soils are the residuum from a wide variety of bedrock types subjected to hot and humid climates and form in tropical to subtropical regions with seasonal rainfall, good drainage, and a fluctuating water table. With few exceptions, such environments are acidic and oxidizing, with rainfall exceeding 102 cm/yr and mean annual temperature above 16°C. Present locations of ferricrete and bauxite range from 70°N to 70°S latitude, reflecting the patterns of continental fragmentation and drift during Phanerozoic time (Nicolas and Bildgen, 1979).

Ferricretes and bauxites are relatively easy to recognize in outcrop because of their exotic compositions and commonly pisolitic texture. These soils are simply the end members of a podzolic weathering regime, which depletes the soil in alkali and alkaline earth elements with resultant enrichment in silica, alumina, and ferric iron. (Many laterites contain partially dissolved quartz grains.) Less easily recognizable in the field are the more common paleopodzols that did not progress to the stage of lateritization. They may be recognized by characteristics such as an obviously altered and reddened upper part; the presence of root tubules or even *in situ* tree trunks;

abundant kaolinized (white color) feldspar grains; assorted insect, earthworm, and larger animal burrows; and other structures peculiar to soils. Detailed descriptions and photographs of many of these features are given by Brewer (1964), Retallack (1976), and Bown and Kraus (1981).

Silcrete

Silcrete is a highly indurated soil composed mainly of quartz grains cemented by silica in any of its various forms: megaquartz, chert, chalcedony, or opal. Most silcretes are developed on quartzose sandstones or gravels and contain well over 90% silica; many have more than 98%. As a result, they can be difficult to distinguish from silica-cemented quartz sandstones. Fresh, unweathered silcrete is usually gray, but the color varies, depending on impurities. It is extremely brittle and shatters readily into sharp, angular pieces with a conchoidal fracture and semivitreous sheen. Beds of silcrete are usually 1–2 m thick but on occasion may be 5 m or more. Vertical jointing is almost invariably present and usually gives rise to a columnar structure. Subhorizontal jointing is also common but tends to be discontinuous. When a massive sheet of silcrete is exposed, such as a cliff face toward the summit of a mesa, vertical columns are particularly pronounced. The columns have fluting features and highly agglutinated ropy surfaces resembling candle-wax drippings (Fig. 7-25).

A detailed description and discussion of the petrography and peculiar diagenetic features of silcretes has been provided by Summerfield (1983).

There is very little reference to silcrete in most geomorphology or geology texts, probably because silcrete is not conspicuous in the parts of the world where most earth science texts are written—North America and Europe. Although these soils occur worldwide, they are most prominent in Australia and South Africa, where they are of early Tertiary age and reflect a climate different from today's. However, silcretes of probable Tertiary age have been reported from several locations in western Europe, northwestern Canada, and the United States.

Four processes have been suggested to explain the occurrence of silcrete soils: (1) evaporation of silica-rich lake or stream water, (2) evaporation of silica-rich groundwater, (3) extensive leaching and dissolution of minerals less stable than quartz, and (4) precipitation of silica from solutions containing colloidal silica. There is some evidence and there are probable examples of silcretes formed by each of these mechanisms. The common factor among these mechanisms is a large infusion of either dissolved or colloidal silica into the soil zone, an infusion that can occur either by decomposition of volcanic detritus or by tropical weathering of the type that also produces laterites. Field observations support the common (although not invariable) association of silcrete with laterite. The silica enters the groundwater during the later stages of podzolization and early stages of lateritization, as silicate minerals and quartz are dissolved. The groundwater then migrates to a drier region, or else the climate changes at the site of dissolution, and the silica is precipitated as overgrowths and colloform siliceous pore fillings in quartz debris in the soil zone.

FIG. 7-25 • Silcrete capping of a small residual west of Dolo Creek, 64 km from Wilcannia on the road to Broken Hill, New South Wales. This residual shows a variety of forms of silcrete, including massive, nodular, ragged, vesicular, ropy, and fluted expressions. A large nodule is exposed in the face just above the center of the photograph, and fluted faces are also discernible. The ropy forms lie on the upper surface, and the ragged and vesicular forms comprise the lower part of the mass of silcrete. (Reprinted by permission of the publisher from Stephens, *Geoderma*, v. 5, Elsevier Science Publishers B.V., 1971.)

Calcrete

Calcrete, unlike silcrete, is well known in the United States because of the extensive modern outcrops of caliche in the southwest. Like silcrete, calcrete forms in hot, dry climates. The difference between the two in method of formation is that silcrete requires an initial period of high rainfall and intense leaching, whereas the formation of caliche requires a continually dry climate (semiarid, 20–60 cm/yr). In such a dry climate there is sufficient precipitation to leach calcium from silicates but insufficient moisture for the calcium to be carried away from the site of leaching. Hence the calcium combines with available carbonate ions to precipitate as microcrystalline calcite in the B-horizon at and near the ground surface.

As with other soils, modern caliche is vertically zoned and is composed of four rock types, in ascending order: (1) massive chalky carbonate, (2) nodular and crumbly carbonate, (3) irregular plates and sheets, and (4) a compact crust or hardpan (Fig. 7-26). The position and development of these lithologies in a vertical sequence are highly variable (Fig. 7-27), but massive chalky carbonate most commonly is found at the base grading downward into underlying rocks or sediment. The detailed character of each of these four segments of caliche profiles is discussed by James and Choquette (1984), and a detailed field study of a Paleozoic caliche is provided by Ettensohn et al. (1988).

Ancient calcretes currently occur at latitudes between 70°S and 45°N, the wide range reflecting postformational continental drift. Because of the relatively high solubility of calcite, many of the paleocaliches are partly recrystallized and lack one or more of the characteristics of modern caliches, such as lamination, textural inhomogeneity, and pisolitic and brecciated texture and structure. Also, calcretes are easily mistaken for stromatolites (Read, 1976). Nevertheless, many of these ancient soils have been recognized. For example, Steel (1974) examined more than 200 calcretes in the New Red Sandstone (Permian) of Scotland, most of them less than 2 m thick and capping floodplain deposits. The presence of the laterally extensive calcrete profiles at the top of alluvial fining-upward sequences indicates that there was a prolonged period of exposure and soil development after the construction of each

FIG. 7-26 • Caliche breccia formed over caliche crusts. Isla Mujeres, Yucatan, Mexico. (Reprinted by permission of the author from Reeves, *Caliche*, Estacado Books, 1976.)

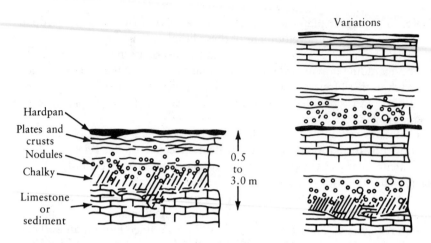

Variations

Hardpan
Plates and crusts
Nodules
Chalky
Limestone or sediment

0.5 to 3.0 m

FIG. 7-27 • Sketch of a caliche (calcrete) profile showing all the elements (left) and some of the observed variations from various modern examples (right). (Reproduced with the permission of the Geological Association of Canada from James and Choquette, *Facies Models,* 1984.)

sequence. This, in turn, implies that precipitation and fluvial discharge were the major factors in the initiation of any new floodplain constructional event. Studies of modern caliches suggest that calcrete profiles develop at rates of 10–50 mm/1000 years with a maximum of 1000 mm/1000 years. Hence each of the caliche soils in the New Red Sandstone represents between 2000 and 200,000 years of subaerial exposure during Permian time.

Summary

The term **facies** is variously defined by stratigraphers and sedimentologists and all definitions accord with the word "aspect," either lithologic, biologic, chemical, or environmental. Some facies are very local, such as an aragonitic facies; others may be wide-ranging, such as shale facies. Some facies require X-ray diffraction for recognition, such as an illitic facies; others are easily recognizable in the field, such as evaporitic facies. Any aspect of a stratigraphic section that can be usefully identified to aid in stratigraphic interpretation should be cited, and the term *facies* is typically used as a general category. Facies form the basis of environmental interpretation and are, therefore, critical to paleogeographic reconstruction.

The principal continental depositional environments are alluvial fans, braided and meandering streams, eolian regimes, lakes, swamps, glacial deposits, and soils. Perhaps the most significant of these is the soil because of the great abundance of soils in the stratigraphic record and their climatic and geomorphic implications. Unfortunately, few geologists have had any training in soil science, so the field recognition of soils has lagged behind our recognition and understanding of other terrestrial environments.

References
Primary

Davis, R. A., Jr., 1983. Depositional Systems. Englewood Cliffs, N.J., Prentice-Hall, 669 pp.

Davis R. A. (ed.), 1985. Coastal Sedimentary Environments, 2nd ed. New York, Springer-Verlag, 716 pp.

Ethridge, F. G. and Flores, R. M., 1981. Recent and Ancient Nonmarine Depositional Environments: Models for Exploration. Tulsa, Society of Economic Paleontologists and Mineralogists, Spec. Pub. No. 31, 349 pp.

Fraser, G. S., 1989. Clastic Depositional Sequences. Englewood Cliffs, N.J., Prentice-Hall, 459 pp.

Reading, H. G. (ed.), 1986. Sedimentary Environments and Facies, 2nd ed. Oxford, Blackwell Scientific Publications, 615 pp.

Scholle, P. A., Bebout, D. G., and Moore, C. H. (eds.), 1983. Carbonate Depositional Environments. Tulsa, American Association of Petroleum Geologists, Mem. 33, 708 pp.

Scholle, P. A. and Spearing, D. (eds.), 1982. Sandstone Depositional Environments. Tulsa, American Association of Petroleum Geologists, Mem. 31, 410 pp.

Walker, R. G. (ed.), 1984. Facies Models, 2nd ed. Toronto, Geological Association of Canada, 317 pp.

Facies

Blatt, H., Middleton, G. V., and Murray, R. C., 1980. Origin of Sedimentary Rocks, 2nd ed. Englewood Cliffs, N.J., Prentice-Hall, 782 pp.

Drever, J. I., 1988. The Geochemistry of Natural Waters, 2nd ed. Englewood Cliffs, N.J., Prentice-Hall, 437 pp.

Middleton, G. V., 1973. Johannes Walther's law of the correlation of facies. Geol. Soc. Amer. Bull., v. 84, pp. 979–988.

Middleton, G. V. and Southard, J. B., 1978. Mechanics of sediment transport. Tulsa, Society of Economic Paleontologists and Mineralogists, Short Course No. 3, Lecture Notes.

Pettijohn, F. J., 1975. Sedimentary Rocks, 3rd ed. New York, Harper & Row, 628 pp.

Stumm, W. and Morgan, J. J., 1981. Aquatic Chemistry, 2nd ed. New York, John Wiley & Sons, 780 pp.

Alluvial Fans

Bull, W. B., 1972. Recognition of alluvial-fan deposits in the stratigraphic record. *In* J. K. Rigby and W. K. Hamblin (eds.), Recognition of Ancient Sedimentary Environments. Tulsa, Society of Economic Paleontologists and Mineralogists, Spec. Pub. No. 16, pp. 63–83.

Elmore, R. D., 1984. The Copper Harbor Conglomerate: A late Precambrian fining-upward alluvial fan sequence in northern Michigan. Geol. Soc. Amer. Bull., v. 95, pp. 610–617.

Heward, A. P., 1978. Alluvial fan sequence and megasequence models. *In* A. D. Miall (ed.), Fluvial Sedimentology. Calgary, Canadian Society of Petroleum Geologists, pp. 669–702.

Klein, G. deV., 1985. Sandstone Depositional Models for Exploration for Fossil Fuels, 3rd ed. Boston, International Human Resources Development Corporation, 209 pp.

Koster, E. H. and Steel, R. J. (eds.), 1984. Sedimentology of Gravels and Conglomerates. Calgary, Canadian Society of Petroleum Geologists, Mem. 10, 441 pp.

Mack, G. H. and Rasmussen, K. A., 1984. Alluvial-fan sedimentation of the Cutler Formation (Permo-Pennsylvanian) near Gateway, Colorado. Geol. Soc. Amer. Bull., v. 95, pp. 109–116.

Steel, R. J., Maehle, S., Nilsen, H., Roe, S. L., and Spinnangr, A., 1977. Coarsening-upward cycles in the alluvium of Hornelen Basin (Devonian), Norway: Sedimentary response to tectonic events. Geol. Soc. Amer. Bull., v. 88, pp. 1124–1134.

Fluvial

Allen, J. R. L., 1964. Studies in fluvial sedimentation: Six cyclothems from the Lower Old Red Sandstone, Anglo-Welsh Basin. Sedimentology, v. 3, pp. 163–198.

Cant, D. J. and Walker, R. G., 1976. Development of a braided-fluvial facies model for the Devonian Battery Point Sandstone, Quebec. Canadian J. Earth Sci., v. 13, pp. 102–119.

Collinson, J. D. and Lewin, J., 1983. Modern and ancient fluvial systems. International Association of Sedimentologists, Spec. Pub. No. 6. Oxford, Blackwell Scientific Publications, 575 pp.

Eschner, T. R., Hadley, R. F., and Crowley, K. D., 1983. Hydrologic and Morphologic Changes in Channels of the Platte River Basin in Colorado, Wyoming, and Nebraska: A Historical Perspective. Washington, D.C., U.S. Geological Survey, Prof. Paper 1277-A, 39 pp.

Laming, D. J. C., 1966. Imbrication, paleocurrents and other sedimentary features in the Lower New Red Sandstone, Devonshire, England. J. Sed. Petrology, v. 36, pp. 940–959.

Miall, A. D. (ed.), 1978. Fluvial Sedimentology. Calgary, Canadian Society of Petroleum Geologists, Mem. 5, 859 pp.

Miall, A. D. (ed.), 1981. Sedimentation and Tectonics in Alluvial Basins. Toronto, Geological Association of Canada, Spec. Paper No. 23, 272 pp.

Eolian

Frostick, L. E. and Reid, I. (eds.), 1987. Desert Sediments: Ancient and Modern. Oxford, Blackwell Scientific Publications, 401 pp.

Fryberger, S. G. and Schenk, C. J., 1988. Pin stripe lamination: A distinctive feature of modern and ancient eolian sediments. Sedimentary Geol., v. 55, pp. 1–15.

Kocurek, G. (ed.), 1988. Late Paleozoic and Mesozoic eolian deposits of the Western Interior of the United States. Sedimentary Geol., v. 56, 413 pp.

McKee, E. D. (ed.), 1979. A Study of Global Sand Seas. Washington, D.C., U.S. Geological Survey, Prof. Paper 1052, 429 pp.

Nurmi, R. D., 1978. Use of well logs in evaporite sequences. *In* W. E. Dean and B. C. Schreiber (eds.), Marine Evaporites. Tulsa, Society of Economic Paleontologists and Mineralogists, Short Course No. 4, pp. 144–176.

Porter, M. L., 1987. Sedimentology of an ancient erg margin: The Lower Jurassic Aztec Sandstone, southern Nevada and southern California. Sedimentology, v. 34, pp. 661–680.

Lacustrine

Anderson, R. Y. and Kirkland, D. W., 1960. Origin, varves and cycles of Jurassic Todilto Formation, New Mexico. Amer. Assoc. Petroleum Geol. Bull., v. 44, pp. 37–52.

Hakanson, L. and Jansson, M., 1983. Principles of Lake Sedimentology. New York, Springer-Verlag, 316 pp.

Lerman, A. (ed.), 1978. Lakes: Chemistry, Geology, Physics. New York, Springer-Verlag, 363 pp.

Matter, A. and Tucker, M. E. (eds.), 1978. Modern and Ancient Lake Sediments. International Association of Sedimentologists, Spec. Pub. No. 2. Oxford, Blackwell Scientific Publications, 290 pp.

Picard, M. D. and High, L. R., Jr., 1972. Criteria for recognizing lacustrine rocks. *In* J. K. Rigby and W. K. Hamblin (eds.), Recognition of Ancient Sedimentary Environments. Tulsa, Society of Economic Paleontologists and Mineralogists, Spec. Pub. No. 16, pp. 108–145.

Picard, M. D. and High, L. R., Jr., 1981. Physical stratigraphy of ancient lacustrine deposits. *In* F. G. Ethridge (ed.), Recent and Ancient Nonmarine Depositional Environments. Society of Economic Paleontologists and Mineralogists, Spec. Pub. No. 31, pp. 233–259.

Swamps

Rahmani, R. A. and Flores, R. M. (eds.), 1984. Sedimentology of Coal and Coal-bearing Sequences. International Association of Sedimentologists, Spec. Pub. No. 7. Oxford, Blackwell Scientific Publications, 412 pp.

Glacial

Frakes, L. A. and Francis, J. E., 1988. A guide to Phanerozoic cold polar climates from high-latitude ice-rafting in the Cretaceous. Nature, v. 333, pp. 547–549.

Gamundi, O. L. and Amos, A. J., 1982. Criteria for identifying old glacigenic deposits. *In* E. B. Evenson, C. Schluchter, and J. Rabassa (eds.), Tills and Related Deposits. Rotterdam, A. A. Balkema, pp. 279–285.

Hambrey, M. J. and Harland, W. B., 1981. Earth's Pre-Pleistocene Glacial Record. New York, Cambridge University Press, 1004 pp.

Mustard, P. S. and Donaldson, J. A., 1987. Early Proterozoic ice-proximal glaciomarine deposition: The lower Gowganda Formation at Cobalt, Ontario, Canada. Geol. Soc. Amer. Bull., v. 98, pp. 373–387.

Ojakangas, R. W., 1988. Glaciation: An uncommon "mega-event" as a key to intracontinental and intercontinental correlation of early Proterozoic basin fill, North American and Baltic cratons. *In* K. L. Kleinspehn and C. Paola (eds.), New Perspectives in Basin Analysis. New York, Springer-Verlag, pp. 431–434.

Veevers, J. J. and Powell, C. McA., 1987. Late Paleozoic glacial episodes in Gondwanaland reflected in transgressive-regressive depositional sequences in Euramerica. Geol. Soc. Amer. Bull., v. 98, pp. 475–496.

Soils

Bown, T. M. and Kraus, M. J., 1981. Lower Eocene alluvial paleosols (Willwood Formation, northwest Wyoming, U.S.A.) and their significance for paleoecology, paleoclimatology, and basin analysis. Paleogeog., Paleoclim., Paleoecol., v. 34, pp. 1–30.

Bown, T. M. and Kraus, M. J., 1987. Integration of channel and floodplain suites. I: Developmental sequence and lateral relations of alluvial paleosols. J. Sed. Petrology, v. 57, pp. 587–601.

Brewer, R., 1964. Fabric and Mineral Analysis of Soils. New York, John Wiley & Sons, 470 pp.

Ettensohn, F. R., Dever, G. R., Jr., and Grow, J. S., 1988. A paleosol interpretation for profiles exhibiting exposure "crusts" from the Mississippian of the Appalachian Basin. *In* Reinhardt, J. and Sigleo, W. R. (eds.). Paleosols and Weathering through Geologic Time: Principles and Applications. Boulder, Geological Society of America, Spec. Paper 316, pp. 49–79.

Fitzpatrick, E. A., 1984. Micromorphology of Soils. New York, Chapman and Hall, 433 pp.

James, N. P. and Choquette, P. W., 1984. Limestones—The meteoric diagenetic environment. Geoscience Canada, v. 11, pp. 161–194.

Kraus, M. J., 1987. Integration of channel and floodplain suites. II: Vertical relations of alluvial paleosols. J. Sed. Petrology, v. 57, pp. 602–612.

Langford-Smith, T. (ed.), 1978. Silcrete in Australia. Armidale, New South Wales, Australia, University of New England, 304 pp.

Lehman, T. M., 1989. Upper Cretaceous (Maastrichtian) paleosols in Trans-Pecos, Texas. Geol. Soc. Amer. Bull., v. 101, pp. 188–203.

Nicolas, J. and Bildgen, P., 1979. Relations between the location of the karst bauxites in the northern hemisphere, the global tectonics and the climatic variations during geological time. Paleogeog., Paleoclim., Paleoecol., v. 28, pp. 205–239.

Read, J. R., 1976. Calcretes and their distinction from stromatolites. *In* M. R. Walter (ed.), Stromatolites. New York, Elsevier Science Publishers, pp. 55–71.

Reeves, C. C., Jr., 1976. Caliche. Lubbock, Texas, Estacado Books, 233 pp.

Reinhardt, J. and Sigleo, W. R. (eds.), 1988. Paleosols and Weathering through Geologic Time: Principles and Applications. Boulder, Geological Society of America, Spec. Paper 216, 181 pp.

Retallack, G. J., 1976. Triassic paleosols in the upper Narrabeen group of New South Wales. I: Features of the paleosols. J. Geol. Soc. Australia, v. 23, pp. 383–399.

Retallack, G. J., 1981. Fossil soils; indicators of ancient terrestrial environments. *In* K. J. Niklas (ed.), Paleobotany, Paleoecology, and Evolution. Vol. I. New York, Praeger Publishers, pp. 55–102.

Retallack, G. J., 1983. Late Eocene and Oligocene paleosols from Badlands National Park, South Dakota. Boulder, Geological Society of America, Spec. Paper 193, 82 pp.

Retallack, G. J. (ed.), 1986a. Special issue of Precambrian Research, v. 32, nos. 2–3, devoted to Precambrian paleopedology.

Retallack, G. J., 1986b. Fossil soils as grounds for interpreting long-term controls on ancient rivers. J. Sed. Petrology, v. 57, pp. 1–18.

Steel, R. J., 1974. Cornstone (fossil caliche)—Its origin, stratigraphic and sedimentological importance in the New Red Sandstone, western Scotland. J. Geol., v. 82, pp. 351–369.

Stephens, C. G., 1971. Laterite and silcrete in Australia: A study of the genetic relationships of laterite and silcrete and their companion materials, and their collective significance in the formation of the weathered mantle, soils, relief and drainage of the Australian continent. Geoderma, v. 5, pp. 5–52.

Summerfield, M. A., 1983. Petrography and diagenesis of silcrete from the Kalihari Basin and Cape coastal zone, southern Africa. J. Sed. Petrology, v. 53, pp. 859–909.

Wright, P. V. (ed.), 1986. Paleosols. Oxford, Blackwell Scientific Publications, 315 pp.

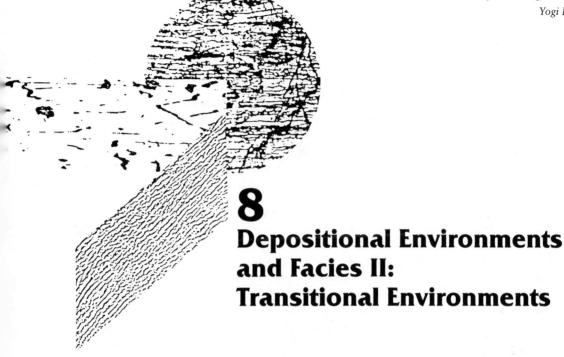

You can observe a lot by watching.

Yogi Berra

8
Depositional Environments and Facies II: Transitional Environments

Transitional environments are conveniently defined as those whose existence depends on nearness to the junction of land and ocean, environments associated with marine shorelines. In this group clearly belong deltas, beaches and barrier bars, and intertidal deposits because each of them is partly above water and partly below. Other environments that might be included as transitional are reefal and nonreefal limestone settings and areas of evaporite deposition because these environments are in contact with both air and seawater. But reefs do not thrive when exposed to air, although they can tolerate exposure for limited periods of time. Evaporites form mostly from highly concentrated seawater solutions, so although they form within a stone's throw of the land-sea interface, it is more convenient to consider them with marine environments in the next chapter. In this chapter we will restrict our concerns to deltas, beaches and barrier bars, and siliciclastic intertidal deposits.

Deltaic Environments

The essential condition for the formation of a large fluviomarine delta is a large or uplifting drainage basin that funnels water and sediment for a long period of time to a coastal site where a large body of sediment can be accommodated, generally by subsidence (Audley-Charles et al., 1977). The tectonic settings of major deltas are varied, but the most common are passive continental margins (such as the Niger delta), marginal basins or foreland basins associated with mountain belts (such as the Ganges-Brahmaputra delta), and cratonic basins (such as the Rhine delta). Deltas are often major depocenters and thus produce exceptionally thick stratigraphic sequences. Commonly, deltaic muds are important source rocks for major petroleum accumulations in deltaic sand units (such as in the Mississippi delta). Major coal deposits can form in swamps on delta surfaces.

For the deltas that persist over several geologic periods, a major river system and stable paleoslope of subcontinental dimensions are required; for example, the Mississippi River system and its ancestors have fed the delta at the entrance to the Gulf of Mexico since at least early Tertiary time. Such major river systems on cratons largely follow structural lows such as deep-seated rifts, aulacogens, and geofracture systems (Potter, 1978), so major deltas reflect major tears in the continental blocks. The less organized and immature drainage systems that lack major structural control produce either small, short-lived deltas or numerous closely spaced rivers that induce uniform progradation of the entire coastal plain rather than point-concentrated progradation.

The physiographic and sedimentologic feature termed a delta is a complex mixture of many very different types of sediment deposits such as a lacustrine, distributary stream channel and a distributary mouth bar, interdistributary bay, and possibly a swamp above mean sea level and delta-front sheet sand and prodelta sands and muds grading into bottomset turbidites below mean sea level. Although the depositional environments of these sediments are linked by their association with a deltaic sediment mass, many of them can and do exist in nondeltaic settings as well—for example, lakes, swamps, and the deep-water turbidite setting. Turbidite units of the major deltas developed at continental margins are parts of extensive deep-water fan facies that usually exceed in volume that of the shallow-water deltaic prism. For example, the subaqueous part of the Magdalena delta in Colombia covers 15 times the area of the subaerial part, and although some of the subaqueous part is not composed of turbidites, it is clear that this part dominates the delta accumulation. Commonly, the original geomorphic and sedimentologic characteristics of the subaqueous beds are modified by deep-water geostrophic currents, resulting in stratigraphic units termed **contourites**.

Delta Models

The distribution, orientation, and internal geometry of deltaic deposits are controlled by a variety of factors, including climate, water discharge, sediment load, river-mouth processes, waves, tides, currents, winds, shelf width and slope, and the tectonics and geometry of the receiving basin. Two currently accepted classifications of deltas based on the relative importances of fluvial processes, wave processes, and tidal processes are shown in Figures 8-1 and 8-2, which are based on many studies of modern deltas. Extensive and detailed subsurface and surface investigations are required to establish the delta type in an ancient deposit. The detail possible in studies of modern deltas cannot be attained for ancient rocks because of limitations of exposure and well control, but it is possible in many cases to establish which of the three major controls is dominant.

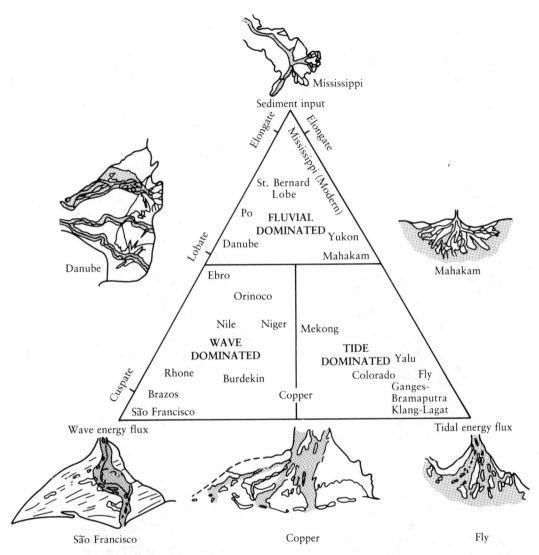

FIG. 8-1 • Ternary classification of deltas according to dominant mode of deposition. The relative importance of sediment input, wave energy flux, and tidal energy flux determine the morphology and internal stratigraphy of the delta. (Reprinted by permission of the publisher from Galloway, *in* Broussard, *Deltas, Models for Exploration*, Houston Geological Society, 1975.)

1. Low wave energy; low littoral drift; high suspended load

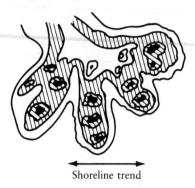

Shoreline trend

2. Low wave energy; low littoral drift; high tide

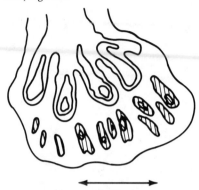

3. Intermediate wave energy; high tide; low littoral drift

4. Intermediate wave energy; low tide

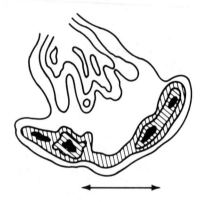

5. High wave energy; low littoral drift; steep offshore slope

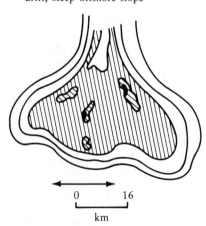

0   16
km

6. High wave energy; high littoral drift; steep offshore slope

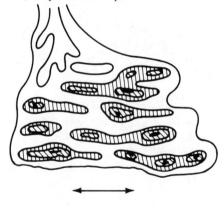

FIG. 8-2 • Patterns of sand deposition as determined by the interaction among river input, wave energy, wind direction (longshore drift), and tidal strength. (Reprinted by permission of the publisher from Coleman and Wright, *in* Broussard, *Deltas, Models for Exploration*, Houston Geological Society, 1975.)

River-dominated deltas.   When the trunk river system brings to the depocenter more sediment than can be distributed by marine processes, rapid seaward progradation of the delta front occurs (Fig. 8-3). Many sand bodies develop as "fingers" within distributary streams on the delta plain surface and are called bar finger sands. These bar finger sands result from migration of the distributary mouth bar and channel system. The sand bodies are oriented at high angles to the regional shoreline and have scoured bases and abundant cross-bedding. The Mississippi delta is the best modern example of a river-dominated delta, the bar fingers causing the effect of bird's foot morphology. It is noteworthy that the sediment deposited on the delta by the Mississippi is only 20% sand, so that domination by sand is not necessary for bar fingers to be an important stratigraphic feature of a fluvial-dominated delta.

Between the sand bar fingers are interdistributary bays, areas of low-energy, muddy sedimentation and organic activity with abundant bioturbation. Breaching of levees and consequent formation of crevasse splays of fine sand and mud eventually fill these bays, developing marshes, swamps, and ultimately coal seams.

In areas where sand dominates the delta sediment, as in arid or glacial regions, a different type of delta may develop. The outline tends to be lobate, and mouth bars merge laterally into a sand sheet. Channel switching (avulsion), which is characteristic of braided streams, serves to spread the sand laterally. The radiating pattern of distributaries is similar to that of alluvial fans, so the sediment accumulation is termed a **fan delta**. Such deltas were probably the common river-dominated delta in pre-Devonian time because, until the advent of land vegetation, which tends to store rainfall (clay formation) and regulate runoff, braided channel networks were probably the main fluvial style.

Wave-dominated deltas.   In areas where wave activity dominates fluvial input into the basin, the sand deposited at the land-sea interface is continually reworked into a series of curved beach ridges whose strike parallels the shoreline. If the wind direction is predominantly onshore, the wind may redistribute much of the beach sand as a coastal dune field on the delta plain (Fig. 8-4). If the wind blows at an angle to the coastline, longshore drift may redistribute the beach sand laterally, producing a lopsided delta morphology reflecting current direction.

Tide-dominated deltas.   In areas where the tidal range is great and is the dominant near-shore marine process, the fluvial sediment is reworked into a series of parallel bars. These sand bodies are approximately normal to the shoreline orientation and are separated from each other by linear channels scoured into the intervening mud. The sands are well sorted and contain bipolar cross-bedding because of the reversing character of tidal flow (Fig. 8-5).

Neritic zone sediments.   Seaward of the subaerial part of the delta plain lies its subaqueous portion, which may be difficult or impossible to distinguish from the subaerial part. It is simply a continuation of the topset delta beds. The chief differences are as follows.

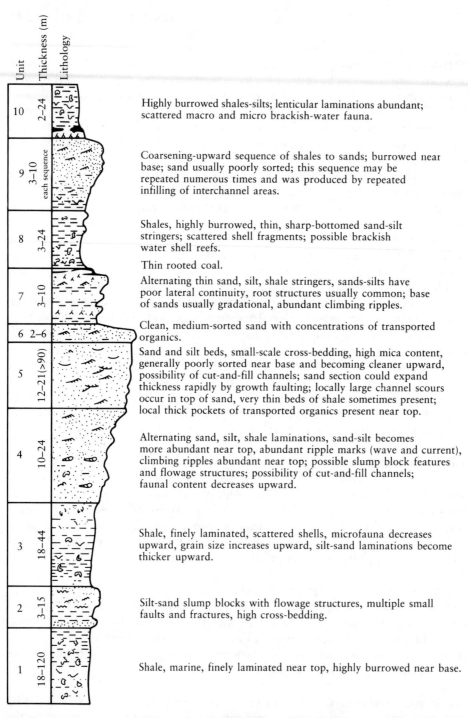

Unit	Thickness (m)	Lithology	Description
10	2–24		Highly burrowed shales-silts; lenticular laminations abundant; scattered macro and micro brackish-water fauna.
9	3–10 each sequence		Coarsening-upward sequence of shales to sands; burrowed near base; sand usually poorly sorted; this sequence may be repeated numerous times and was produced by repeated infilling of interchannel areas.
8	3–24		Shales, highly burrowed, thin, sharp-bottomed sand-silt stringers; scattered shell fragments; possible brackish water shell reefs. Thin rooted coal.
7	3–10		Alternating thin sand, silt, shale stringers, sands-silts have poor lateral continuity, root structures usually common; base of sands usually gradational, abundant climbing ripples.
6	2–6		Clean, medium-sorted sand with concentrations of transported organics.
5	12–21(>90)		Sand and silt beds, small-scale cross-bedding, high mica content, generally poorly sorted near base and becoming cleaner upward, possibility of cut-and-fill channels; sand section could expand thickness rapidly by growth faulting; locally large channel scours occur in top of sand, very thin beds of shale sometimes present; local thick pockets of transported organics present near top.
4	10–24		Alternating sand, silt, shale laminations, sand-silt becomes more abundant near top, abundant ripple marks (wave and current), climbing ripples abundant near top; possible slump block features and flowage structures; possibility of cut-and-fill channels; faunal content decreases upward.
3	18–44		Shale, finely laminated, scattered shells, microfauna decreases upward, grain size increases upward, silt-sand laminations become thicker upward.
2	3–15		Silt-sand slump blocks with flowage structures, multiple small faults and fractures, high cross-bedding.
1	18–120		Shale, marine, finely laminated near top, highly burrowed near base.

FIG. 8-3 • Composite, idealized stratigraphic section of Mississippi delta sediments reflecting progradation resulting from dominance of river input over marine dispersal capability. (Reprinted by permission of the publisher from Coleman and Wright, *in* Broussard, *Deltas, Models for Exploration*, Houston Geological Society, 1975.)

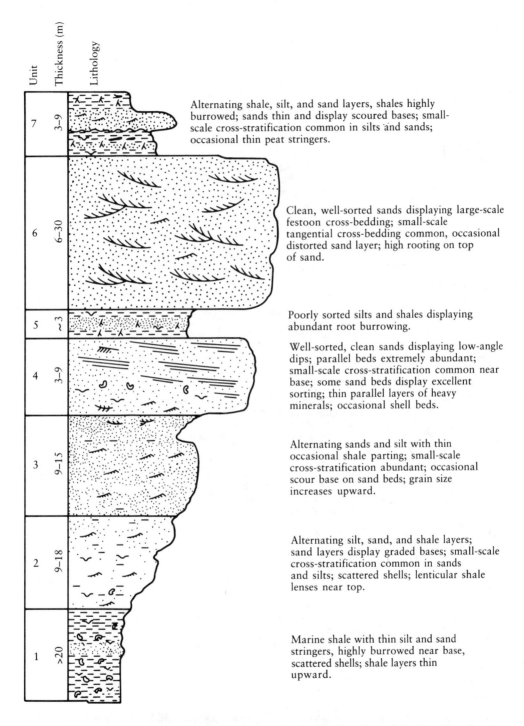

Unit / **Thickness (m)** / **Lithology**

7 — 3–9: Alternating shale, silt, and sand layers, shales highly burrowed; sands thin and display scoured bases; small-scale cross-stratification common in silts and sands; occasional thin peat stringers.

6 — 6–30: Clean, well-sorted sands displaying large-scale festoon cross-bedding; small-scale tangential cross-bedding common, occasional distorted sand layer; high rooting on top of sand.

5 — ~3: Poorly sorted silts and shales displaying abundant root burrowing.

4 — 3–9: Well-sorted, clean sands displaying low-angle dips; parallel beds extremely abundant; small-scale cross-stratification common near base; some sand beds display excellent sorting; thin parallel layers of heavy minerals; occasional shell beds.

3 — 9–15: Alternating sands and silt with thin occasional shale parting; small-scale cross-stratification abundant; occasional scour base on sand beds; grain size increases upward.

2 — 9–18: Alternating silt, sand, and shale layers; sand layers display graded bases; small-scale cross-stratification common in sands and silts; scattered shells; lenticular shale lenses near top.

1 — >20: Marine shale with thin silt and sand stringers, highly burrowed near base, scattered shells; shale layers thin upward.

FIG. 8-4 ● Composite, idealized stratigraphic section in São Francisco delta (Brazil), reflecting the domination of wave influence on the delta plain. Note the abundance of thick sands with unidirectional cross-bedding and progradation from marine shale at the base to mixed lithology with peat at the top. (Reprinted by permission of the publisher from Coleman and Wright, *in* Broussard, *Deltas, Models for Exploration*, Houston Geological Society, 1975.)

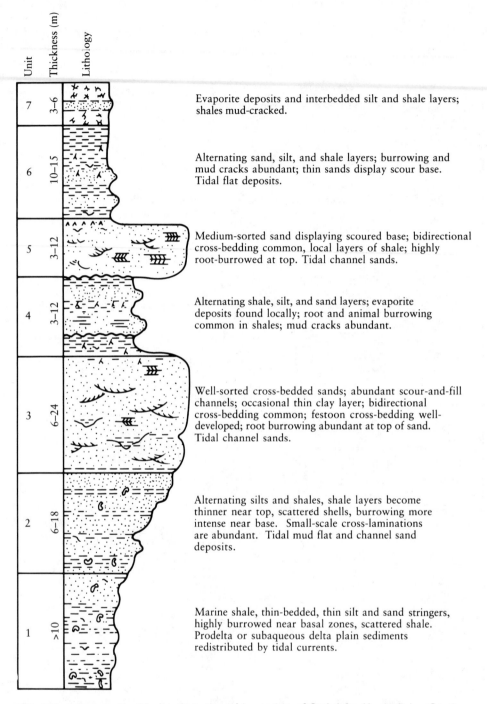

Unit	Thickness (m)	Lithology	
7	3–6		Evaporite deposits and interbedded silt and shale layers; shales mud-cracked.
6	10–15		Alternating sand, silt, and shale layers; burrowing and mud cracks abundant; thin sands display scour base. Tidal flat deposits.
5	3–12		Medium-sorted sand displaying scoured base; bidirectional cross-bedding common, local layers of shale; highly root-burrowed at top. Tidal channel sands.
4	3–12		Alternating shale, silt, and sand layers; evaporite deposits found locally; root and animal burrowing common in shales; mud cracks abundant.
3	6–24		Well-sorted cross-bedded sands; abundant scour-and-fill channels; occasional thin clay layer; bidirectional cross-bedding common; festoon cross-bedding well-developed; root burrowing abundant at top of sand. Tidal channel sands.
2	6–18		Alternating silts and shales, shale layers become thinner near top, scattered shells, burrowing more intense near base. Small-scale cross-laminations are abundant. Tidal mud flat and channel sand deposits.
1	>10		Marine shale, thin-bedded, thin silt and sand stringers, highly burrowed near basal zones, scattered shale. Prodelta or subaqueous delta plain sediments redistributed by tidal currents.

FIG. 8-5 • Composite, idealized stratigraphic section of Ord delta (Australia), reflecting the domination of tidal influence on the delta plain and progradation. (Reprinted by permission of the publisher from Coleman and Wright *in* Broussard, *Deltas, Models for Exploration*, Houston Geological Society, 1975.)

1. There is higher sand content on the subaqueous surface because of pervasive strong current activity. However, sand size decreases seaward, and muddiness occurs (decrease in textural maturity) within a few kilometers of the shoreline.
2. Marine fauna may occur, although brackishness of the water in very near-shore areas may limit its full development.

As is emphasized in many publications concerned with the Mississippi delta plain, the major streams of which the lower Mississippi River is composed shift their positions laterally and construct new delta lobes within very brief periods of time, periods on the order of only thousands of years (Fig. 8-6). Hence the sequence of beds that a stratigrapher sees in an outcrop of, say, an Ordovician delta is an amalgamation of probably hundreds of small deltas. During the period of time represented by a small outcrop, perhaps ten million years, events in the near-shore area may have changed repeatedly from river dominance to wave dominance to tidal

FIG. 8-6 • Distinguishable delta complexes and lobes of the premodern and modern Mississippi delta. (Reprinted by permission of the publisher from Elliott, *in* Reading, *Sedimentary Environments and Facies*, Blackwell Scientific Publications, 1986.)

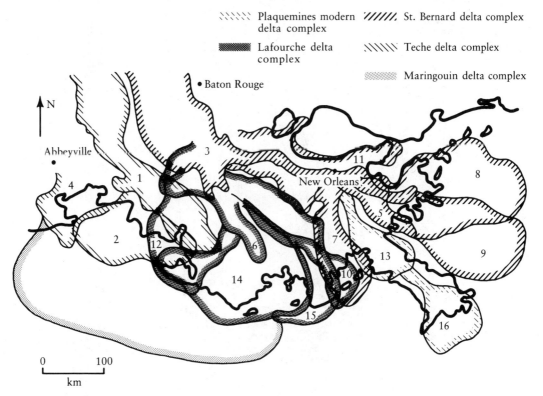

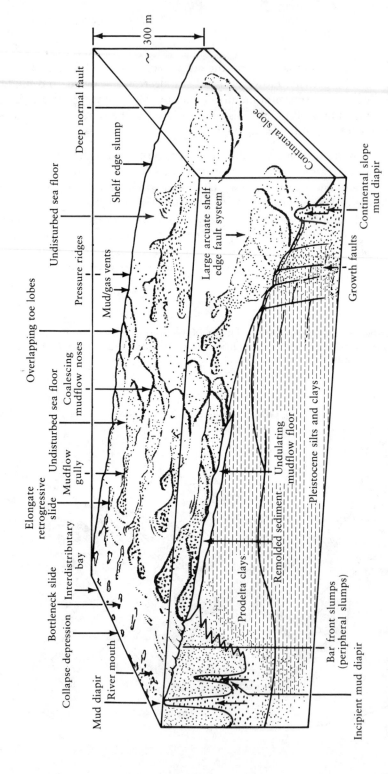

FIG. 8-7 • Schematic block-diagram showing the relationship of the various types of subaqueous sediment instabilities off the Mississippi River delta. (Reprinted by permission of the publisher from Coleman et al., *in* Stanley and Moore, *Society of Economic Paleontologists and Mineralogists Spec. Pub. 33*, 1983.)

Deep normal fault

Shelf edge slump

Undisturbed sea floor

Pressure ridges

Mud/gas vents

Overlapping toe lobes

Undisturbed sea floor

Coalescing mudflow noses

Elongate retrogressive slide

Mudflow gully

Bottleneck slide

Interdistributary bay

Collapse depression

Mud diapir

River mouth

Continental slope

Large arcuate shelf edge fault system

Continental slope mud diapir

Growth faults

Undulating mudflow floor

Pleistocene silts and clays

Remolded sediment

Prodelta clays

Bar front slumps (peripheral slumps)

Incipient mud diapir

300 m

dominance. In other words, it is easier to approach perfect descriptions of present-day events, processes, and their results than it is to describe long-past events by studying ancient rocks. What we see in outcrops of ancient rocks might bear little resemblance to the effects of day-to-day "normal" events reported by the weather-person on the *Today Show*, *Good Morning America*, or *CBS This Morning*. The present may be a very imperfect key to the past. Nevertheless, it is the only key we have, and for this reason, "process sedimentology" is an essential component of the training of a modern stratigrapher.

Prodelta slope sediments.   The inclined (clinoform) sediments deposited in front of the subaqueous part of the delta plain form the bulk of the continental slope in deltaic areas. Rather homogeneous mud dominates these slope sediments, although thin stringers of sand may occur. Deposition is from suspension and gives rise to thin laminations that might not be visible to the unaided eye but might show up well on X-radiographs. Burrowing organisms may destroy the laminations if rates of deposition are low and benthonic organisms are abundant. Both benthonic and planktonic shells may occur in slope muds.

Prodelta sediments are notoriously unstable and deformable. Differential compaction of prodelta clays underlying the distributary-mouth bar sands leads to localized increases in vertical sand accumulation, creating the series of locally thick sand pods of Figure 8-2. In the modern Mississippi delta these sand pods can attain local thicknesses as great as 120 m.

Penecontemporaneous slumps and slides are frequent (Fig. 8-7) because of the characteristics of prodelta muds. On a geologic time scale, such mass movements are always occurring somewhere on the delta. Density contrasts among superjacent beds are common, as is liquefaction, so detached semiconsolidated masses of sediment can rise upward for appreciable distances through the sediment mass, and "down-to-the-coast" listric normal faulting is abundant in progradational delta sequences. In the subsurface of the Mississippi delta these faults frequently serve as the trapping mechanism for petroleum accumulations. The causes and results of sediment instability on continental slopes are discussed by Doyle and Pilkey (1979). Examples of penecontemporaneous deformation are numerous in ancient deltaic sequences.

Deltas in the Rock Record

As we noted earlier, deltaic masses vary greatly in size (areal extent and volume) from masses so small that they would be difficult to recognize in the rock record to those so large they would be difficult to miss. The Mississippi River currently delivers about 0.5 km^3 of uncompacted sediment each year to the delta, and the delta has existed at its present location for 60–70 m.y., so it would be impossible to miss it in the rock record. But many other, much smaller deltas have also been described from the ancient record, deltas having areal extents of only 10^3 km^2. The great interest in ancient deltas reflects not only their commonness as loci of sediment deposition but

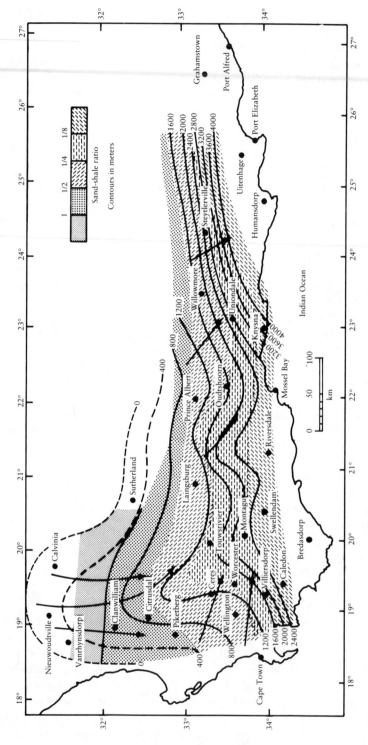

FIG. 8-8 ● Isopach, paleocurrent, and lithofacies map of the Bokkeveld Group (Series). (Adapted from Theron, 1970.)

also the fact that much of the oil found in detrital rocks and a significant proportion of the coal are associated with deltaic accumulations.

As an example of the field appearance of ancient deltaic sediments we can examine the Devonian wave-dominated sequence exposed along the southern tip of South Africa: the Bokkeveld Basin (Tankard and Barwis, 1982; Theron, 1970). The Bokkeveld Group has a maximum thickness of about 4000 m and consists typically of five superimposed upward-coarsening sequences that form a prominent hogback topography. Based on a fauna of shallow-water bivalves, brachiopods, and trilobites, the unit is of Emsian-Givetian age, deposited during a period of about 20 m.y. From north to south (Fig. 8-8) the Bokkeveld shows an increase in total thickness and a decrease in both sand/shale ratio and number of discrete sandstone beds. Paleocurrent directions based on cross-bedding vary from southerly in the western part of the outcrop to easterly in the eastern part.

On the basis of distinctive rock types and assemblages of physical and biogenic sedimentary structures, the Bokkeveld Group is divisible into four or five facies assemblages (Fig. 8-9), analogous to facies of modern wave-dominated deltas: (1) shelf-prodelta, (2) distributary mouth bar, (3) tidal flat with beach-shoreface complex, and (4) interdistributary bay.

Shelf-prodelta muds consist of laminated gray to black shales and siltstones with upward-coarsening textural trends over thicknesses up to 100 m. Some massive units occur among the laminated beds, and the randomly oriented micas and frequent presence of mottled textures in them indicate either bioturbation or extensive soft-sediment deformation. Both graded and ripple cross-laminated siltstones are present, and thin lenses of fine-grained cross-bedded clayey sandstones appear higher in the section. The sequence is interpreted as fine-grained prodelta muds at the base that grade upward into nearer-shore shallower-water fine sands.

The beds that gradationally overlie the shelf-prodelta sequence consist of upward-coarsening siltstones and sandstones, the finer-grained clastics dominating the lower part of the section. The lower part also contains mudstone interbeds. The lithic sandstones are 0.3–1.0 m thick and contain soft-sediment deformational structures reflecting the instability of the underlying muds (Fig. 8-10). In the upper part of the section the sandstones become less clayey and increase in thickness, and the ripples become increasingly asymmetric; all of these changes reflect progressively shallowing water. U-shaped *Arenicolites* burrows (Fig. 8-11) and plant debris, which are characteristic of modern mouth-bar environments, also occur. Some cross-lamination has clay-draped foresets and flaser-bedded bottomsets, indicating alternating traction and suspension sedimentation, which is characteristic of fluctuating river discharges or tidal currents.

Locally overlying the mouth-bar sediments are sequences of gray-brown mudstone up to 17 m thick, which gradationally coarsens upward to texturally immature lithic arenites. The sandstones are lenses 5–30 cm thick and can be graded or cross-bedded. Articulated plant fossils, occasional rooted zones, and vertical burrow types all suggest rapid deposition into a dominantly muddy environment. As a result, these

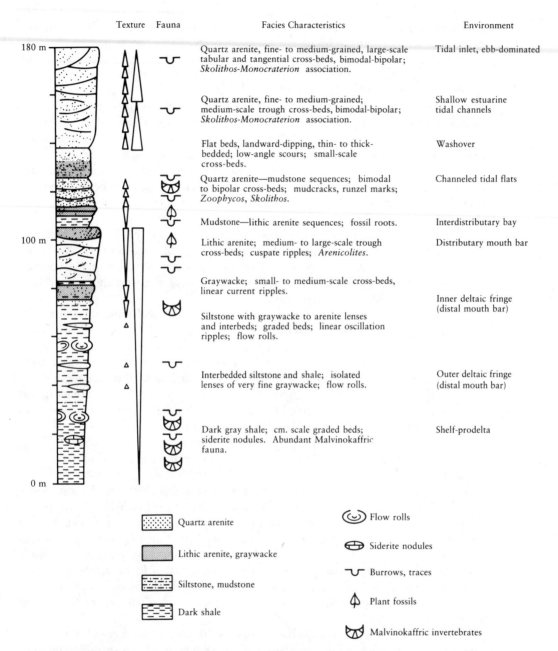

Texture	Fauna	Facies Characteristics	Environment
		Quartz arenite, fine- to medium-grained, large-scale tabular and tangential cross-beds, bimodal-bipolar; *Skolithos-Monocraterion* association.	Tidal inlet, ebb-dominated
		Quartz arenite, fine- to medium-grained; medium-scale trough cross-beds, bimodal-bipolar; *Skolithos-Monocraterion* association.	Shallow estuarine tidal channels
		Flat beds, landward-dipping, thin- to thick-bedded; low-angle scours; small-scale cross-beds.	Washover
		Quartz arenite—mudstone sequences; bimodal to bipolar cross-beds; mudcracks, runzel marks; *Zoophycos, Skolithos.*	Channeled tidal flats
		Mudstone—lithic arenite sequences; fossil roots.	Interdistributary bay
		Lithic arenite; medium- to large-scale trough cross-beds; cuspate ripples; *Arenicolites.*	Distributary mouth bar
		Graywacke; small- to medium-scale cross-beds, linear current ripples.	Inner deltaic fringe (distal mouth bar)
		Siltstone with graywacke to arenite lenses and interbeds; graded beds; linear oscillation ripples; flow rolls.	
		Interbedded siltstone and shale; isolated lenses of very fine graywacke; flow rolls.	Outer deltaic fringe (distal mouth bar)
		Dark gray shale; cm. scale graded beds; siderite nodules. Abundant Malvinokaffric fauna.	Shelf-prodelta

Quartz arenite

Lithic arenite, graywacke

Siltstone, mudstone

Dark shale

Flow rolls

Siderite nodules

Burrows, traces

Plant fossils

Malvinokaffric invertebrates

FIG. 8-9 • Facies characteristics of a generalized upward-coarsening Bokkeveld sequence. (Reprinted by permission of the publisher from Tankard and Barwis, *Journal of Sedimentary Petrology*, v. 52, Society of Economic Paleontologists and Mineralogists, 1982.)

FIG. 8-10 • Disturbed bedding in lower portions of distributary mouth bar sequences range from the centimeter-scale convolutions shown here to meter-scale flow rolls. The scale is 10 cm long. (Reprinted by permission of the publisher from Tankard and Barwis, *Journal of Sedimentary Petrology*, v. 52, Society of Economic Paleontologists and Mineralogists, 1982.)

mudstones are interpreted as fills of interdistributary bays. The graded units represent suspension settling during "normal" bay filling; the cross-bedded units represent more rapid water movement associated with occasional crevasse-splays into the bay.

At most outcrops the distributary mouth-bar sequence is overlain by a series of quartz arenites, mudstones, and shales arranged in upward-fining cycles that range in thickness from 2 to 10 meters. The total thickness of these beds ranges to 170 m. The sandstones are erosively based, lenticular, up to 1.4 m thick, and separated by mudstone partings. Flaser bedding and bipolar cross-lamination are common. Bioturbation is extensive, but body fossils are rare. Shales are mud-cracked, and sandstones contain rill marks and other evidence of periodic emergence. The upward-fining cycles, internal sedimentary structures, and mineralogic purity of the sandstones all suggest repeated and intense sand movement in a moderate environment.

Overlying the tidal flat sequence are up to 70 m of quartz sandstones interpreted as a beach-shoreface complex. These sandstones are sheetlike and both erosively overlie and interfinger with the progradational deltaic sediments, suggesting contemporaneous reworking of delta lobes by marine processes. It is the tidal-inlet and

FIG. 8-11 • Common biogenic sedimentary structures. (a) Base of a flaggy sandstone bed from the upper prodelta, with horizontal trails criss-crossing a network of much smaller, vertical features. These smaller features may represent casts of burrow tops in the underlying shale. (b) U-shaped burrow in tidal-flat sandstone. The scale is 10 cm long. (Reprinted by permission of the publisher from Tankard and Barwis, *Journal of Sedimentary Petrology*, v. 52, Society of Economic Paleontologists and Mineralogists, 1982.)

tidal-channel deposits that are reworked, the mud having been removed offshore in the process. Apparently, the sheet sandstones were formed by shore-zone reworking during temporary periods of sea level stability.

Nondeltaic Coastlines

Modern coastlines are subdivided by sedimentologists on the basis of whether coastline processes are dominated by waves or by tides, and three categories have been established: microtidal (<2 m difference in elevation between normal low and high tides), mesotidal (2–4 m), and macrotidal (>4 m). In microtidal and mesotidal settings, wave activity dominates tidal influences, and the largest volumes of sediment deposits are barrier island sands and landward lagoonal muds. In macrotidal environments the dominant influence on sediment accumulation is tidal forces, and the major deposits are large intertidal and subtidal sand ridges that trend roughly parallel to the tidal currents (normal to the coastline), and these are separated by major tidal channel deposits. The sediment accumulations form typically in funnel-shaped estuaries or bays. Waves are important only in coastal erosion and alongshore movement of sand and gravel.

Microtidal and Mesotidal Deposits

The barrier island–lagoon depositional system is a complex of marine and brackish-water environments that can appear in surface outcrops of ancient rocks either as pure sand or as an association of sand and mud (Fig. 8-12). Offshore deposits consist of intensely bioturbated sandy muds (slow deposition) with occasional beds of graded, laminated storm deposits. Shell fragments and ichnofossils are common. As the water shallows toward the beach, bioturbation becomes less common, and the sediment gradually coarsens to sand (Fig. 8-13). Grain sorting and rounding increase, and low-angle ripple cross-lamination becomes more important and may dip either landward or seaward. Large-scale, low-angle, wedge-shaped sets of planar laminae also may be prominent, a reflection of the offshore bars that develop along low-gradient shorefaces with an abundant sediment supply. The presence of hummocky cross-bedding testifies to the influence of storm-generated waves, and this structure is so common in the geologic record that it is now thought that storm-related deposits may constitute much of the ancient record of inner shelf to middle shoreface environments (Reinson, 1984). Washover fan deposits occur when wind-generated storm surges overflow and cut through barriers, creating lobate or sheet deposits of fine and medium sand in the lagoons. Such deposits may be very abundant in ancient barrier island settings.

The backshore dune environment is characterized by subaerial, predominantly wind-generated depositional processes and consists of sand blown shoreward from adjacent beaches. Most coastal dunes are vegetated, and trapping of sand by vegetation produces a large amount of low-angle cross-bedding along with higher-angle cross-bedding that results from avalanching down the slip-face of migrating dunes. In pre-Devonian rocks the low-angle structure may not be present because of the absence of vegetation.

Tidal inlet channels are relatively narrow areas that separate barrier bars (Fig. 8-12) and are zones of swift currents moving into and out of the lagoonal area. Rock sequences formed in the inlet setting are characterized by erosional bases often marked by a coarse lag deposit, and this is overlain by bidirectional cross-bedded sand of various scales in a fining-upward textural trend (Fig. 8-14). A dominance of either flood or ebb currents may cause one of the two directions of cross-bedding to be much more prominent than the other. Immediately landward and seaward of the inlet channel deposits flood and ebb tidal deltas occur, which are often difficult to

Map

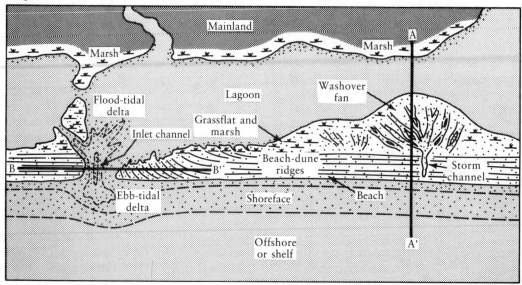

Section perpendicular to shore

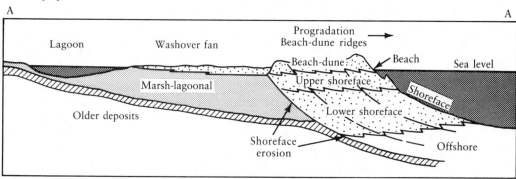

Section parallel to shore

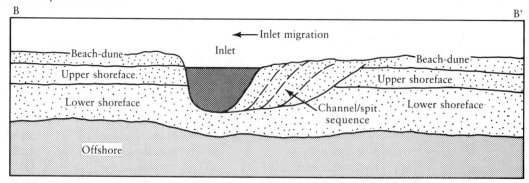

FIG. 8-12 • Generalized map and cross-sections showing major environments and facies of a barrier island–lagoonal system. Similar environments are associated with a strand-plain system, except that the progradational beach-ridge plain generally is wider and inlet-lagoonal environments are less well developed on strand plains. (Adapted from McCubbin, *in* Scholle and Spearing, 1982.)

FIG. 8-13 • Alternating laminated-to-burrowed (*Ophiomorpha*) beds in lower to middle shoreface fine sand deposits of the Upper Cretaceous Blood Reserve Sandstone, southern Alberta. (Reproduced with the permission of the Geological Association of Canada from Reinson, *in* Walker, *Facies Models,* 1984.)

separate in ancient deposits on the basis of sedimentary structures. Perhaps the major difference between the two is the occurrence of multidirectional cross-beds in ebb delta sequences, in contrast to the predominantly flood-oriented or bidirectional cross-beds of flood-tidal delta sequences. Both types of delta deposits have textures and structures similar to inlet fill sequences, and so the identification of delta sand bodies may depend largely on their geometry and stratigraphic position relative to surrounding facies.

Lagoonal deposits generally consist of interbedded and interfingering sandstone, shale, siltstone, and coal facies that are characteristic of a number of overlapping subenvironments (Fig. 8-15). Lagoonal sequences are in marked contrast to the predominantly clean sandstone sequences of the barrier-beach, tidal inlet, and tidal delta environments. And although interfingering occurs, the lateral facies change from sandstone to mud-rich sediments is relatively abrupt. The fine-grained facies include those of (1) subaqueous lagoon, characterized by brackish-water macro-

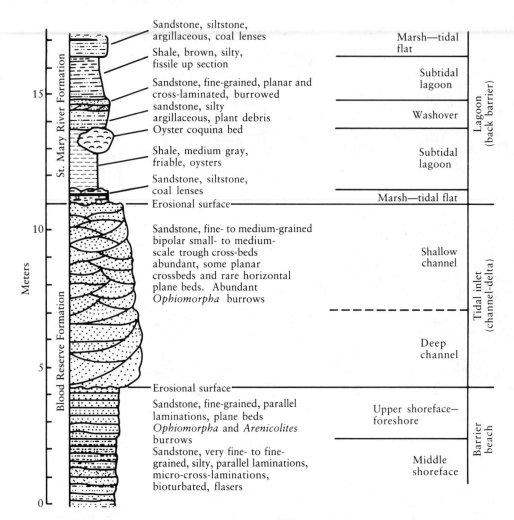

FIG. 8-14 • Composite stratigraphic section of the Blood Reserve–St. Mary River Formations (Upper Cretaceous), southern Alberta. (Reproduced with the permission of the Geological Association of Canada from Reinson, *in* Walker, *Facies Models*, 1984.)

invertebrate shells and carbonaceous plant remains; (2) marsh and swamp flatlands landward of the lagoon, recognized by the presence of thin peat or coal beds formed on mud flats of the lagoonal margin; and (3) tidal flat deposits formed on the lagoonal margin. Tidal flat deposits are minor in microtidal settings and increase in importance as tidal range increases. The common sedimentary structures in tidal flat accumulations are ripple-laminated fine sand and interbedded sand and mud containing flaser bedding and lenticular layers. (See the following section on macrotidal deposits.)

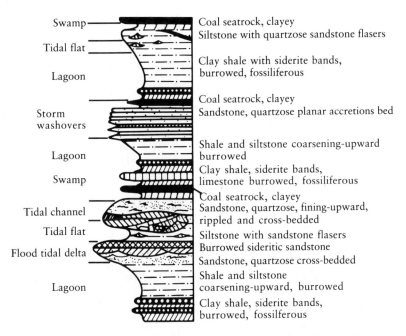

Swamp
Tidal flat
Lagoon

Storm
washovers

Lagoon
Swamp

Tidal channel
Tidal flat
Flood tidal delta

Lagoon

Coal seatrock, clayey
Siltstone with quartzose sandstone flasers

Clay shale with siderite bands,
burrowed, fossiliferous

Coal seatrock, clayey
Sandstone, quartzose planar accretions bed

Shale and siltstone coarsening-upward
burrowed
Clay shale, siderite bands,
limestone burrowed, fossiliferous
Coal seatrock, clayey
Sandstone, quartzose, fining-upward,
rippled and cross-bedded
Siltstone with sandstone flasers
Burrowed sideritic sandstone
Sandstone, quartzose cross-bedded
Shale and siltstone
coarsening-upward, burrowed
Clay shale, siderite bands,
burrowed, fossilferous

FIG. 8-15 ● Generalized lagoonal sequence through back-barrier deposits in the Carboniferous of eastern Kentucky and southern West Virginia. Such sequences range from 7.5 to 24 m thick. (Reprinted by permission of the American Association of Petroleum Geologists from Horne et al., *American Association of Petroleum Geologists Bulletin*, v. 62, 1978.)

As an example of the field appearance of a microtidal sequence that contains all the environments shown in Figure 8-12, we can examine the Vryheid Formation (Permian) near Durban, South Africa (Tavener-Smith, 1982). The formation is exposed in three large quarries, each of which shows a different part of the total stratigraphic thickness of 70 m (Fig. 8-16). The sequence is conveniently divisible into two parts of approximately equal thickness, the lower one a beach-barrier association, the upper one a back barrier–lagoonal complex. The 14 sedimentary facies present in the Vryheid Formation (Table 8-1, Fig. 8-17) illustrate a range in depositional environment including open-water shelf silts, sandy shoreface and littoral deposits, organic-rich muds and peats of lagoonal origin, a tidal inlet, washover fans, and a fluvial channel sand.

The 70 m of section described by Tavener-Smith represents perhaps 10 million years of sedimentation, a period of time so long that we must regard the resulting sedimentary model (Fig. 8-18) almost as a cartoon. It is the best we can do, given the incompleteness of the stratigraphic record (Chapter 3), but we should always be aware of how generalized it is. The same statement is true of all sedimentary packages that have accumulated over extended periods of time. The only really accurate

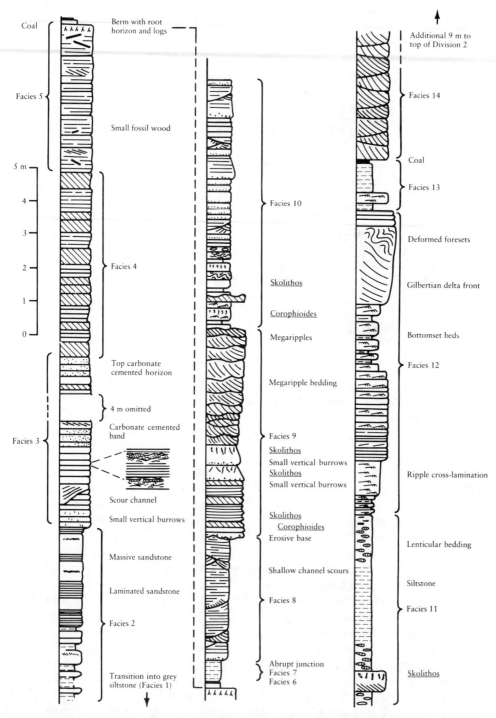

Top of Division 1

Coal

Berm with root
horizon and logs

Facies 5

Small fossil wood

5 m

4

3

2

1

0

Facies 4

Top carbonate
cemented horizon

4 m omitted

Carbonate cemented
band

Facies 3

Scour channel

Small vertical burrows

Massive sandstone

Laminated sandstone

Facies 2

Transition into grey
siltstone (Facies 1)

Facies 10

Skolithos

Corophioides

Megaripples

Megaripple bedding

Facies 9
Skolithos
Small vertical burrows
Skolithos
Small vertical burrows

Skolithos
Corophioides
Erosive base

Shallow channel scours

Facies 8

Abrupt junction
Facies 7
Facies 6

Additional 9 m to
top of Division 2

Facies 14

Coal

Facies 13

Deformed foresets

Gilbertian delta front

Bottomset beds

Facies 12

Ripple cross-lamination

Lenticular bedding

Siltstone

Facies 11

Skolithos

FIG. 8-16 • Columnar section of the Vryheid Formation as exposed in quarries near Durban, South Africa. (Reprinted by permission of the author and the publisher from Tavener-Smith, *Sedimentary Geology*, v. 32, Elsevier Science Publishers B.V., 1982.)

facies
8, 9 →

facies
6, 7 →

facies
4, 5 →

facies
3 →

FIG. 8-17 • Main face of the old quarry. The thin and regular beds of facies 3 pass up into the thicker sandstones of facies 4 and 5. A red-brown stained horizon of carbonate-cemented sandstone crosses the rock face just below the top of facies 3. The base of the prominent thin, dark siltstone (facies 6 and 7) marks the position of the berm. The upper part of the quarry face is in facies 8 and 9. Figure for scale. (Reprinted by permission of the author and the publisher from Tavener-Smith, *Sedimentary Geology*, v. 32, Elsevier Science Publishers B.V., 1982.)

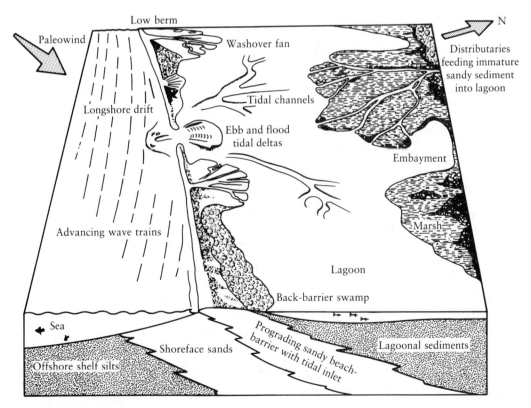

FIG. 8-18 • A sedimentary model for facies 1–12 in beach-barrier and lagoon environments. The marsh-delta complex on the right is inferred. (Reprinted by permission of the publisher from Tavener-Smith, *Sedimentary Geology*, v. 32, Elsevier Science Publishers B.V., 1982.)

TABLE 8-1 • Sedimentary facies in the Vryheid Formation

Facies Number	Description	Interpretation
	Lower Division, 38 m Thick	
1	Silty, black micaceous shale and structureless silt-stone, only occasionally flat-laminated with lenticles of fine sand, 1–2 mm thick. Some burrowing. No body fossils.	Shelf sediment in a reducing environment of moderate to "deep" water.
2	Thin sandstones with dark siltstones in lower part; 8.5 m thick. Sandstone fine-grained, micaceous, regularly laminated in bands 3–10 mm thick; some graded. Some scours and cross-bedding a few centimeters deep. In upper part, sandstone beds up to 70 cm thick and with erosive bases; low-angle cross-bed sets up to 5 cm thick.	Coastally derived sand carried offshore by storms. Micaceous matrix indicates deposition below normal wave base. Grading resulted from settling of storm-generated suspensions. Small-scale cross-beds from reworking between storm events.
3	Well-sorted, thinly flat-bedded, medium-grained sandstone beds, 12–30 cm thick. Commonly flat lamination in lower ⅔ of a bed passing upward into cross-lamination. Transitional or mildly erosive bed contacts with scour-and-fill up to 15 cm deep. Primary current lineation indicates traction transport. Vertical burrows up to 5 cm long and 3 mm wide, filled with carbonaceous material.	Lower shoreface deposit at depths of 5–10 m. In the absence of evidence of sea level fluctuation, contact between facies 2 and 3 is interpreted as normal wave base, about ½ wave length.
4	Texturally mature, thick bedded quartz arenite, 5.5 m thick. Beds 30–60 cm thick. Abundant shallow trough cross-beds in sets up to 50 cm thick. Planar sets up to 120 cm thick. Some flat lamination with primary current lineation.	Upper shoreface environment.
5	Mature to supermature quartz arenite 4.5 m thick; bed thickness 60–150 cm, thickest at top. Prominent lamination with primary current lineation; minor wood near base of facies. Loss of lamination in upper 50 cm; upper 30 cm has ramifying system of carbonized rootlets. Top 1 m has branches and logs.	Swash zone or beach foreshore, completing transition from offshore facies of facies 1 to the barrier beach. Subaerial exposure caps facies.
6	12 cm of structureless, carbonaceous, sandy silt-stone capped by 2–8 cm of anthracite coal. Seam thin and discontinuous.	Organic accumulation on landward side of a beach-barrier bar.
	Upper Division, 32 m Thick	
7	Carbonaceous siltstones with unoriented plant fragments up to 4 cm long.	Slightly deeper-water equivalent of facies 6. Suspension settling in reducing environment of back barrier lagoon.

continued

TABLE 8-1 • continued

Facies Number	Description	Interpretation
8	3.5 m of texturally submature fine sandstone with scattered coarse grains in basal 5 cm. Laminated and with some low-angle cross-beds a few centimeters thick. Shallow scour channels up to 5 cm wide and 20 cm deep. *Skolithos* burrows in top 50 cm.	Sudden change from underlying facies; greater current strength and oxygenation, possibly because of development of a new tidal inlet nearby.
9	5 m of medium- to coarse-grained, mature feldspathic sandstone; first appearance of this mineral. Base erosive with 6 mm feldspar pebbles. Overlying basal sediment are 12 couplets, 12–20 cm thick, of laminated or cross-bedded medium sandstone capped by a few millimeters of dark siltstone. Also present are bioturbated fine sandstone with 25-cm *Skolithos* burrows. Facies culminates in megarippled surface with wave lengths up to 110 cm and heights up to 14 cm.	Feldspar indicates terrestrial sediment influx from rivers, which reworked lagoonal muds meandering through nearshore area (dark siltstone). Swift currents (pebbles and megaripples). Burrows suggest seaward facies of the beach barrier.
10	Distinguished by heterogeneity. 8 m thick, mostly fine sandstone with burrowed, organic-rich siltstone beds in lower 2.5 m. Sandstones are flaser-bedded, cross-laminated, with conglomeratic feldspathic lenses with megaripples and bipolar current directions. Mud cracks and rill marks.	Subaqueous tidal flats within lagoon, not far from tidal inlet. Not intertidal (mud cracks).
11	5 m of dark siltstone with laminae and lenses of fine sand. Much bioturbation.	Lagoon with some water circulation. Sand may be blown in from fronting barrier bar.
12	Fine sand lenses of facies 11 grow larger and are replaced by flaser bedding and small ripples. Feldspar rare; cross-beds show transport from seaward, not landward. Three foreset-bottomset sequences of minideltaic origin.	Carbonaceous siltstone partings between storm-generated washover fan increments formed by normal lagoon margin sedimentation.
13	2 m thick carbonaceous mud grading upward from facies 12 as the amount of sand decreases. Mudstone grades upward into coal 15 cm thick. Also present is a 30-cm bed of rippled, cross-bedded, medium sand.	Decrease in washover activity and reestablishment of lagoon margin environment. Coal swamps behind barrier.
14	9 m of highly cross-bedded, texturally immature, feldspathic and lithic pebbly sandstone with erosive basal contact on facies 13. Unimodal currents, trough and planar cross-beds.	Fluvial channel, perhaps on deltaic distributary.

(Data from Tavener-Smith, 1982.)

pictures of geography are those that represent very brief periods, as is clearly shown by the example of the Mississippi River delta (Fig. 8-6). As the length of time represented by sedimentary rock increases, our interpretations must be more generalized.

Macrotidal Deposits

In modern coastal settings, macrotidal deposits are associated with wide continental shelves or embayed coasts on narrower shelves. The reason for this is that broad continental shelves or V-shaped gulfs or estuaries can cause resonant amplification of the tidal motion, with the result that the tidal range of about 0.3 m in the open ocean can be increased near coastlines to values of 15 m or more. The change from wave dominance (microtidal coasts) to tide dominance is, of course, gradational, and many modern coastlines show effects of both processes. No doubt the same was true of ancient coastlines, although textbook discussions may leave the impression that all coastlines are of either one type or the other. Along modern coasts, polar end members might be the high-relief California coastline for wave dominance and the Gulf of California (12-m range) or the Bay of Fundy (16.3-m range) for tidal dominance.

The importance of macrotidal deposits in ancient rocks has been argued by sedimentologists for about 20 years without a clear result. Certainly, many intertidal deposits have been recognized on the basis of examples from the North Sea. The concept is that progradation should give rise to a fining-upward sequence from sand to mud, the thickness of which should give an indication of mean tidal range. Klein's (1971) model assumes a stillstand of sea level, but without an independent time control there is no way to determine how long the prograding sequence took to form. And as Hallam (1981) has pointed out, it is difficult to unravel the effects of subsidence, compaction, and erosion. When ancient intertidal fining-upward sequences are traced laterally, they vary in thickness. How is this to be treated? The problem is even more difficult with carbonate deposits because of their biologic origin and ease of recrystallization.

Another important but unresolved controversy concerning macrotidal deposits in ancient rocks is their presence in epeiric seas, such as those that covered much of the interior of cratonic North America during the Paleozoic Era. Shaw (1964) has estimated that submarine slopes in these seas averaged about 2 cm/km, so water depth 1000 km from shore would have been only 20 m. For comparison we can consider the slope along the periphery of eastern United States off New Jersey or Virginia, a divergent continental margin generally thought of as sloping very gently offshore. Its slope is about 100 cm/km, 50 times as steep as Shaw's picture of the intracratonic sea floor slopes. Because of the exceedingly low slope of epeiric sea floors, Shaw has argued that the effect of friction would have had a dampening effect on tidal range, so macrotidal deposits would be rare. However, the presence of mesotidal or macrotidal deposits is well documented from many epeiric sea sequences, both siliciclastic and carbonate. For example, numerous workers have identified probable tidal sandstones and clastic carbonate rocks in the epeiric middle and upper Devonian Catskill Sea of the central Appalachians. Tidal ridges and tidal flat deposits have been identified. Slingerland (1986) has used numerical simulation to estimate paleotides in the Catskill Sea, incorporating hydrodynamic theory and basin geometry and paleogeography. His calculations indicate high mesotidal to low macrotidal ranges on the Catskill shelf.

Criteria for the recognition of tidal processes in ancient shoreline successions include (Elliott, *in* Reading, 1986) (1) a close spatial and temporal association of current-formed structures indicating bipolar or bimodal flow; (2) an abundance of reactivation surfaces in cross-beds; and (3) the presence of structures such as flaser bedding and mud-draped foresets, which reflect small-scale, repeated alternations in sediment transport conditions. Additional criteria that apply only to intertidal settings include tidal bundles (Fig. 8-19), indications of repeated emergence, and the presence of surface runoff features formed as the tide falls (Klein, 1971). An estuarine setting for intertidal deposits, rather than a broad, open coastal setting, would be indicated when fluvial channel sands with unidirectional cross-bedding directions underlie intertidal sediments.

FIG. 8-19 ● Tidal bundles near the base of the McMurray Formation (Cretaceous), near the confluence of the Athabaska and Clearwater rivers, Alberta, Canada. The thicker sand units were deposited by the dominant flood-tide current. The thinner units are mud deposited during flow stoppage as the tide reverses. (Reprinted by permission of the publisher from Smith, *Bulletin of Canadian Petroleum Geology*, v. 36, 1988.)

TABLE 8-2 • Associations of sedimentary structures, Middle Member, Wood Canyon Formation, California and Nevada (Precambrian), and interpreted flow models

Structures	Flow Model
1. Herringbone cross-stratification, parallel laminae	Tidal current bedload transport with bipolar reversals of flow directions
2. Reactivation surfaces, multimodal distribution of set thicknesses of cross-strata	Time-velocity asymmetry of tidal current bedload transport
3. Current ripples, current ripples superimposed at 90° on current ripples, current ripples superimposed at 90° on cross-strata, interference ripples, B-C sequences of micro-cross-laminae overlying cross-strata	Late-stage emergence outflow and emergence with sudden changes in flow directions at extremely shallow water depths (less than 2.0 m)
4. Cross-strata with flasers Simple flaser bedding, wavy bedding, isolated, thin lenticular bedding, tidal bedding	Alternation of tidal current bedload transport with suspension deposition during slack-water periods
5. Flaser and lenticular bedding	Tidal slack-water mud deposition
6. Mudchip conglomerates at base of washouts	Tidal scour
7. Mud cracks, runzel marks	Exposure
8. Tracks and trails, *Monocraterion* escape burrows Burrows	Burrowing
9. Load casts, pseudonodules, convolute bedding	Differential compaction and loading due to rapid deposition
10. Paleotidal range sequences	High rate of prograding tidal flat sedimentation

(Reprinted by permission of the publisher from Klein *in* Ginsburg, *Tidal Deposits*, Springer-Verlag New York, Publishers, 1975.)

The interactions between fluid flow and sediment movement on modern tidal flats have been intensely studied by process-oriented sedimentologists for about 20 years, so quite detailed interpretations of sedimentary structures are possible (Weimer et al., 1982; Elliott, *in* Reading, 1986). Among the events that can be recognized in ancient tidal flat sediments are stage in the tidal cycle, velocity of water on the tidal flat, and the relative competencies of flood and ebb currents (Table 8-2).

An excellent field example of a sequence whose deposition was dominated by tidal influence (tidal range unknown) has been given by Sellwood (1972; *in* Ginsburg, 1975) for Lower Jurassic sediments on the Danish island of Bornholm on the south flank of the Baltic Shield. The two-dimensional outcrop is about 400 m in length and 15 m thick and consists of a lower sandy unit that is succeeded upward by wavy- and flaser-bedded sands and clays that are capped by rootlet beds with thin coals.

The basal sand is fine- to medium-grained and is dominated by planar and tabular cross-bedding. Although some herringbone cross-bedding is present, most opposed foreset dips are found in units several beds apart. Megaripples (subaqueous dune forms) occur as reactivation surfaces (surfaces formed by slight erosion of the lee side of a ripple form), and *Skolithos* burrows sometimes descend from them. Both the megaripple surfaces and foreset laminae within them are often draped with clay formed during deposition from slack water, possibly during tidal reversals. Channels up to 1 m deep are cut into the basal sand unit and contain clay-draped foresets (Fig. 8-20) formed by lateral accretion of sediment in the tidal channel. Occasional burrows descend at right angles to the gently inclined foreset laminae, which are well defined by small fragments of plant debris.

Above the basal sand unit the clay content increases, and the bulk of the sequence consists of discretely alternating laminae of rippled sands draped by clays. Flaser, wavy, and lenticular bedding are very common. The sand layers are either current- or oscillation-rippled or show climbing-ripple features that are characteristic of rapid deposition. Plant debris is abundant, but body fossils are absent, and ichnofossils are uncommon. Channel fills cut through the flaser-bedded sediments and exhibit a range of sedimentary structures similar to those in the basal sand unit. Water-escape structures occur at many levels where water-filled sands were sealed by clay bands and later became thixotropic (Fig. 8-21).

FIG. 8-20 ● Channel-fill sediments of sand laminae draped with clay. Burrows descended perpendicular to individual foreset laminae. Tidal-bedded unit. The length of the knife is 15 cm. (Reprinted by permission of the author and the publisher from Sellwood, *Paleogeog., Paleoclim., Paleoecol.,* v. 11, Elsevier Science Publishers B.V., 1972.)

FIG. 8-21 • Water-escape structures. Tidal-bedded unit. The length of the knife is 15 cm. (Reprinted by permission of the author and the publisher from Sellwood, *Paleogeog., Paleoclim., Paleoecol.,* v. 11, Elsevier Science Publishers B.V., 1972.)

Capping the tidal sequence are several coal beds up to 30 cm thick, each resting on rootlet beds. The rootlets descend about 60 cm into the underlying sediments, which consist of bioturbated clays with sand lenticles, remnants of the underlying wavy- and flaser-bedded unit. Above each coal the facies immediately reverts to flaser-bedded clays and sands, which again may pass up into similar sediments bearing rootlets and a capping of coal.

The sequence on Bornholm is a clear example of a regressive tidal flat succession containing sands in the lower part (lower tidal flat) that becomes more clay-rich upward (high tidal flat) and culminates in supratidal coals deposited in a salt-marsh environment.

Summary

Depositional environments located at the land-sea boundary include deltas, beaches, barrier bars, and assorted types of intertidal deposits. The most complex stratigraphy is found in the deltaic sequences, in part because of the geologically rapid shifting of the locus of deltaic sedimentation and in part because of the great variety of environments within the deltaic sediment mass. Deltas are categorized as wave-dominated, river-dominated, or tidally dominated, and numerous modern examples of each type are known. The type of dominant current may change during the life of the delta, greatly increasing interpretive difficulties.

Nondeltaic coastlines can be either wave- or tidally dominated. In wave domination, lagoonal muds and barrier islands parallel to the coast are abundant. When the tidal range exceeds about 4 m, tidal current effects prevail and the major detrital accumulations are sand ridges normal to the coastline.

References

Audley-Charles, M. G., Curray, J. R., and Evans, G., 1977. Location of major deltas. Geology, v. 5, pp. 341–344.

Broussard, M. L. (ed.), 1975. Deltas, Models for Exploration. Houston, Houston Geological Society, 555 pp.

Doyle, L. J. and Pilkey, O. H. (eds.), 1979. Geology of Continental Slopes. Tulsa, Society of Economic Paleontologists and Mineralogists, Spec. Pub. No. 27, 374 pp.

Ginsburg, R. N. (ed.), 1975. Tidal Deposits. New York, Springer-Verlag, 428 pp.

Greenwood, B. and Davis, R. A. (eds.), 1984. Hydrodynamics and Sedimentation in Wave-Dominated Coastal Environments. New York, Elsevier Science Publishers, 473 pp.

Hallam, A., 1981. Facies Interpretation and the Stratigraphic Record. New York, W. H. Freeman and Co., 291 pp.

Horne, J. C., Ferm, J. C., Caruccio, F. T., and Baganz, B. P., 1978. Depositional models in coal exploration and mine planning in Appalachian region. Amer. Assoc. Petroleum Geol. Bull., v. 62, pp. 2379–2411.

Klein, G. deV., 1971. A sedimentary model for determining a paleotidal range. Geol. Soc. Amer. Bull., v. 82, pp. 2585–2592. Discussion and reply in Geol. Soc. Amer. Bull., v. 83, pp. 539–546.

McCubbin, D. G., 1981. Barrier-island and strand-plain facies. In P. A. Scholle and D. Spearing (eds.), Sandstone Depositional Environments. Tulsa, American Association of Petroleum Geologists, Mem. 31, pp. 247–279.

Potter, P. E., 1978. Significance and origin of big rivers. J. Geol., v. 86, pp. 13–33.

Reading, H. G. (ed.), 1986. Sedimentary Environments and Facies, 2nd ed. Oxford, Blackwell Scientific Publications, 628 pp.

Reineck, H. E. and Singh, I. B., 1980. Depositional Sedimentary Environments, with Reference to Terrigenous Clastics, 2nd ed. New York, Springer-Verlag, 549 pp.

Reinson, G. E., 1984. Barrier-island and associated strand-plain systems. In R. G. Walker (ed.), Facies Models. Toronto, Geological Association of Canada, pp. 119–140.

Sellwood, B. W., 1972. Tidal-flat sedimentation in the Lower Jurassic of Bornholm, Denmark. Paleogeog., Paleoclim., Paleoecol., v. 11, pp. 93–106.

Shaw, A. B., 1964. Time in Stratigraphy. New York, McGraw-Hill, 365 pp.

Slingerland, R., 1986. Numerical computation of co-oscillating paleotides in the Catskill epeiric sea of eastern North America. Sedimentology, v. 33, pp. 487–497.

Smith, D. G., 1988. Tidal bundles and mud couplets in the McMurray Formation, northeastern Alberta, Canada. Bull. Canadian Petroleum Geol., v. 36, pp. 216–219.

Stanley, D. J. and Moore, G. T. (eds.), 1983. The Shelfbreak: Critical Interface on Continental Margins. Tulsa, Society of Economic Paleontologists and Mineralogists, Spec. Pub. No. 33, 467 pp.

Tankard, A. J. and Barwis, J. H., 1982. Wave-dominated deltaic sedimentation in the Devonian Bokkeveld Basin of South Africa. J. Sed. Petrology, v. 52, pp. 959–974.

Tavener-Smith, R., 1982. Prograding coastal facies associations in the Vryheid Formation

(Permian) at Effingham quarries near Durban, South Africa. Sedimentary Geol., v. 32, pp. 111–140.

Theron, J. N., 1970. A stratigraphical study of the Bokkeveld Group (Series). *In* 2nd Gondwana Symposium, International Union of Geological Sciences, South Africa, pp. 197–204.

Weimer, R. J., Howard, J. D., and Lindsay, D. R., 1982. Tidal flats and associated tidal channels. *In* P. A. Scholle and D. Spearing (eds.), Sandstone Depositional Environments. Tulsa, American Association of Petroleum Geologists, Mem. 31, pp. 191–245.

Woodrow, D. L. and Sevon, W. D. (eds.), 1985. The Catskill Delta. Boulder, Geological Society of America, Spec. Paper 201, 246 pp.

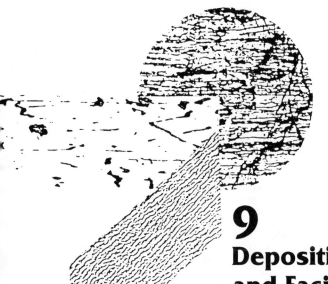

9
Depositional Environments and Facies III: Marine Environments

About 70% of the Earth's surface is covered by seawater, and this great areal extent is matched by the variety of sedimentary processes and sediment types that accumulate in the sea. Away from the shoreline on ancient cratonic shelves we find elongate and tabular sand bodies that grade into muddy facies with increasing distance offshore, although this simple relationship is commonly obscured on modern shelves by the effects of Pleistocene glaciation. On shelves in low latitudes, abundant carbonate deposits occur as reefs (bioherms) or as fragmental accumulations (biostromes).

If we traverse perhaps several hundred kilometers of shelf normal to the shoreline, we reach the shelfbreak (Stanley and Moore, 1983) at an average depth of 124 m, where the shelf gradient of perhaps 0.1° steepens to 3–6° as the shelf sea floor becomes the continental slope (Doyle and Pilkey, 1979) and descends to the continental rise and to abyssal depths. Sediment deposition occurs on modern continental slopes, but it is not clear how much of it is preserved in the geological record, as contrasted to slope sediment that is in transit (on a geological time scale).

At the base of the slope occur deep-sea fans formed by turbidity currents and other types of gravity flows, which deposit their sediment where the slope gradient flattens to the near-planar surface of abyssal plains. Further into the deep ocean basin we find sediment deposits characteristic of the abyssal environment: brown clay, siliceous ooze, and calcareous ooze.

Terrigenous Shelf Sediments

Ancient terrigenous shelf deposits (10- to 200-m depth) formed along the margin of the craton are abundant in stratigraphic sections but cannot always be distinguished from intracratonic marine sands. For example, during mid- and late Mesozoic time on the North American craton a north-south seaway with an average width on the order of 1000 km bisected the continent, and the sand bodies formed in it appear much like those on the modern cratonic shelf margin off the eastern United States. Another intracratonic depositional setting is the epeiric sea. Sand bodies formed tens or hundreds of kilometers from its shoreline may be indistinguishable from those of either the linear intracratonic seaway or the cratonic margin. In many ancient rock sections the location of and distance to the shoreline are uncertain, and the processes that formed the sand and mud accumulations may be similar. Regional paleogeographic studies may be the only way to achieve a distinction among the three types of depositional settings.

Terrigenous shelves can be either storm-dominated or tide-dominated, analogous to the shoreline environments considered in Chapter 8. The criteria for distinguishing between the tidal, normal shelf wave, and shelf storm influence are the same as were used in that chapter. For tidal sand bodies, herringbone cross-beds, mudstone clay drapes over cross-beds, and bioturbation features are diagnostic. Accumulations formed by normal shelf waves have the internal stratification pattern shown in Figure 9-1 (de Raaf et al., 1977) with ripple-form wave lengths less than 30–40 cm. For storm-generated sand bodies the wave lengths exceed 1 m, and large swells (hummocks) and swales dominate the bedding surface. One hummock appears in the lower left of Figure 9-2. Hummocks are characterized by lower bounding erosional set surfaces sloping in all directions at low angles away from the hummock crest, laminae overlying these surfaces in parallel arrangement, and lateral thickening of laminae. Associated with the hummocky stratification (Fig. 9-2) are overlying flat laminae, micro-cross-laminae, and a cap of bioturbated layer reflecting interstorm periods when the sea floor was colonized by burrowing organisms (Dott and Bourgeois, 1982).

Mud-dominated offshore facies are the most abundant of all ancient siliciclastic offshore deposits but are difficult to describe and characterize in outcrop. Deposition of modern shelf muds is highly episodic with maximum rates immediately following storms and minimum rates during prolonged fair weather periods. Although it appears paradoxical at first, fine-grained muds are best developed in association with unusually high-energy shallow-water events, as are the coarser clastic deposits. The two lithologies do not form at the same place or necessarily at the same time as the high-energy event, but both are closely related to it. Evidence of storms is provided by the concentration of reworked and abraded skeletal material (coquinas) and by muddy storm layers with their diagnostic bioturbated tops developed during interstorm periods. Some shelf shale sequences may contain unbioturbated units whose bases cut sharply through underlying bioturbation and show only minor poststorm

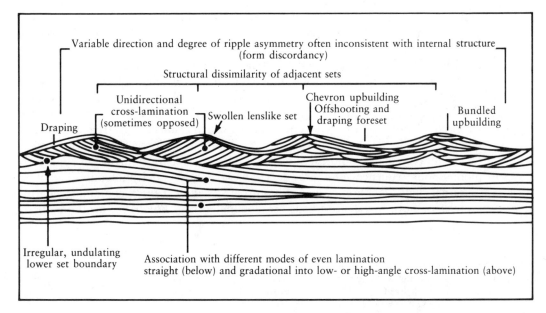

Variable direction and degree of ripple asymmetry often inconsistent with internal structure (form discordancy)

Structural dissimilarity of adjacent sets

Draping

Unidirectional cross-lamination (sometimes opposed)

Swollen lenslike set

Chevron upbuilding
Offshooting and draping foreset

Bundled upbuilding

Irregular, undulating lower set boundary

Association with different modes of even lamination
straight (below) and gradational into low- or high-angle cross-lamination (above)

FIG. 9-1 • Diagnostic features of wave-generated cross-stratification. (Reprinted by permission of the publisher from de Raaf et al., *Sedimentology*, v. 24, Blackwell Scientific Publications, 1977.)

FIG. 9-2 • Hummocky cross-stratification, Cardium Formation (Turonian), Seebe, Alberta, Canada, showing interbedded bioturbated mudstones. (Photo courtesy of R. G. Walker.)

bioturbation toward their tops. It is possible that the internal fabric or grain size distribution of muddy storm deposits is diagnostic, but such laboratory studies of ancient shelf muds are rare.

Most studies of ancient shelf muds have been paleontologic, with emphasis on trace fossil assemblages (Johnson and Baldwin, *in* Reading, 1986). Research has concentrated on whether the water was oxygenated, which results in a diverse benthonic assemblage of burrowing and bottom-dwelling organisms, or has restricted water circulation and lowered oxygen supply, which results in a restricted fauna and few trace fossils. An example of this approach is that of Morris (1979), who studied Lower Jurassic shales in northern England. He was able to subdivide an apparently monotonous sequence into three facies based largely on ichnofossils (Fig. 9-3). "Normal shale" is a homogeneous bioturbated sediment often containing sideritic nodules or horizons. Trace fossils are dominated by *Chondrites*; benthonic body

FIG. 9-3 • Triangular plot showing distinct distribution of bivalves for each facies in Lower Jurassic shales of Yorkshire, England. (Reprinted by permission of the publisher from Morris, *Paleogeog., Paleoclim., Paleoecol.,* v. 26, Elsevier Science Publishers B.V., 1979.)

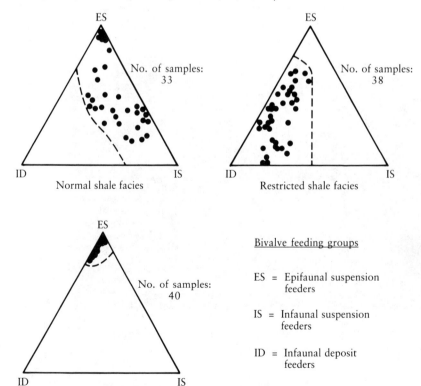

Normal shale facies

Restricted shale facies

Bituminous shale facies

Bivalve feeding groups

ES = Epifaunal suspension feeders

IS = Infaunal suspension feeders

ID = Infaunal deposit feeders

fossils are abundant and diverse. "Restricted shale" is a poorly laminated sediment with scattered calcareous concretions. Bioturbation is sparse, thin pyritic burrows are present, and the benthonic fauna is dominated by deposit-feeding protobranch valves. "Bituminous shale" is finely laminated with pyritic calcareous concretions, little or no bioturbation, and a benthonic fauna that is sparse and entirely epifaunal. These three types of shale are not necessarily related to water depth but rather to the presence of topographic barriers to the circulation of seawater.

An influx of siliciclastic sediment has a pronounced inhibiting effect on carbonate-secreting organisms, so sediments composed of a mixture of carbonate and siliciclastic debris are uncommon. They do occur, however, because of the sporadic nature of detrital influx. Mount (1984) recognized four processes that can be responsible for carbonate-silicate mixing:

1. punctuated mixing, in which sporadic storms and other periodic events transfer sediments from one depositional environment to another, for example, from the beach to an offshore position;
2. facies mixing, in which sediments are mixed along the diffuse boundaries between contrasting facies;
3. *in situ* mixing, in which the carbonate fraction consists of body fossils that accumulated on or within siliciclastic substrates; and
4. source mixing, in which admixtures are formed by the uplift and erosion of nearby carbonate source terranes.

Examples of each of these four causes of mixing have been recognized in the stratigraphic record.

The Shannon Sandstone is a petroliferous unit of Late Cretaceous age that has been studied extensively in the Powder River Basin of northeastern Wyoming in both outcrop and the subsurface (Tillman and Martinsen, *in* Tillman and Siemers, 1984). It occurs as a time-transgressive series of elongate north-south-trending lithic sandstone bodies up to 25 m in thickness within a much thicker sequence of marine sandstones and shales. From examination of 14 surface sections and abundant subsurface electric log data, 11 facies were identified on the basis of sedimentary structures, mineralogy, and paleontologic criteria (Table 9-1, Fig. 9-4). All facies appear to have formed from southerly flowing shore-parallel currents intensified periodically and frequently by storms.

The Shannon intertongues both landward and seaward with fossiliferous marine shales, and there is no evidence of associated continental or shoreline facies such as fluvial deposits, soil zones, or lagoons. No low-angle, swash-generated cross-bedding is present in the sands, indicating an absence of beach deposits. In addition, clay laminae occur in the Shannon, a feature not normally present in the high-energy breaker zone on a shoreline. And glauconite is present throughout the Shannon, in amounts greater than 15% in some outcrops. Glauconite forms only in the marine environment, not in the transitional setting of a beach. For these reasons the Shannon is interpreted to be an offshore shelf deposit.

The many sandstone facies in the Shannon represent gradational changes in

TABLE 9-1 ● Facies of the Shannon shelf-ridge complexes in the area of Salt Creek anticline, Wyoming

	Central Marine Bar Facies	Central Bar (Planar Laminated) Facies	Bar Margin Facies (Type 1)	Bar Margin Facies (Type 2)	Interbar Facies	Interbar Sandstone Facies	Shelf Sandstone Facies	Bioturbated Shelf Sandstone Facies	Shelf Siltstone Facies	Bioturbated Shelf Siltstone Facies	Shelf Silty Shale Facies
Lithology	Predominantly medium-grained quartzose sandstone, moderately glauconitic; local siderite clasts	Fine- to medium-grained quartzose sandstone	Fine- to medium-grained sandstone, shale, and limonite rip-up clasts and lenses, very glauconitic	Fine- to medium-grained sandstone with only rare shale interbeds; fewer clasts and lenses and less glauconitic than Type 1	Thinly interbedded fine- to very fine-grained silty sandstone and silty shale, slightly glauconitic	Fine-grained sandstone, virtual absence of silty shale	Very fine-grained sandstone; trace of laminated shale	Shaly, slightly sandy dark gray siltstone, traces to moderate amounts of glauconite	Siltstone and very fine grained sandstone and shale	Siltstone, very fine-grained sandstone and shale; some glauconite and limonite	Silty shale; rare thin (3 mm), silty sandstone lenses
Sedimentary Structures	Predominantly moderate-angle trough and planar-tangential cross-bedding; trough sets commonly horizontally truncated	Mostly subhorizontal plane-parallel laminated sandstone, 15 cm thick laminasets; minor shale and sandstone ripples	Mostly moderate angle troughs, some current ripples, shale clasts rarely show preferred orientations	Interbedded sequences of several beds of troughs overlain by several beds of ripples	Predominantly horizontal ripple-form bedding surfaces marked by interbedded shales; trace of wave ripples; current ripples predominate	Predominantly horizontal ripple-form bedding surfaces; bedding commonly indistinct; trace of wave ripples; current ripples predominate	Subhorizontally to low angle laminated; rare troughs	Few physical structures preserved; scattered thin rippled sand and horizontal laminasets; bedding commonly destroyed	Somewhat mottled to massive-appearing; some low-angle bedding and lamination	Highly mottled; few physical structures preserved; trace of horizontal diffuse bedding	Current ripples and subhorizontal laminae; bedding surfaces indistinct horizontal
Burrowing	Sparse	Sparse	Sparse	Sparse	Moderate to locally high	Low to moderate	Sparse	More than 75% burrowed	Moderate	Abundant, more than 75% burrowed	Low to moderate
Reservoir Potential	Excellent	Limited?	Good	Moderate to good	Limited	Limited	Limited?	None	None	None	None
Outcrop Occurrences	Moderately common	Uncommon	Common	Common	Uncommon	Very common	Uncommon	Common	Uncommon	Common	Poorly preserved

(Reprinted by permission of the publisher from Tillman and Martinsen, *in* Tillman and Siemers, *Society of Economic Paleontologists and Mineralogists, Spec. Pub. 34,* 1984.)

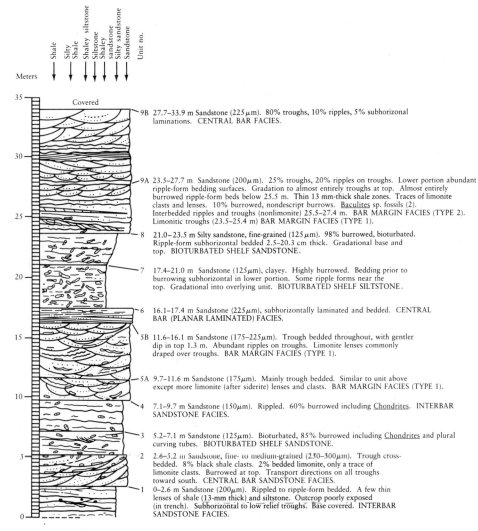

FIG. 9-4 • Measured section of Shannon Sandstone, Natrona County, Wyoming, containing many of the facies present in the formation. The lower Shannon sandstone extends up to 18 m. The upper Shannon (10 m thick) includes units 9A and 9B. (Reprinted by permission of the publisher from Tillman and Martinsen, *in* Tillman and Siemers, *Society of Economic Paleontologists and Mineralogists, Spec. Pub. 34,* 1984.)

sedimentary processes related to variations in wave and current energies on the wide depositional shelf, and as a result, the 11 discernable facies tend to occur in a preferred sequence (Fig. 9-5). The most likely facies to overlie bioturbated shelf siltstone is bioturbated shelf sandstone, which in turn probably will be overlain by interbar sandstone, the central bar, and either a bar margin facies or a shelf sandstone facies. These various facies differ greatly in their reservoir potential, so the rather detailed

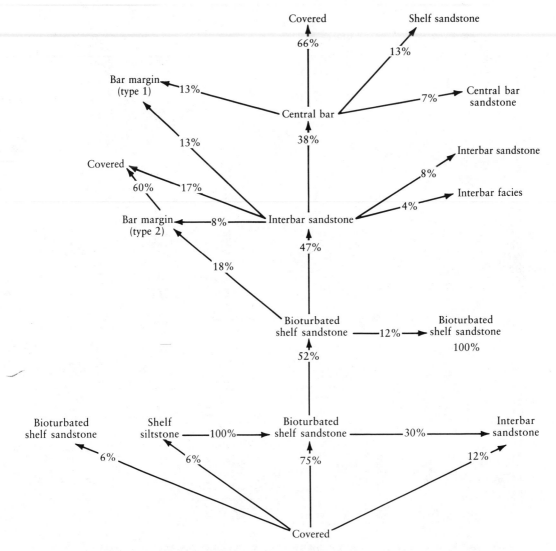

FIG. 9-5 ● Vertical sequence of facies probabilities in order of deposition (bottom to top). Individual percentage values may be considered as the probability of finding the overlying facies indicated. The percentages are compiled from 127 facies successions. (Reprinted by permission of the publisher from Tillman and Martinsen, *in* Tillman and Siemers, *Society of Economic Paleontologists and Mineralogists, Spec. Pub. 34*, 1984.)

facies distinctions made by Tillman and Martinsen are of great economic as well as academic importance.

Isopach maps of the Shannon Sandstone were constructed on the basis of outcrop and subsurface well data, and the map for the lower of the two sandstone sequences shows a distinctly linear, north-south orientation, suggestive of ridge and swale depositional patterns (Fig. 9-6). The most prominent ridge (ridge 2) exceeds

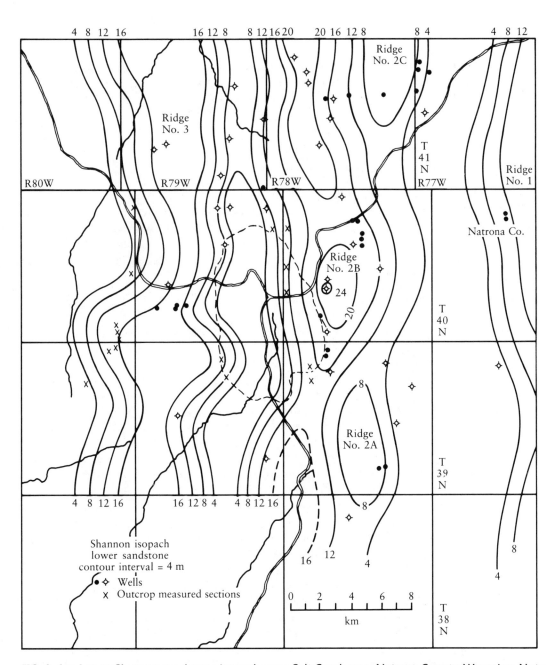

FIG. 9-6 • Lower Shannon sandstone isopach map. Salt Creek area. Natrona County, Wyoming. Note that the grain of map is north-south. Ridges are numbered from right to left. Narrow areas between ridges are designated as swales. The maximum thickness of sandstone encountered in the lower Shannon interval is 24 m. All control points and interpreted thicknesses are shown. (Reprinted by permission of the publisher from Tillman and Martinsen, *in* Tillman and Siemers, *Society of Economic Paleontologists and Mineralogists, Spec. Pub. 34*, 1984.)

30 km in length, is about 11 km wide at its widest point, and is up to 24 m thick. It appears to be a coalescence of several smaller ridges and is multicrested. Regional paleogeography indicates that the Shannon sand bars were located at least 115 km from the shoreline and were oriented at about a 30° angle to it, an orientation also found on the modern Atlantic coastal shelf of the United States.

Continental Slope Sediments

The continental slope is taken to begin at 200 m and descends to perhaps 1000 m at an initial gradient of 3°–6°; the characteristic width is a few tens of kilometers. Modern continental slopes occupy 8% of the marine environment. There have been relatively few studies of ancient slope deposits (Doyle and Pilkey, 1979; Pickering, 1982; Buck and Bottjer, 1985), but on the basis of these few there are several criteria for their recognition.

1. Slope deposits are thin in comparison to the thickness of time-equivalent shelf deposits in shallower depths or deep-sea fan deposits at abyssal depths. This is to be expected because of the relatively high gradient of the continental slope compared to the gradient of the neritic zone or continental borderland–deep-sea environment.

2. Because of the high gradient, sediment slumps, intraformational breccias, and assorted types of gravity flow deposits are very common in continental slope deposits. Load structures and evidence of thixotropic sediment behavior are common. Perhaps half the total deposit may consist of detached sediments of various kinds (Fig. 9-7). Transported coarse-grained sediments also occur in the submarine channels that are cut into the upper parts of the slope at sites controlled by the location of major rivers along the paleocoastline. Several submarine canyon deposits have been identified in the geologic record.

3. Excluding the secondarily produced translational blocks and coarse debris transported toward deeper water from the shelf environment, slope sediment is fine-grained. Typical lithologies are mud-size, either siliciclastic or carbonate. Bedding, when undisturbed, is typically finely laminated or graded, and amalgamation surfaces and cross-bedding may occur, indicating the presence of traction currents in addition to suspension settling. On a microscopic scale, quartz grain orientation and imbrication may be present.

4. The paleontology of slope sediments consists of planktonic microfossils, indigenous benthonic organisms, and carbonate-shelled shallow-water faunas transported onto the slope from the shelf environment. The indigenous benthonic faunas indicate bathyl depths. Ichnofossils may occur, but the unstable character of slope sediments makes preservation difficult.

At the base of the slope, gravity deposits may be reworked and transported by ocean currents termed contour currents. These currents flow parallel to the conti-

FIG. 9-7 • Completely conglomeratic texture developed at the base of translational slide. Clasts are set within a pervasive lime-mud matrix. The texture of the conglomerate is virtually identical to mass-flow deposits commonly inferred to be products of debris flows. The tape is 84 cm long. (Reprinted by permission of the publisher and author from Cook, *in* Doyle and Pilkey, *Society of Economic Paleontologists and Mineralogists, Spec. Pub. No. 27*, 1979.)

nental margin, and their deposits (contourites) are identified in ancient rocks largely by sedimentary structures, which indicate transport perpendicular to the direction indicated by turbidity currents carrying sediment down the continental slope.

Deep-Sea Fans

Ancient deep-sea fan deposits are very common at the base of the continental slope in the modern world ocean, have been recognized in great abundance in ancient sediments (e.g., Barnes, 1988), and are commonly very petroliferous; well-known examples are the productive Tertiary sandstones in southern California. Because these fans are formed generally at depths greater than 10^3 m, it is at first surprising to find them in such great abundance on the land surface, but the explanation is found in their location. They occur in ancient rocks near former convergent plate margins where intense tectonic compression has raised them to a position often much above present sea level.

The distinguishing features of deep-sea fans in outcrop are as follows.

1. The regional geologic setting can be suggestive. Structural deformation is intense, as in southern California, where vertical displacements of 1000 meters or more have been documented using foraminifera.

2. Although structural deformation is intense on a regional or basinal scale, outcrops tens of kilometers in length often are largely undeformed; individual beds have been traced laterally for as much as 175 km. Such beds are part of apparently monotonous sequences of shale and sandstone many hundreds of meters in thickness (Fig. 9-8).

3. The sandstone units of deep-sea fans are deposited mostly by turbidity currents that flow down the continental slope in submarine canyons and deposit their sediment load as they leave the canyon mouth at the lower part of the slope (continental rise). Turbidity currents typically deposit graded beds and many of the sandstone beds in fan sequences are graded, although not all of the Bouma divisions (A through E) need be present. Some of the shale beds may also be graded, but this grading usually is not detected in outcrop.

4. Body fossils normally are rare, although some can be introduced into the deep-water environment by the turbidity currents that originate on the continental shelf. These translocated fossils should, however, be highly broken and abraded and may be confined to channels on the fan, the sites of deposition of the coarsest sediment. Ichnofossils can be helpful, if not diagnostic of deep-water facies (Curran, 1985; D'Alessandro et al., 1986).

The morphology of modern deep-sea sediment fans is much like a combination of alluvial fans and wave-dominated deltas. Differences between the alluvial deposits and those in the deep marine setting result from (1) the greater sediment concentration in the grain-water mixture flowing downslope in the marine environment and (2) the difference in viscosity between air and seawater.

The basic process constructing submarine fans is the deposition of sediment from turbidity currents. As these currents and other gravity-flow phenomena leave the confines of submarine canyons and spread out over the lower-gradient surface of the continental rise or abyssal plain, there is a loss of both lateral confinement and competency. Distributaries form, commonly with levees, and sediment is deposited within the channel. As distance from the canyon mouth increases, the sediments generally become finer-grained; shale replaces sand as the dominant lithology. However, the use of grain size alone as an indicator of distance from the canyon mouth is dangerous because of lobe switching, differences in amount, grain size, and current velocity leaving the canyon mouth, and normal progradation of the fan with time.

The dominant theme during development of a submarine fan is progradation as additional sediment is intermittently supplied by turbidity currents originating on the continental shelf. Ideally, the sedimentary sequence will have a coarsening-upward and thickening-upward bedding trend. The basal part of the sequence will

FIG. 9-8 • Distal part of Paleocene submarine fan, Point San Pedro, California. The continuity of beds and absence of a massive sandstone facies suggests that the sequence is not a channel fill but may result from lateral switching of lobe positions on the fan. Alternatively, the beds may be interchannel, the thinning-upward sequence representing channel migration away from this area. (Reprinted by the permission of the Geological Association of Canada from Walker, *Facies Models*, 1984.)

be shale representing the most distal part of the fan, which will be succeeded upward by increasing amounts of sandy beds with increasingly complete Bouma sequences and graded bedding. Massive and pebbly sandstones or conglomerates cap the sequence. Within the overall grain-size trend there will be smaller trends of fining-upward sequences when avulsion occurs or as channels fill and are abandoned and different channels become dominant on the fan surface. As with most dominant themes, however, other melodies are interwoven with the main theme. Thicknesses of lobes vary from a few meters to tens of meters, depending on their position on the suprafan. Channel and lobe switching may result in removal of some part of the underlying lobe sequence. The differing extent of progradation of different lobes into the deeper ocean basin can cause confusing bedding sequences when outcrops are limited. Many descriptions of fan sequences are detailed by Pickering et al. (1986), Walker (1984), and Howell and Normark (*in* Scholle and Spearing, 1982).

Substantial accumulations of deep-sea fans and turbidite sandstone facies are associated with periods of lowstands of sea level and are most common in basins where slope instability is enhanced by tectonic uplift. Both eustatic lowering of sea level and uplift enhance erosion and delivery of coarse sediment to marine basins. Conversely, most predominantly fine-grained turbidite fans are deposited during periods of highstand and are generally associated with active seaward progradation of deltas on adjacent shelves.

Although a large number of descriptions have been published of ancient deep-sea fans, very few papers have been able to provide three-dimensional facies analyses. This has resulted from the large size of many fans, pervasive and intense structural deformation subsequent to fan deposition, and limitations of outcrop. One of the few three-dimensional studies that analyze a prograding fan sequence from basin-plain turbidites upward through fan deposits into silty mudstones of a continental slope is an investigation of the Shale Grit Formation in the Pennine Basin, northern England (Walker, 1978, 1984).

The Shale Grit Formation is one of a group of five formations of Namurian age (Table 9-2), based on goniatitic cephalopod zonation. The units total 730–890 m in thickness. The Edale Shales are unfossiliferous, very thinly laminated, black basinal mudstones deposited in the unoxygenated distal part of the depositional basin, and they are gradationally overlain by a series of unchanneled turbidites termed Mam Tor Sandstone. The Mam Tor from base to top becomes sandier, coarser-grained, and thicker-bedded and passes upward into the Shale Grit. The Mam Tor is interpreted as basin-plain and smooth lower fan deposits landward of the basinal mudstones. The base of the Shale Grit is defined by the first appearance of massive sandstones, beds consistently thicker than 60 cm, with many composite amalgamated sandstone beds without interbedded shales. The sandstones are granular and pebbly. Also present are turbidites with Bouma sequences and sole marks. Thinly laminated dark mudstones occur between the sandstone beds. In the lower part of the Shale

TABLE 9-2 • Stratigraphy of Shale Grit fan and associated rocks

Formation	Approximate Thickness (m)	Lithology and Interpretation
Kinderscout Grit	150	Mainly coarse sandstones, some shales; shallow-water deltaic complex
Grindslow Shales	100–120	Massive and laminated mudstones and shales; mainly prograding slope deposits; upper fan channels at base
Shale Grit	130–240	Sandstones and shales; the upper part was mainly deposited on the braided suprafan; the lower part was deposited on smooth suprafan lobes
Mam Tor Sandstones	100–130	Classical turbidites deposited on lower fan or basin plain
Edale Shales	250	Black basinal mudstones

(Adapted from Walker, 1978.)

Grit, classic turbidites are more abundant than the massive sandstones, and there are several thickening-upward and coarsening-upward sequences that begin with dark mudstones, pass upward into classic turbidites, and finally pass into massive sandstones. These sequences may be up to 60 m thick. Because of the rarity of channels, the lower Shale Grit is assigned to the same facies as the underlying Mam Tor beds, the smooth outer parts of suprafan lobes. The dark mudstones represent blanket covering of an abandoned fan lobe as a channel elsewhere on the fan became a more important distributary. An average fan lobe, the thickness of sandstone between two dark mudstones, is 30 m.

In the upper part of the Shale Grit, massive and pebbly sandstones are much more abundant than in the lower part, and numerous channels up to 20 m in depth are present. The association of these facies with channels strongly suggests a braided suprafan environment. Thickening-upward and coarsening-upward sequences are present, up to 120 m thick, but contain coarser-grained facies than the equivalent sequences in the lower part of the Shale Grit. Nonlaminated pebbly sandstones are common, as is bed amalgamation. Rapidly moving currents on the steep, upper part of the suprafan are responsible for deposition of this very coarse facies, and channel braiding is inferred. Uppermost Shale Grit channels have widths of at least 1 km and depths up to 50 m. The upper fan channels are stratigraphically surrounded by well-laminated mudstones formed by overbank spillover, analogous to crevasse-splay deposits in continental fluvial deposits.

Overlying the Shale Grit is the Grindslow Shale, composed largely of sandy mudstone, commonly burrowed, and carbonaceous sandstone. Few turbidite beds or massive and pebbly sandstones are present, and deposition is inferred to have taken place high on the continental slope immediately below the shallow-water platform represented by the facies of the Kinderscout Grit (Table 9-2). The Kinderscout is a shallow-water deltaic complex of sandstones and shales that grade from marine at the base to brackish interdeltaic and fluvial in the upper part. Collinson (1969) has recognized 14 facies in this formation, facies characterized by the sedimentologic features described for fluvial and deltaic deposits in Chapters 7 and 8. The three-dimensional paleogeographic picture represented by the Namurian stratigraphic section in the Pennine Basin is shown in Figure 9-9.

Abyssal Sediments

The abyssal parts of the world ocean are defined by depth rather than by relationship to tectonic features or topographic isolation. Abyssal areas are at depths greater than 1000 m and, as a result, are typically located far from terrigenous sediment sources. Despite this fact, the bulk of sediments in some parts of the abyssal realm are not distinctive of sedimentologic isolation. This is the case, for example, in the western Atlantic Ocean, where turbidity current deposits emanating from the divergent margin of eastern North and South America have been voluminous during Late Tertiary time.

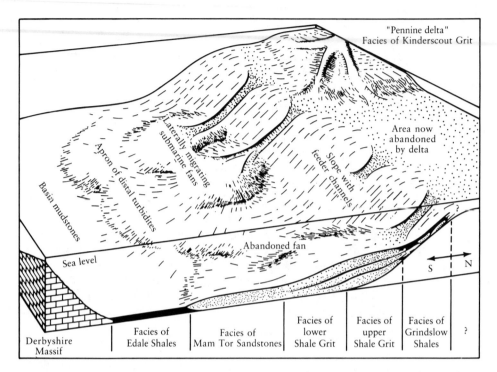

FIG. 9-9 ● Hypothetical reconstruction of the Namurian facies of North Derbyshire, arranging the formations laterally rather than vertically and giving interpretations of their environments of deposition. The observed vertical sequence can thus be visualized in terms of a southward advance of facies belts. (Reprinted by permission of the publisher from Walker, *Journal of Sedimentary Petrology*, v. 36, Society of Economic Paleontologists and Mineralogists, 1966.)

These nondiagnostic sediment types may contain some planktonic microfossil material but not in large amounts because of the great difference in emplacement rate between, say, a turbidity current deposit and a diatom skeleton settling from surface waters. The lateral facies change from terrigenous debris to pelagic sediment should be rather sharp. Pelagic units can also grade into reef talus or into volcanogenic sediment aprons, for example, in the Pacific Ocean. There also are unusual circumstances in which pelagic sediments can occur at depths much shallower than 1000 m, for example, the famous Late Cretaceous chalks of western Europe, which probably accumulated at depths of only a few hundred meters (Scholle, 1974).

Characteristic types of stratigraphic units formed in abyssal depths are microcoquinas composed of the remains of calcareous organisms such as globigerinid foraminifera (Jurassic-Holocene), and coccolithophorids (Jurassic-Holocene) and siliceous organisms such as radiolaria (Ordovician-Holocene) and diatoms (Jurassic-Holocene). Because of the importance of these pelagic microorganisms in abyssal sediments, the term *pelagic sediment* is commonly used as a synonym for *abyssal sediment*, even though microcoquinas are not the only sediment type present at abyssal depth. The sediment called brown clay covers about one-third of the modern ocean floor, mostly at depths greater than 3500 m (Fig. 9-10). It is composed largely of clay minerals transported from the continents by either wind or oceanic currents. Settling of these very fine particles through the ocean mass is accelerated by organic pelletization by filter feeders and by chemical flocculation. The brown clay deposits also contain relatively large proportions of meteoric "dust," shark teeth, whale earbones, and manganese nodules.

On the modern ocean floor the areal distributions of the biologically generated oozes are controlled by a variety of factors, including variation in water temperature as functions of latitude and oceanic depth, state of saturation of seawater with respect to calcium carbonate and silica, increase in hydrostatic pressure with increasing depth in the ocean, and current patterns both at the ocean surface and at depth. The influence of each of these variables on sediment distribution at abyssal depths is discussed in books and articles on oceanography, sedimentology, and marine biology (e.g., Barron and Whitman, 1981). What can a stratigraphic outcrop of an ancient pelagic microcoquina or brown clay tell us about the paleo-oceanographic depositional environment?

Calcareous Pelagic Rocks

The two chief pieces of information obtainable from an ancient calcareous, pelagic microfossil assemblage are geologic age and the minimum water depth at the site of deposition. The determination of water depth derives from two facts.

1. The solubility of calcium carbonate increases with increasing hydrostatic pressure and with decreasing temperature, and because oceanic temperatures always decrease with depth (although the present rate of decrease may not be typical of the geologic nonglacial past), calcareous shells dissolve as they sink through the oceanic water column. Below a certain depth defined by chemical calculation, the rate of dissolution in the water column and at the sea floor exceeds the rate of influx of carbonate particles, so carbonate accumulation should not occur. This depth is termed the **carbonate compensation depth** (CCD) (Fig. 9-11). The *actual* loss of carbonate particles, however, occurs in a depth range termed the **lysocline**, and it is the paleolysocline that we hope to determine in an ancient pelagic sediment. The base or lower boundary of the lysocline will generally be at greater depth than the base of the CCD because of the existence on the carbonate particles of metabolically produced organic coatings or because of pelletization of the particles in fecal matter as they settle through the water column. As a result of this and other factors, the disappearance of the shells of pelagic organisms with depth is most accurately visualized as a vertical facies change, not as a sharply defined surface.
2. According to the tenets of sea floor spreading, sediments are deposited on mov-

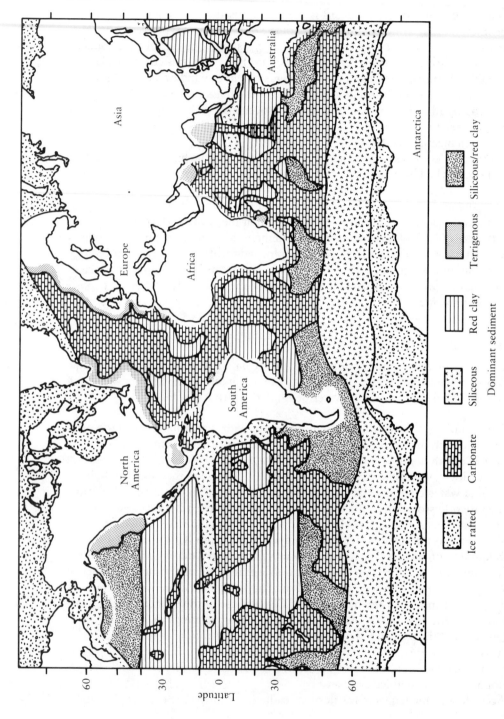

FIG. 9-10 • Modern oceanic sediments. The distribution of dominant sediments in the world ocean. (Reprinted by permission of John Wiley & Sons, Inc., from Barron and Whitman, *The Sea*, v. 7, copyright © 1981.)

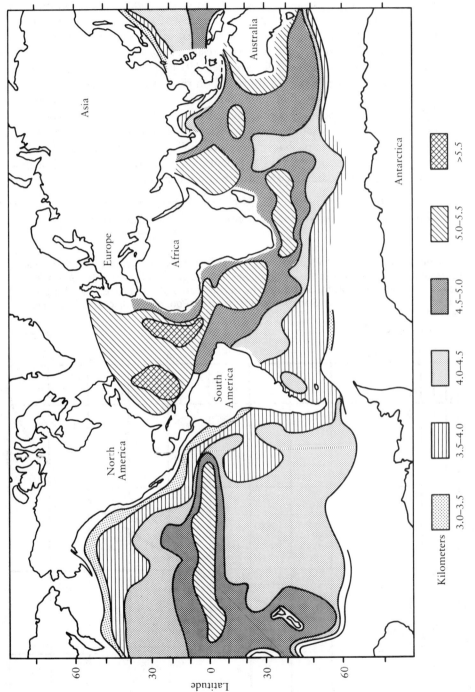

FIG. 9-11 • Calcium carbonate compensation depth mapped throughout the world ocean. (Reprinted by permission of John Wiley & Sons, Inc., from Barron and Whitman, *The Sea*, v. 7, copyright © 1981.)

ing subsiding plates, so the sediments must change both their geographical location and depth during geological time (Fig. 9-12). Thus the present locations and depths of facies boundaries in the modern ocean, such as between calcareous ooze and brown clay, are in part a result of sea floor subsidence and horizontal plate motions (Berger and Winterer, 1974). These relationships have been termed **plate stratigraphy** and are based on the existing age-depth relationship of the sediments on midocean ridge flanks (Fig. 9-12). It is assumed in interpreting ancient pelagic sediments that the same age-depth relationship was present, an assumption that is satisfactory as a first approximation but whose precision as a depth indicator is probably no better than ± 1 km. But the assumption of an age-depth constancy allows the depth of deposition of a dated sediment sample overlying a dated basement to be established (Fig. 9-13).

The volumetrically abundant groups of microfossils (globigerinid foraminifera) and nannofossils (coccoliths) that form modern oceanic pelagic oozes did not evolve

FIG. 9-12 • Generalized age-depth relationship for a modern actively spreading sea floor, based on current average spreading rates in the world ocean. The ridge crest is arbitrarily set at the present average elevation of 2700 m. (Reprinted by permission of the publisher from Berger and Winterer, *International Association of Sedimentologists, Spec. Pub. No. 1*, Blackwell Scientific Publications, 1974.)

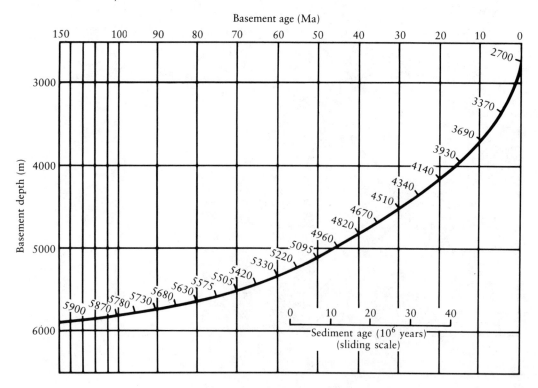

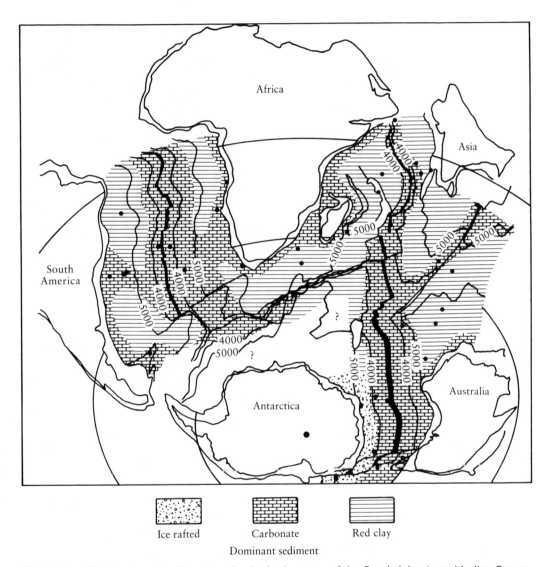

FIG. 9-13 • Distribution of sediment and paleobathymetry of the South Atlantic and Indian Ocean at 20 Ma (Lower Miocene). (Reprinted by permission of John Wiley & Sons, Inc., from Barron and Whitman, *The Sea*, v. 7, copyright © 1981.)

until mid-Mesozoic time and did not become abundant until the Cretaceous. Calcareous planktonic organisms did not exist in Paleozoic and Early Mesozoic seas. Thus ancient analogs of modern deposits can be found only in oceanic rocks less than about 150 m.y. old. We are aware of no papers in English that deal with stratigraphic relations in outcrop of such rocks, although supposed "deep-water" finely laminated micritic limestones of Paleozoic age have been described (Bissell and Barker, 1977). These "flyschlike" rocks contain 25–30% quartz silt and are believed to have an origin related to tectonic activity in adjacent highlands.

Although outcrop data are lacking, Barron and Whitman (1981) and Berger and Winterer (1974) provide extended discussions of the position of the lysocline in the world ocean during Cretaceous and Tertiary time.

Siliceous Pelagic Rocks

In ancient rocks, siliceous pelagic sediments are represented by bedded cherts, such as occur in the Ouachita Mountains of Arkansas (Arkansas Novaculite, Devonian), Sublette Range of Idaho (Rex Chert, Permian), and Coast Range of California (Monterey Chert, Miocene). Sometimes, such rocks are intimately associated with ophiolites; other times, they are not. The reason for the difference is uncertain, but the chemical purity of bedded cherts leaves no doubt that they were deposited far from terrigenous sources and in "deep" waters.

The precise paleobathymetric interpretation of bedded cherts is vague because the dominance of siliceous forms in a pelagic sediment depends in part on the unavailability of calcareous forms, and the calcite compensation depth has fluctuated in time and space. Presumably, the CCD was generally at greater depth in the absence of carbonate-shelled plankton, prior to mid-Cretaceous time. The appearance of calcareous plankton must have marked a significant change in the distribution of biogenic pelagic sediments (Garrison, 1974). Whereas prior to mid-Mesozoic time, pelagic sedimentation in regions of high plankton productivity was largely radiolarian ooze regardless of water depth, radiolarian oozes most commonly have been restricted to abyssal depths below the CCD from the Cretaceous onward, the shallower depths being dominated by calcareous forms.

The behavior of the opaline tests of diatoms and radiolaria as they settle through seawater differs from that of calcareous shells. The difference results from two factors:

1. Seawater is undersaturated with respect to silica at all depths, whereas seawater is supersaturated with respect to calcium carbonate in the upper few thousand meters. Thus dissolution of opal starts immediately on the death of the organism, most dissolution occurring in the upper 1000 meters.
2. Silica becomes less soluble at lower temperature, in contrast to calcium carbonate, for which the reverse is true. Thus accumulations of siliceous tests occur most commonly in deep waters below the lysocline and/or in shallow but cold waters such as those around the Antarctic continent. In the vicinity of oceanic ridge crests, the opal/calcium carbonate ratio increases as the moving plate descends into deeper and colder waters.

Miocene time saw a great explosion of diatomaceous pelagic facies that lasted 6–8 m.y. around the rim of the North Pacific Ocean. In California these sediments are known as the Monterey Formation and form a sequence with an estimated volume of 100,000 km^3. Monterey sediments accumulated in a series of extensional basins of variable size that formed as a consequence of the collision of the spreading ridge separating the Pacific and Farallon plates with the western margin of the North American plate (Pisciotto and Garrison, 1981). These basins subsided rapidly at different times, depending on the local tectonic situation. Thus considerable facies variations exist among basins, depending on the time of basin development and subsidence.

The Monterey Formation is a basinal sequence normally composed of three facies:

1. A lower foraminiferal-coccolith calcareous sequence of shales, siltstones, and mudstones of hemipelagic origin; commonly, the sediments were subsequently displaced by turbidity currents and slumps.
2. A middle phosphatic facies formed by phosphatization of foraminiferal shales and mudstones. This facies is distinguished from the underlying calcareous facies only by mineral composition and is simply a penecontemporaneous replacement of it.
3. The upper unit is the siliceous facies by which the Monterey Formation is characterized. It is the thickest and most widespread of the three facies and consists essentially of diatomites and diatomaceous mudrocks and their diagenetic equivalents—chert, porcelanite, and siliceous mudrocks (Fig. 9-14). Other constituents in the siliceous facies include sandstone, mudstone, shale, breccia, phosphorite, volcanic ash, and diagenetic carbonates.

FIG. 9-14 • Hand specimen of laminated diatomite from the Lompoc area. Note variability of type and spacing of laminae. Foraminifera are common. (Reprinted by permission of the publisher from Pisciotto and Garrison, *Society of Economic Paleontologists and Mineralogists, Spec. Pub. Book 15*, 1981.)

FIG. 9-15 • Outcrop in the Santa Monica Mountains showing second-order clastic-biogenic cycles. Light-colored layers are laminated porcelanites, dark layers are siltstones and thin turbiditic sandstones. The pen is 13 cm long. (Reprinted by permission of the publisher from Pisciotto and Garrison, *Society of Economic Paleontologists and Mineralogists, Spec. Pub. Book 15,* 1981.)

Lamination and rhythmic bedding with cycles ranging from 1 mm to 1 m in thickness are conspicuous in the siliceous facies (Fig. 9-15). The undisturbed thin laminae indicate deposition either on an anoxic basin floor or on an outer shelf-slope environment intersected by the oxygen-minimum zone. Pisciotto and Garrison (1981) have described two types of cycles, each containing two or three orders, with time scales ranging from years to thousands of years. For example, one common kind of cycle consists of massive, burrowed layers about 1 m thick, interbedded with laminated unburrowed layers of equal thickness. This cycle probably records fluctuations in the intensity and/or position of the oxygen-minimum zone.

Another very common cycle is rhythmic interbedding of siliceous beds 1–10 mm thick, intervening beds being separated either by bedding-plane partings or by paper-thin shale layers. Bases of the siliceous beds are in sharp contact with underlying shale, and some tops are gradational to overlying shale. In a few cases the uppermost parts of the cherts show curved laminae that represent low-amplitude ripples. In thin sections it is apparent that sand- and silt-size impurities in the cherts are graded. This type of cycle is interpreted as distal turbidites composed of a diatomaceous A-division and rare, rippled C-divisions. The reworked deposit is capped by a Bouma E-division, the paper-thin shales. Many of the chert beds are structureless because of recrystallization of the opaline diatom skeletons from opal-A (totally amorphous as seen in X-ray diffraction patterns) to opal-CT (tridymite or cristobalite crystallites detected in the X-ray diffractogram) to chert (only well-crystallized quartz detected by the X-rays).

Abyssal Shale

Rarely has an ancient pelagic brown clay deposit been recognized in a surface outcrop or even during subsurface exploration for natural resources. Nearly all supposed bathyl-abyssal siliciclastic deposits identified in the stratigraphic record are turbidites, sediments transported downslope from nearby terrigenous sources. Some of the shales in such sequences, however, may indeed be of pelagic origin, and a method of recognition has been described by O'Brien et al. (1980). When examined by using a scanning electron microscope, the pelagic clays are found to have a fabric of preferred orientation, while the clay units of turbiditic origin (Bouma E-division) have a random orientation of clay mineral flakes.

Another method of recognition of pelagic shale units, one more useful in the field setting, is the presence in the shale of granule-size ferromanganese nodules and shark teeth, as occurs in the Late Cretaceous Wai Bua Formation on the island of Timor in Indonesia (Audley-Charles, 1968). Supporting the interpretation of a pelagic origin for the shale is its association in outcrop with a bedded radiolarian chert facies (Fig. 9-16) and a very finely laminated microcrystalline limestone composed

FIG. 9-16 • Outcrop photo of the Wai Bua Formation at its type locality in southeastern Timor. The thick, light-colored beds are radiolarian cherts; the thin, dark layers are pelagic shales. The hammer in the left center is for scale. (Reproduced by permission of the Geological Society of London from Audley-Charles, *The Geology of Portuguese Timor*, 1968.)

largely of pelagic foraminifera. The presence of these other two types of pelagic sediment in the Wai Bua strengthens the inference of a pelagic origin for the thin shale units.

Carbonate Shelf Sediments

The occurrence of shallow-water carbonate deposits is controlled largely by biologic and chemical variables, hydrodynamic considerations being of distinctly secondary importance (Scholle et al., 1983). The main determinants of sedimentation patterns are water temperature, depth, light penetration, and water turbidity, because they control biologic productivity in the marine environment. All of these factors can be affected by rates of eustatic sea level change and latitudinal plate movements. Autochthonous carbonate deposits are formed almost entirely in tectonically quiescent areas such as the epeiric seas of Paleozoic time on the North American craton and the continental shelf of Florida during Tertiary time, or on isolated shallow-water banks that rise from oceanic depths such as the modern Bahama Platform. These restrictions occur because of the deleterious effects on marine organisms of siliciclastic mud influx and the fresh waters that transport them to the coastline.

Shallow-water carbonate deposits can be conveniently considered as being of two types: (1) those composed of particulate carbonate sediment unrelated to reefs (biostromes) and (2) those carbonate deposits whose presence is related to reefal buildups (bioherms), including the reef core, lagoon, and talus cone that surrounds the reef. Reefs have topographic relief and thus influence their own development because of the effect of the relief on current patterns and water depth. But for both biostromes and bioherms, intrabasinal factors control facies development. Because of this, it is possible to recognize a sequence of facies belts in carbonate basins and to use an idealized model as an aid in determining the position of local outcrops within the larger depositional basin (Fig. 9-17). The extent of each facies will vary in response to changing water depths, marine currents, and paleobiology. Rarely will all nine facies be developed in a single area.

Biostromes

Perhaps the most commonly encountered carbonates are laterally persistent, evenly bedded limestones and dolomites of apparent shallow-water origin, as demonstrated by abundant mud cracks, stromatolites, cross-bedding, and other features (James, *in* Walker, 1984). The vertical stratigraphic section in such sequences typically reflects shallowing-upward, because biologic productivity is generally much greater than the rate of basin or platform subsidence. This shallowing-upward packet is repeated many times in thick limestone sequences, reflecting gradual decrease in water depth followed by relatively rapid submergence. The packets are regressive sequences because of the high rate of biologic growth, not because of encroachment of siliciclastic debris as occurs in sandstone-mudrock basins.

Each shoaling-upward packet usually consists of three carbonate facies: (1) subtidal (open marine or lagoon), (2) intertidal, and (3) supratidal. Each facies has distinctive lithologic and paleontologic characteristics that permit it to be recognized in outcrop, so rather detailed stratigraphic analyses can be made.

One of the clearest examples of sequential development of nonreefal platform carbonate facies is provided by the Lower Devonian rocks of the northeastern United States (Laporte, 1967; 1971a, b; 1975). The Helderberg Group of western New York State and surrounding areas to the west and south ranges in thickness to 110 m (Fig. 9-18) and is divided into four formations. In ascending order they are Manlius (7–15 m), Coeymans (3–30 m), Kalkberg (15–30 m), and New Scotland (15–50 m). Paleontologic evidence indicates that all four units are of Gedinnian age and thus accumulated during a period of about 7 m.y., an apparent accumulation rate of 1 m/70,000 yrs.

Four major facies have been recognized in the Helderberg Group:

1. Supratidal-intertidal, best developed in the lower Manlius Formation.
2. Shallow subtidal, above wave base, as seen in the upper Manlius and Coeymans.
3. Subtidal, below wave base, without terrigenous clastic influx, in the lower Kalkberg.
4. Subtidal, below wave base, with intermittent terrigenous influx, in the upper Kalkberg and New Scotland Formations.

This interpretation of progressively farther offshore environments from Manlius to New Scotland Formations is indicated by both sedimentary structures and changing taxonomic diversity of shelly marine invertebrates. The Manlius contains 5–10 species, Coeymans contains 50–80, and Kalkberg and New Scotland contain 300. The overall stratigraphic progression is thus from a few meters above normal high tide (lower Manlius) to maximum depths of perhaps a few tens of meters (New Scotland). No formation is composed entirely of a single facies, of course, because shorelines will always have initial irregularities on which are superimposed local variations in carbonate production, subsidence, and hydrography. The description of a formation as subtidal, with or without terrigenous influx, is a generalization; some intertidal facies may be present as well.

The facies of the Helderbergian platform carbonates in the central Appalachians have been studied in greatest detail in the Manlius Formation, where supratidal, intertidal, and subtidal facies are all well developed.

1. **Supratidal facies.** Characterized by irregular laminar stratification, mud cracks, birdseye structures, and scattered ostracod valves in a pelletal carbonate mudstone. Often the laminations, about 1 mm thick, are alternately calcite/dolomite and are separated by thin, wispy, bituminous films. Laminae seem to be of algal origin (organic matter). Ichnofossils are rare.
2. **Intertidal facies.** Characterized by typically thin-bedded, pelletal calcite mudstone and skeletal grainstones with cross-bedding and scour-and-fill structures. Abundant fossils include ostracods, tentaculitids, brachiopods, bryozoans, stro-

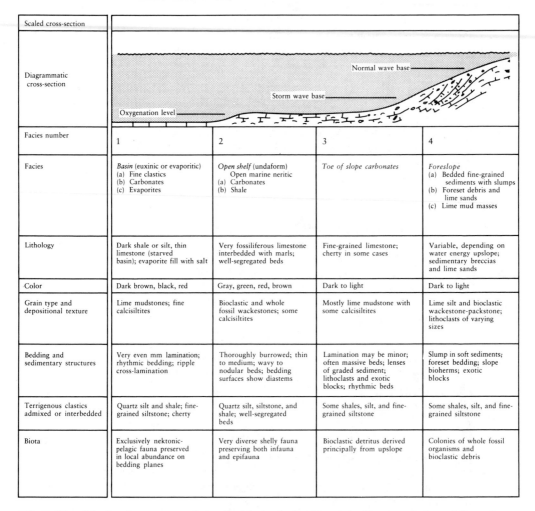

	1	2	3	4
Facies number	1	2	3	4
Facies	*Basin* (euxinic or evaporitic) (a) Fine clastics (b) Carbonates (c) Evaporites	*Open shelf* (undaform) Open marine neritic (a) Carbonates (b) Shale	*Toe of slope carbonates*	*Foreslope* (a) Bedded fine-grained sediments with slumps (b) Foreset debris and lime sands (c) Lime mud masses
Lithology	Dark shale or silt, thin limestone (starved basin); evaporite fill with salt	Very fossiliferous limestone interbedded with marls; well-segregated beds	Fine-grained limestone; cherty in some cases	Variable, depending on water energy upslope; sedimentary breccias and lime sands
Color	Dark brown, black, red	Gray, green, red, brown	Dark to light	Dark to light
Grain type and depositional texture	Lime mudstones; fine calcisiltites	Bioclastic and whole fossil wackestones; some calcisiltites	Mostly lime mudstone with some calcisiltites	Lime silt and bioclastic wackestone-packstone; lithoclasts of varying sizes
Bedding and sedimentary structures	Very even mm lamination; rhythmic bedding; ripple cross-lamination	Thoroughly burrowed; thin to medium; wavy to nodular beds; bedding surfaces show diastems	Lamination may be minor; often massive beds; lenses of graded sediment; lithoclasts and exotic blocks; rhythmic beds	Slump in soft sediments; foreset bedding; slope bioherms; exotic blocks
Terrigenous clastics admixed or interbedded	Quartz silt and shale; fine-grained siltstone; cherty	Quartz silt, siltstone, and shale; well-segregated beds	Some shales, silt, and fine-grained siltstone	Some shales, silt, and fine-grained siltstone
Biota	Exclusively nektonic-pelagic fauna preserved in local abundance on bedding planes	Very diverse shelly fauna preserving both infauna and epifauna	Bioclastic detritus derived principally from upslope	Colonies of whole fossil organisms and bioclastic debris

FIG. 9-17 • Idealized sequence of standard facies belts. (Reprinted by permission of the publisher from Wilson, *Carbonate Facies in Geologic History*, Springer-Verlag New York, Publishers, 1975.)

matolites, algal oncolites, and small U-shaped burrows. Many fossil occurrences have abundant individuals, but few taxa are represented. Mudcracks and minor erosional relief on carbonate mudstone beds indicate intermittent subaerial exposure.

3. **Subtidal facies.** Characterized by burrowed, calcitic pelletal mudstone with stromatoporoids, rugose corals, brachiopods, bryozoans, codiacian algae, and gastropods. The relatively diverse biota of this facies required continuous marine

Normal wave base

⟶ Salinity increases ⟶

37–45 ppm | >45 ppm

5	6	7	8	9
Organic (ecologic) reef (a) Boundstone mass (b) Crust on accumulation of organic debris and lime mud; bindstone (c) Bafflestone	*Sands on edge of platform* (a) Shoal lime sands (b) Islands with dune sands	*Open platform* (normal marine, limited fauna) (a) Lime-sand bodies (b) Wackestone-mudstone areas, bioherms (c) Areas of clastics	*Restricted platforms* (a) Bioclastic wackestone, lagoons and bays (b) Lithobioclastic sands in tidal channels (c) Lime mud-tide flats (d) Fine clastic units	*Platform evaporites* (a) Nodular anhydrite and dolomite on salt flats (b) Laminated evaporite in ponds
Massive limestone-dolomite	Calcarenitic-oolitic lime sand or dolomite	Variable carbonates and clastics	Generally dolomite and dolomitic limestone	Irregularly laminated dolomite and anhydrite, may grade to red beds
Light	Light	Dark to light	Light	Red, yellow, brown
Boundstones and pockets of grainstone; packstone	Grainstones well-sorted; rounded	Great variety of textures; grainstone to mudstone	Clotted, pelleted mudstone and grainstone; laminated mudstone; coarse litho-clastic wackestone in channels	
Massive organic structure or open framework with roofed cavities; lamination contrary to gravity	Medium- to large-scale cross-bedding; festoons common	Burrowing traces very prominent	Birdseye, stromatolites, mm lamination, graded bedding, dolomite crusts on flats. Cross-bedded sand in channels	Anhydrite after gypsum; nodular. rosettes, chickenwire, and blades; irregular lamination; carbonate caliche
None	Only some quartz sand admixed	Clastics and carbonates in well-segregated beds	Clastics and carbonates in well-segregated beds	Windblown, land-derived admixtures; clastics may be important units
Major frame building colonies with ramose forms in pockets; in situ communities dwelling in certain niches	Worn and abraded coquinas of forms living at or on slope; few indigenous organisms	Open marine fauna lacking (e.g., echinoderms, cephalopods, brachiopods); mollusca, sponges, forams, algae abundant; patch reefs present	Very limited fauna, mainly gastropods, algae, certain foraminifera (e.g., miliolids) and ostracods	Almost no indigenous fauna, except for stromatolitic algae

submergence; its close association and juxtaposition with rocks that clearly indicate periodic subaerial emergence further suggest that water depths were very shallow. The stromatoporoids often form tabular beds 2–3 m thick and several tens of meters long (Fig. 9-19). These sponges are interpreted as having grown either as encrusting masses within tidal creeks or channels or as closely crowded, individual heads in front of the tidal flats of the intertidal facies. A summary of Manlius biofacies in relation to shoreline position is shown in Figure 9-20.

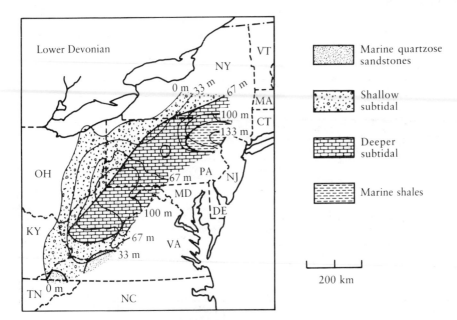

Lower Devonian

Marine quartzose sandstones

Shallow subtidal

Deeper subtidal

Marine shales

200 km

FIG. 9-18 • Generalized facies and isopach map for Lower Devonian (Helderberg) rocks in the central Appalachians. (Reprinted by permission of the publisher from Laporte, *Journal of Sedimentary Petrology*, v. 41, Society of Economic Paleontologists and Mineralogists, 1971.)

FIG. 9-19 • Weathered outcrop showing lateral interfingering of encrusting stromatoporoid masses with skeletal carbonate mudstones. Truncated stratification of nonstromatoporoid beds suggests lateral erosion of sediments followed by accumulation of stromatoporoid masses. Subtidal facies. (Reprinted by permission of the American Association of Petroleum Geologists from Laporte, *American Association of Petroleum Geologists Bulletin*, v. 51, 1967. Photo courtesy of L. V. Rickard.)

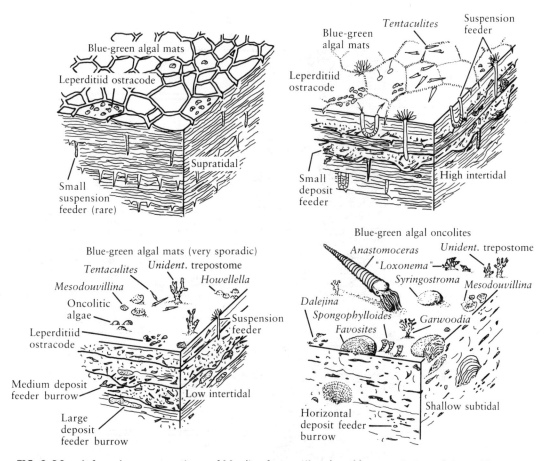

FIG. 9-20 • Inferred reconstructions of Manlius facies. (Reprinted by permission of the publisher from Laporte, *Tidal Deposits*, Springer-Verlag New York, Publishers, 1975.)

Bioherms

Bioherms vary in size from small "patch reefs" only a few meters or tens of meters in diameter (Fig. 9-19) to elongate masses hundreds of kilometers in length, much like the modern Great Barrier Reef of eastern Australia. The dominant type of fauna in a reef tends to vary with the age of the structure (Fig. 9-21), and the relative proportions of plant and animal species are variable within wide limits. Because all plants require a wide spectrum of wave-lengths in the visible spectrum to thrive and because carbonate removal from seawater is easier in warmer conditions, reefs are good indicators of temperate, shallow-water deposition.

The first-order facies of the reef environment are as follows.

1. Reef core, a massive, unbedded limestone mass consisting of unoriented skeletons of reef-building organisms and a matrix of calcite mud. Many ancient reefs

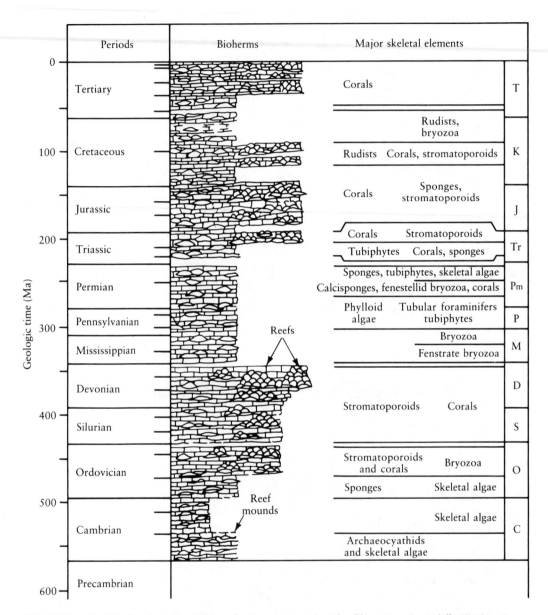

FIG. 9-21 • An idealized stratigraphic column representing the Phanerozoic and illustrating times when there appear to be no reefs or bioherms (gaps), times when there were only reef mounds and times when there were both reefs and reef mounds. (Reproduced with the permission of the Geological Association of Canada from James, *in* Walker, *Facies Models*, 1984.)

exhibit striking community successions within reef core facies, reflecting the sequential replacement of biotic assemblages during the growth of the complex. For example, in Oligocene reefs, four major phases of growth can be recognized (Sellwood, *in* Reading, 1986), each typified by a characteristic community style, diversity quality, and sediment type (Fig. 9-22).

Initially a set of lime-sand shoals may become abandoned and *stabilized* by opportunistic organisms possessing "holdfasts" or roots (e.g., pelmatozoans, calcareous green algae, sea-grass) and once established provide a refuge for other immigrants. *Colonization* now begins with the growth of low diversity thickets of tolerant groups often able to cope with high turbidities. Branching colonizers provide new niches for encrusters and the accumulation evolves into the *diversification* stage which comprises the bulk of the reef mass, with high diversities of framework and binding organisms accompanied by nestlers and borers. Ultimately, as growth proceeds to the surf zone, the *domination* stage commences, often abruptly, and diversities decrease dramatically with most organisms exhibiting encrusting or laminated growth styles and rubbled zones becoming common (Sellwood, *in* Reading, 1986, pp. 329–330).

2. Reef talus, consisting of bedded, poorly sorted limestone conglomerates and calcarenites that flank the reef core and that originated from biologic and oceanographic fragmentation of the core. This flank debris, which may exceed the reef core in volume, dips away from the core and becomes thinner-bedded and finer-grained as distance from the core increases. Fragments that compose the talus clearly reflect their origin and are composed of the same organisms that form the core.

FIG. 9-22 • Schematic representation of the four divisions of the reef-core facies with a tabulation of the most common types of limestone, relative diversity and shape of reef-builders found in each stage. (Reprinted by permission of the publisher from Sellwood, *in* Reading, *Sedimentary Environments and Facies*, Blackwell Scientific Publications, 1986.)

Stages of reef growth

	Stage	Type of limestone	Species diversity	Shape of reef builders
	Domination	Bindstone to framestone	Low to moderate	Laminate, encrusting
	Diversification	Framestone (bindstone) mudstone to wackestone matrix	High	Domal, massive, lamellar, branching, encrusting
	Colonization	Bafflestone to floatstone (bindstone) with a mudstone to wackestone matrix	Low	Branching, lamellar, encrusting
	Stabilization	Grainstone to rudstone (packstone to wackestone)	Low	Skeletal debris

A second type of bioherm is the mud mound, sometimes termed a Waulsortian reef after a village in Belgium where these carbonate mud mounds are particularly prominent. The mounds occur in rocks ranging from Cambrian to Jurassic in age, in locations farther offshore than reefs formed by frame-building organisms (Pratt, 1982; Wilson, 1975). A location at the shelf edge is typical. Mud mounds commonly have simple domal shapes (Fig. 9-23) that are circular in plan with thicknesses up to several hundred meters and diameters of more than 1 km. They are composed mainly of calcite mud and show no evidence of current activity, although they are often flanked by pelmatozoan grainstones that contain no mud matrix. The mounds are not erosional remnants but are accretional in origin, as indicated by steep depositional slopes of up to 50° with sheets of stromatactoid and rare bioclastic layers paralleling the external configuration. Also, most mounds lack detrital fans or breccias at their peripheries.

The internal structure of mud mounds sometimes contains abundant stromatactis structures and vague laminar structures interpreted to be of cryptalgal origin. The syndepositional stromatactis cavities are larger than spaces that could normally be supported by the host particles, and they may be filled with internal sediment or cement. This fenestral fabric is strongly suggestive of sediment binding by organic mats, analogous to binding by blue-green algal mats in shallow-water cryptalgal structures.

One of the most spectacular known examples of an ancient barrier reef complex crops out along the northern margin of the Canning Basin in Western Australia (Playford and Lowry, 1966; Playford, 1980). The reefs occur in a northwesterly trending belt about 350 km long and up to 50 km wide and grew for about 20 m.y. in Middle and Late Devonian time along the shoreline bordering the Precambrian basement rocks and around islands of Precambrian rocks. An interval of block faulting is believed to have immediately preceded reef growth in Middle Devonian time, giving rise to rugged topography with local relief of several hundred meters on which the reefs were established. The barrier reef belt may originally have extended another 1000 km beyond its present extent around the Precambrian craton. Because of Cenozoic erosion and the present semiarid climate in the area, the reef facies have been almost perfectly exhumed and exist today much as they appeared 370 m.y. ago (Fig. 9-24).

The reef complexes developed as platform deposits flanked by marginal slopes that descended to the basin floor at maximum depths of perhaps 300–400 m. Three major facies are recognized: platform, marginal slope, and basin. The platform facies was deposited nearly horizontally, and the depth of water over the reef margin and interior shelf lagoon probably did not exceed 10 m. The facies has been divided into three formations, one of which has been studied in detail by Read (1973a, b), who recognized approximately 70 cycles in it. Each cycle consists of distinctive lithofacies that reflect (1) initial rapid submergence, (2) shoaling by sedimentation to intertidal-supratidal levels, and (3) progressive decrease in kinetic energy. Rapid submergence is suggested by the general absence of reversed cycles and by the presence of the most diverse assemblages of organisms in lower parts of cycles. Shoaling was accompanied by increasing biotic restriction and by a decrease in kinetic energy,

(a)

(b)

FIG. 9-23 • Waulsortian mud mounds in the Lodgepole Formation (Mississippian), Montana. (a) Outcrop photo of two mud mounds. Overlying stratified units are limestones. (Photo courtesy of R. C. Murray.) (b) Slabbed hand specimen showing stromatactis structure. Orientation unknown. (Reprinted by permission of the author and the publisher from Cotter, *Journal of Geology*, v. 73, The University of Chicago Press, 1965.)

(a)

FIG. 9-24 • Aerial view looking northwest over Windjana Gorge and Napier
Range, exhumed Frasnian-Famennian reef complex. Gorge is 4 km long and up to

as is evidenced by upward change within cycles from grain-supported to mud-supported rocks. The limited thickness of cycle biostromes indicates that biostrome growth occurred under rather stable sea level conditions rather than under subsiding conditions, which would result in thick biostromes. Each cycle begins with a subtidal assemblage of in-place tabular or subspherical stromatoporoids up to 10 m thick in a fragmental matrix (Fig. 9-25a), and this is followed by a bed of fragments of cylindrical *Stachyoides* and *Amphipora* stromatoporoids (Fig. 9-25b,c). As the depth of water decreased on the platform in the latter part of a cycle, the fragmental stromatoporoid layer was overlain by pellet/intraclast limestone with cryptalgal and fenestral structure. Cryptalgal fabrics indicate the presence of algal mats in intertidal and supratidal habitats kept moist by frequent tidal flooding and/or proximity of tidal groundwaters to the surface. The absence of intraclast pavements and breccias, which form by superficial desiccation and lithification of emergent cryptalgal sediments, also suggests that the Pillara tidal flats were mostly damp and rarely subjected to extended desiccation. Cycles are terminated by horizons containing root-casts and incipient carbonate soils. In supratidal deposits, internal sediments, leach fabrics,

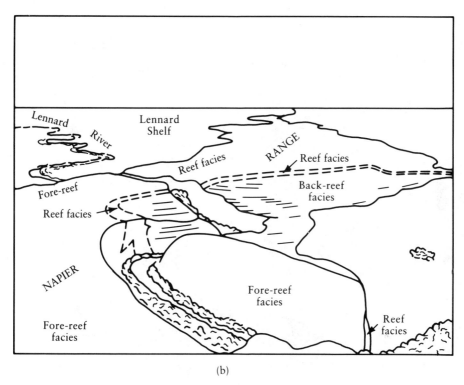

Lennard River

Lennard Shelf

Reef facies

RANGE

Reef facies

Back-reef facies

Fore-reef

Reef facies

NAPIER

Fore-reef facies

Fore-reef facies

Reef facies

(b)

80 m deep. (Reprinted by permission of the publisher from Playford and Lowry, *Geological Survey of Western Australia Bulletin 118*, 1966.)

and color-mottles were developed in sediments under vadose zone conditions. Tubular fenestrae occur as root molds of salt-tolerant land plants that colonize both supratidal and suitable intertidal substrates.

The reef itself is generally only a few tens of meters wide and is discontinuous around the seaward margin of most platforms. It consists of a massive organic framework of stromatoporoids, algae, and corals (Fig. 9-26), but substantial parts of the reef deposits consist of clastic limestones with only minor in-place frame-building organisms. The clastic reef limestones are formed mainly of grain-supported reefal rubble and skeletal sand, with ooids, pellets, micritic intraclasts, and carbonate mud. Relatively structureless reef limestone also occurs, composed of nonskeletal algae, sometimes stromatolitic. The unifying feature of the reef deposits, both in-place and clastic, is that they were lithified penecontemporaneously to form rigid wave-resistant zones around the platform margins. This early rigidity allowed many of the platform margins to be constructed as steep submarine scarps.

The marginal slope deposits consist mostly of platform-derived debris and reef material with variable contributions from indigenous organisms and terrigenous

(a)

FIG. 9-25 • (a) Tabular stromatoporoid biostrome. (b) *Stachyoides* limestone, polished slab (scale in centimeters). (c) *Amphipora* limestone, with oriented *Amphipora* sticks on bedding planes. The ruler is 15 cm long. (Reprinted by permission of the publisher from Read, *Bulletin of Canadian Petroleum Geology*, 1973a.)

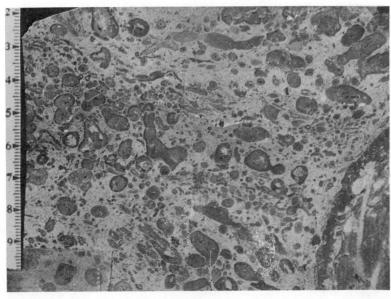

(b)

(c)

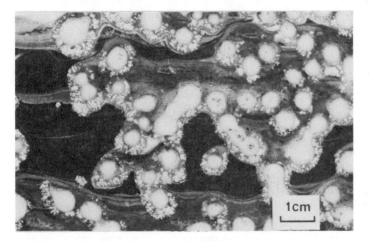

1cm

FIG. 9-26 ● Stromatoporoid-algal limestone, showing dense colonies of *Renalcis* encrusting branches of *Stachyoides*(?). The interstices between the colonies are occupied by very thin-bedded red and yellow calcilutite or sparry calcite. Block of reef facies in conglomerate. (Reprinted by permission of the publisher from Playford and Lowry, *Geological Survey of Western Australia Bulletin 118*, 1966.)

sediments. Depositional dips were as high as 35°–40° in loose debris and even higher where early submarine cementation by algae provided slope stability. Depositional dips shallow to less than 5° where the slope facies interfingers with basinal facies.

The sediments of the basin facies consist largely of terrigenous sediments ranging in size from conglomerate through shale, interbedded with thin beds of pelagic limestone and tongues of platform-derived detrital limestone. Early Frasnian basinal mudrocks are black and rich in organic matter; most late Frasnian and Famennian deposits are red and oxidized. The red coloration is strongest in condensed sequences, in which the bacteria may have destroyed the organic matter and caused precipitation of ferric hydroxide, which subsequently dehydrated to hematite.

Evaporite Sediments

The commonness of evaporite facies in the rock record is typically underestimated by geologists because they do not crop out in proportion to their abundance, a circumstance clearly related to their origin. By definition, evaporite minerals are very soluble in water; the most soluble of them (gypsum/anhydrite) precipitates from seawater when about 80% of the water has been removed. Thus ancient evaporite beds will be easily seen only in semiarid to arid climates. In more humid areas, evaporitic beds will have negligible topographic relief and be covered by modern soils. Subsurface data, however, reveal that at least one-third of the sedimentary rocks in the conterminous United States are underlain at some depth by bedded evaporites. Because of the combined effect of latitudinal continental drift and orogenesis, dry climates have been common in the geologic history of the United States (Blatt, 1982).

There are three major types of evaporite facies: (1) those formed by evaporative concentration of a large, standing body of marine water (early Dead Sea sediments), (2) those formed in supratidal areas at the edges of marine seas (Persian Gulf, Baja California), and (3) those formed from nonmarine waters in intracontinental topographic depressions (southern California; Lake Magadi, Kenya). The first two types are both more abundant and more easily recognized in ancient rocks, but numerous examples of all three types have been described. They can be distinguished from each other in outcrop on the bases of regional paleogeographic setting, sedimentary structures, mineral composition, and perhaps also textural criteria, although laboratory studies are usually required to recognize textural distinctions.

Large Marine Evaporite Basins

Examples of ancient deposits of evaporites deposited from originally large, standing bodies of marine water include the Miocene (Messinian) deposits of the Mediterranean Sea, described most recently by Schreiber et al. (1976); Permian deposits of west Texas (Anderson et al., 1972); Silurian deposits of the Michigan Basin (Nurmi and Friedman, 1977); and the Permian Zechstein evaporites of northwestern Europe (Peryt, 1987). In outcrop the diagnostic criteria for these deposits are (1) regionally circular or oval distribution pattern with distinguishable (but commonly incomplete) concentric belts of carbonate rock, gypsum/anhydrite, halite, and posthalite evaporites from outer ring to inner (Fig. 9-27); (2) thin laminae of limestone, gypsum/anhydrite, and possibly halite that are laterally continuous for distances of many tens of kilometers (Fig. 9-28), which reflects a widespread, undisturbed, flat area of precipitation; mineralogic alternations resembling varves are common; and (3) only

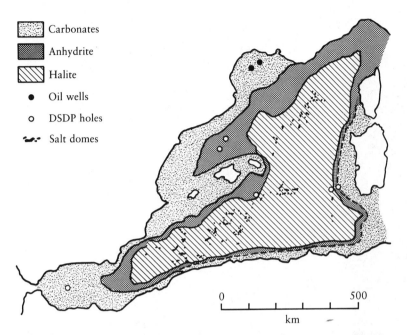

Carbonates

Anhydrite

Halite

● Oil wells

○ DSDP holes

〰 Salt domes

0   500

km

FIG. 9-27 ● Bull's-eye distribution of evaporites in the Balearic Basin of the western Mediterranean Sea. The same pattern occurs in many other thick and areally extensive evaporite sections. (Reprinted by permission of the publisher from Hsü, *Earth-Science Reviews*, v. 8, Elsevier Science Publishers B.V., 1972.)

minerals characteristic of evaporated seawater, such as gypsum/anhydrite, halite, sylvite, and carnallite.

The depth of water in which the basin—central marine evaporites were deposited is the subject of debate among sedimentologists and stratigraphers. Those favoring deep water (probably meaning depths of more than 50 m, although exact depths are rarely specified) use as evidence the evident large topographic relief of the evaporite basin, as in the Permian evaporite basin of west Texas; the great lateral continuity of thin evaporite laminae; the absence of organic structures; and the absence of shallow-water features such as ripple marks, desiccation cracks, and rip-up clasts. Those favoring a shallow-water origin for evaporites deposited in deep basins point out that topographic relief of rocks surrounding a basin does not necessarily mean a great depth of water within the basin; that the density stratification known to exist in evaporite basins can effectively dampen the effects of surface waves at shallow depth; and that in waters with salinities of about 110,000 ppm, the salinity required for gypsum precipitation, we would not expect to find either body fossils, ichnofossils, or many algae. Because there is no modern analog of the large evaporite basins of the past, the disagreement has not been resolved.

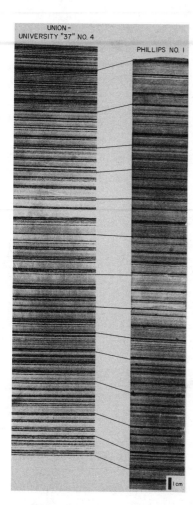

FIG. 9-28 • Correlative couplets in Castile Formation composed of dark organic-rich calcite and anhydrite, core locations are 91 km apart. (Reprinted by permission of the author from Anderson et al., *Geological Society of America Bulletin*, v. 83, 1972.)

Sabkha Evaporites

Examples of evaporites formed in supratidal areas (sabkhas) at the margins of marine basins include the Upper Devonian evaporites of the Williston Basin (Wilson, 1967); Ordovician of Ellesmere Island, Canada (Mossop, 1978); and Carboniferous of the Canadian maritime provinces (Schenk, 1969). Sabkha evaporites are deposited in dry climates along the margin of depositional basins surrounded by topographically low-lying land areas. Typically, basinward deposits are limestones, and landward deposits are fluvial red shales and fine-grained sandstones. The evaporites are deposited in shallow salt pans that lie above normal high tide. Such areas are flooded by marine waters only during events such as storms or the fortnightly spring tides. Between floods, the pan water evaporates to the point of gypsum precipitation,

FIG. 9-29 • Laminated do-lomicrites of probable hyper-saline lagoonal origin overlain by algal mat with nodular and mosaic sabkha anhydrite pseudomorphs after gypsum. The core fits together but was rotated during slabbing. Core width is 10 cm. The anhydrite is overlain by anhydrite-bear-ing laminated micrite of the next lagoonal cycle. Frobisher Evaporite (Mississippian), Saskatchewan, Canada. (Re-produced with the per-mission of the Geological As-sociation of Canada from Kendall, *in* Walker, *Facies Models*, 1984.)

and the mineral precipitates. Occasionally, halite will form, but this is not nearly as common as gypsum in ancient sabkha deposits. Removal of calcium from the pan water increases the magnesium/calcium ratio in the remaining brine and typically causes dolomite to form, so the dolomite-gypsum/anhydrite association is common in ancient carbonate sequences, particularly in back-reef areas (Fig. 9-29). As the sea transgresses over the land surface, the supratidal area moves landward, so extensive evaporite deposits can be formed.

In outcrop the diagnostic criteria for sabkha evaporites are (1) repeated carbonate-evaporite cycles in the stratigraphic sequence, reflecting the numerous transgressions and regressions of the sea that typify low-lying, near-shore areas during geologically significant periods of time (periods greater than perhaps 10^6 years); (2) the dominance of gypsum/anhydrite as the evaporite species in the sequence; (3) the presence of rip-up clasts, intraformational conglomerates unrelated to solution collapse-brecciation, and other evidence of periodic desiccation; and (4) abundant dolomite intercalated with the gypsum/anhydrite.

Intracontinental Lakes and Playas

Examples of ancient nonmarine evaporites are perhaps not as abundant as marine examples, but several have been described, including those in the Green River Formation (Eocene) of the western United States (Smoot, 1983) and Neogene sediments of Lake Magadi, Kenya (Eugster, 1980). Criteria for the recognition of playa evaporites have been described from modern deposits by Lowenstein and Hardie (1985). Diagnostic criteria applicable to outcrops of ancient deposits include (1) paleogeographic setting—it is a definite continental deposit; (2) an abundance of evaporite minerals that do not occur when normal seawater is evaporated, such as the carbonates trona and northupite, borate minerals, and sulfates such as glauberite and mirabilite; and (3) evidence of lacustrine sedimentation in intercolated non-evaporite beds.

Summary

The fully marine depositional environment contains a wide variety of both detrital and intrabasinal sediments. Terrigenous shelves extend to an average depth of 124 m at a gradient of only 0.1°, at which point the slope steepens to 3–6° and descends to the continental rise and abyssal depths. Sediment transport down the continental slope commonly is in turbidity current flows down submarine canyons and deposition occurs as submarine fans prograding outward from the base of the slope.

In deep oceanic areas far from terrigenous sediment sources are found calcareous ooze (globigerinid, coccolith, pteropod), opaline oozes (diatom and radiolarian), and brown clay. Calcareous oozes are restricted to depths above the lysocline which in equatorial regions reaches depths exceeding 5.5 km.

Carbonate sediments are abundant on shallow shelves of epicontinental seas and divergent plate margins with little terrigenous sediment influx. Both reefs (bioherms) and loose carbonate debris (biostromes) are common. The occurrence of limestone is essentially biologically controlled and the most common sites of accumulation are in warm waters with depths less than 100 m, areas where nutrients and sunlight tend to be high.

Most evaporite deposits form in areas of restricted marine waters although non-marine, lacustrine evaporites are not uncommon. The giant evaporite accumulations such as those that underlie the modern Mediterranean Sea or dominate the fill of the Paleozoic Michigan Basin seem to represent areally extensive and possibly topographically deep basins that evaporated to near-dryness because of periodic restrictions to inflow of marine waters. Supratidal (sabkha) evaporites form most commonly near the borders of carbonate basins and may be associated with dolomite.

References

Terrigenous Shelf Sediments

de Raaf, J. F. M., Boersma, J. R., and van Gelder, A., 1977. Wave-generated structures and sequences from a shallow marine succession, Lower Carboniferous, County Cork, Ireland. Sedimentology, v. 24, pp. 451–483.

Dott, R. H., Jr. and Bourgeois, J., 1982. Hummocky stratification: Significance of its variable bedding sequences. Geol. Soc. Amer. Bull., v. 93, pp. 663–680.

Morris, K. A., 1979. A classification of Jurassic marine shale sequences: An example from the Toarcian (Lower Jurassic) of Great Britain. Paleogeog., Paleoclim., Paleoecol., v. 26, pp. 117–126.

Mount, J. F., 1984. Mixing of siliciclastic and carbonate sediments in shallow shelf environments. Geology, v. 12, pp. 432–435.

Pashin, J. C. and Ettensohn, F. R., 1987. An epeiric shelf-to-basin transition: Bedford-Berea sequence, northeastern Kentucky and south-central Ohio. Amer. J. Sci., v. 287, pp. 893–926.

Reading, H. G. (ed.), 1986. Sedimentary Environments and Facies, 2nd ed. Oxford, Blackwell Scientific Publications, 615 pp.

Scholle, P. A. and Spearing, D., 1982. Sandstone Depositional Environments. Tulsa, American Association of Petroleum Geologists, Mem. 31, 410 pp.

Stanley, D. J. and Moore, G. T. (eds.), 1983. The Shelfbreak: Critical Interface on Continental Margins. Tulsa, Society of Economic Paleontologists and Mineralogists, Spec. Pub. No. 33, 467 pp.

Tillman, R. W. and Siemers, C. T., 1984. Siliciclastic Shelf Sediments. Tulsa, Society of Economic Paleontologists and Mineralogists, Spec. Pub. No. 34, 268 pp.

Continental Slope Sediments

Buck, S. P. and Bottjer, D. J., 1985. Continental slope deposits from a Late Cretaceous, tectonically active margin, southern California. J. Sed. Petrology, v. 55, pp. 843–855.

Doyle, L. J. and Pilkey, O. H. (eds.), 1979. Geology of Continental Slopes. Tulsa, Society of Economic Paleontologists and Mineralogists, Spec. Pub. No. 27, 374 pp.

Pickering, K.T., 1982. A Precambrian upper basin-slope and prodelta in northeast Finnmark, north Norway—A possible ancient upper continental slope. J. Sed. Petrology, v. 52, pp. 171–186.

Deep-Sea Fans

Barnes, P. M., 1988. Submarine fan sedimentation at a convergent margin: The Cretaceous Mangapokia Formation, New Zealand. Sedimentary Geol., v. 59, pp. 155–178.

Bouma, A. H., Normark, W. R., and Barnes, N. E. (eds.), 1985. Submarine Fans and Related Turbidite Systems. New York, Springer-Verlag, 350 pp.

Collinson, J. D., 1969. The sedimentology of the Grindslow Shales and the Kinderscout Grit: A deltaic complex in the Namurian of northern England. J. Sed. Petrology, v. 36, pp. 194–221.

Curran, H. A. (ed.), 1985. Biogenic Structures: Their Use in Interpreting Depositional Environments. Tulsa, Society of Economic Paleontologists and Mineralogists, Spec. Pub. No. 35, 347 pp.

D'Alessandro, A., Ekdale, A. A., and Sonnino, M., 1986. Sedimentologic significance of turbidite ichnofacies in the Saraceno Formation (Eocene), southern Italy. J. Sed. Petrology, v. 56, pp. 294–306.

Klein, G. deV., 1984. Relative rates of tectonic uplift as determined from episodic turbidite deposition in marine basins. Geology, v. 12, pp. 48–50.

Pickering, K. T., 1986. Deep-water facies, processes and models: A review and classification scheme for modern and ancient sediments. Earth-Sci. Rev., v. 23, pp. 75–174.

Stow, D. A. V. and Piper, D. J. W. (eds.), 1984. Fine-Grained Sediments: Deep Water Processes and Facies. Oxford, Blackwell Scientific Publications, 668 pp.

Walker, R. G., 1966. Shale Grit and Grindslow Shales: Transition from turbidite to shallow water sediments in the Upper Carboniferous of northern England. J. Sed. Petrology, v. 36, pp. 90–114.

Walker, R. G., 1978. Deep-water sandstone facies and ancient submarine fans: Models for exploration for stratigraphic traps. Amer. Assoc. Petroleum Geol. Bull., v. 62, pp. 932–966. Also discussion in v. 64, pp. 1094–1112.

Walker, R. G. (ed.), 1984. Facies Models, 2nd ed. Toronto, Geological Association of Canada, 317 pp.

Abyssal Sediments

Audley-Charles, M. G., 1968. The Geology of Portuguese Timor. London, Geological Society of London, Mem. No. 4, 76 pp.

Barron, E. J. and Whitman, J. M., 1981. Oceanic sediments in space and time. In C. Emiliani (ed.), The Sea. Vol. 7: The Oceanic Lithosphere. New York, John Wiley & Sons, pp. 689–731.

Berger, W. H. and Winterer, E. L., 1974. Plate stratigraphy and the fluctuating carbonate line. In K. J. Hsü and H. C. Jenkyns (eds.), Pelagic Sediments: On Land and Under the Sea. International Association of Sedimentologists, Spec. Pub. No. 1. Oxford, Blackwell Scientific Publications, pp. 11–48.

Bissell, H. and Barker, H. K., 1977. Deep-water limestones of the Great Blue Formation (Mississippian) in the eastern part of the Cordilleran miogeosyncline in Utah. In H. E. Cook and P. Enos (eds.), DeepWater Carbonate Environments. Tulsa, Society of Economic Paleontologists and Mineralogists, Spec. Pub. No. 25, pp. 171–186.

Garrison, R. E., 1974. Radiolarian cherts, pelagic limestones and igneous rocks in eugeosynclinal assemblages. *In* K. J. Hsü and H. C. Jenkyns (eds.), Pelagic Sediments: On Land and Under the Sea. International Association of Sedimentologists, Spec. Pub. No. 1, Oxford, Blackwell Scientific Publications, pp. 367–399.

Ingle, J., Jr., 1981. Origin of Neogene diatomites around the North Pacific rim. *In* R. E. Garrison and R. G. Douglas (eds.), The Monterey Formation and Related Siliceous Rocks of California. Tulsa, Society of Economic Paleontologists and Mineralogists, Pacific Sec., Spec. Pub. Book 15, pp. 159–179.

Jones, D. L. and Murchey, B., 1986. Geologic significance of Paleozoic and Mesozoic radiolarian chert. Ann. Rev. Earth Planet. Sci., v. 14, pp. 455–492.

O'Brien, N. R., Nakazawa, K., and Tokuhashi, S., 1980. Use of clay fabric to distinguish turbiditic and hemipelagic siltstones and silts. Sedimentology, v. 27, pp. 47–61.

Pisciotto, K. A. and Garrison, R. E., 1981. Lithofacies and depositional environments of the Monterey Formation, California. *In* R. E. Garrison and R. G. Douglas (eds.), The Monterey Formation and Related Siliceous Rocks of California. Tulsa, Society of Economic Paleontologists and Mineralogists, Pacific Sec., Spec. Pub. Book 15, pp. 97–122.

Scholle, P. A., 1974. Diagenesis of upper Cretaceous chalks from England, Northern Ireland, and the North Sea. *In* K. J. Hsü and H. C. Jenkyns (eds.), Pelagic Sediments: On Land and Under the Sea. International Association of Sedimentologists, Spec. Pub. No. 1. Oxford, Blackwell Scientific Publications, pp. 177–210.

Weaver, P. P. E. and Thomson, J., 1987. Geology and Geochemistry of Abyssal Plains. Oxford, Blackwell Scientific Publications, 246 pp.

Carbonate Shelf Sediments

Burchette, T. P., 1988. Tectonic control on carbonate platform facies distribution and sequence development: Miocene, Gulf of Suez. Sedimentary Geol., v. 59, pp. 179–204.

Cotter, E., 1965. Waulsortian-type carbonate banks in the Mississippian Lodgepole Formation of central Montana. J. Geol., v. 73, pp. 881–888.

Laporte, L. F., 1967. Carbonate deposition near mean sea-level and resultant facies mosaic: Manlius Formation (Lower Devonian) of New York State. Amer. Assoc. Petroleum Geol. Bull., v. 51, pp. 73–101.

Laporte, L. F., 1971a. Paleozoic carbonate facies of the central Appalachian shelf. J. Sed. Petrology, v. 41, pp. 724–740.

Laporte, L. F., 1971b. Recognition of a transgressive carbonate sequence within an epeiric sea: Helderberg Group (Lower Devonian) of New York State. *In* G. M. Friedman (ed.), Depositional Environments in Carbonate Rocks. Tulsa, Society of Economic Paleontologists and Mineralogists, Spec. Pub. No. 14, pp. 98–119.

Laporte, L. F. (ed.), 1974. Reefs in Time and Space. Tulsa, Society of Economic Paleontologists and Mineralogists, Spec. Pub. No. 18, 256 pp.

Laporte, L. F., 1975. Carbonate tidal-flat deposits of the Early Devonian Manlius Formation of New York State. *In* R. N. Ginsburg (ed.), Tidal Deposits. New York, Springer-Verlag, pp. 243–250.

Playford, P. E., 1980. Devonian "Great Barrier Reef" of Canning Basin, Western Australia. Amer. Assoc. Petroleum Geol. Bull., v. 64, pp. 814–840.

Playford, P. E. and Lowry, D. C., 1966. Devonian Reef Complexes of the Canning Basin, Western Australia. Perth, Geologic Survey of Western Australia, Bull. No. 118, 150 pp.

Pratt, B. R., 1982. Stromatolitic framework of carbonate mud-mounds. J. Sed. Petrology, v. 52, pp. 1203–1227.

Read, J. F., 1973a. Carbonate cycles, Pillara Formation (Devonian), Canning Basin, Western Australia. Bull. Canadian Petroleum Geol., v. 21, pp. 38–51.

Read, J. F., 1973b. Paleo-environments and paleogeography, Pillara Formation (Devonian), Western Australia. Bull. Canadian Petroleum Geol., v. 21, pp. 344–394.

Scholle, P. A., Bebout, D. G., and Moore, C. H. (eds.), 1983. Carbonate Depositional Environments. Tulsa, American Association of Petroleum Geologists, Mem. 33, 708 pp.

Toomey, D. F. (ed.), 1981. European Fossil Reef Models. Tulsa, Society of Economic Paleontologists and Mineralogists, Spec. Pub. No. 30, 546 pp.

Wilson, J. L., 1975. Carbonate Facies in Geologic History. New York, Springer-Verlag, 471 pp.

Evaporite Sediments

Anderson, R. Y., Dean, W. E., Kirkland, D. W., and Snider, H. I., 1972. Permian Castile varved evaporite sequence, west Texas and New Mexico. Geol. Soc. Amer. Bull., v. 83, pp. 59–85.

Blatt, H., 1982. Sedimentary Petrology. New York, W. H. Freeman and Co., 564 pp.

Dean, W. E. and Schreiber, B. C. (eds.), 1978. Marine Evaporites. Tulsa, Society of Economic Paleontologists and Mineralogists, Short Course No. 4, 188 pp.

Eugster, H. P., 1980. Lake Magadi, Kenya, and its precursors. *In* A. Nissenbaum (ed.), Hypersaline Brines and Evaporitic Environments. New York, Elsevier Science Publishers, pp. 195–232.

Handford, C. R., Loucks, R. G., and Davies, G. R. (eds.), 1982. Depositional and Diagenetic Spectra of Evaporites—A Core Workshop. Tulsa, Society of Economic Paleontologists and Mineralogists, Core Workshop No. 3, 395 pp.

Hsü, K. J., 1972. Origin of the saline giants: A critical review after the discovery of the Mediterranean evaporite. Earth-Sci. Rev., v. 8, pp. 371–396.

Kendall, A. C., 1988. Aspects of evaporite basin stratigraphy. *In* B. C. Schreiber (ed.), Evaporites and Hydrocarbons. New York, Columbia University Press, pp. 11–65.

Lowenstein, T. K. and Hardie, L. A., 1985. Criteria for the recognition of salt-pan evaporites. Sedimentology, v. 32, pp. 627–644.

Mossop, G. D., 1978. The Ordovician Baumann Fiord Formation Evaporites of Ellesmere Island, Arctic Canada. Toronto, Geological Survey of Canada, Bull. 298, 52 pp.

Nurmi, R. D. and Friedman, G. M., 1977. Sedimentology and depositional environments of basin-center evaporites, lower Salina Group (Upper Silurian), Michigan Basin. *In* Reefs and Evaporites—Concepts and Depositional Models. Tulsa, American Association of Petroleum Geologists, Stud. in Geol. No. 5, pp. 23–52.

Peryt, T. M. (ed.), 1987. The Zechstein Facies in Europe. New York, Springer-Verlag, 272 pp.

Schenk, P. E., 1969. Carbonate-sulfate-redbed facies and cyclic sedimentation of the Windsorian Stage (Middle Carboniferous), Maritime Provinces. Canadian J. Earth Sci., v. 6, pp. 1037–1066.

Schreiber, B. C., Friedman, G. M., Decima, A., and Schreiber, E., 1976. Depositional environments of Upper Miocene (Messinian) evaporite deposits of the Sicilian Basin. Sedimentology, v. 23, pp. 729–760.

Smoot, J. F., 1983. Depositional subenvironments in an arid closed basin: The Wilkins Peak Member of the Green River Formation (Eocene), Wyoming. Sedimentology, v. 30, pp. 801–828.

Warren, J. K. and Kendall, C. G. St. C., 1986. Comparison of sequences formed in marine sabkha (subaerial) and salina (subaqueous) settings—Modern and ancient. Amer. Assoc. Petroleum Geol. Bull., v. 69, pp. 1013–1023.

Wilson, J. L., 1967. Carbonate-evaporite cycles in lower Duperow Formation of Williston Basin. Bull. Canadian Petroleum Geol., v. 15, pp. 230–312.

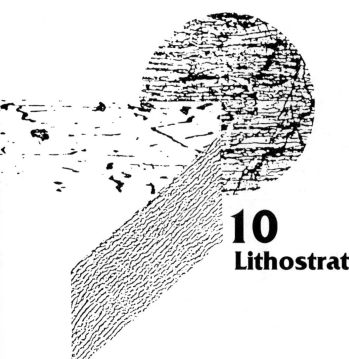

The flight of time is measured by the weaving of composite rhythms—day and night, calm and storm, summer and winter, birth and death—such as these are sensed in the brief life of man. . . .
. . . the stratigraphic series constitutes a record, written on tablets of stone, of the lesser and greater waves of change which have pulsed through geologic time.

Joseph Barrell

10
Lithostratigraphy

Lithostratigraphy, or physical stratigraphy, is concerned with the description and analysis of the areally significant building blocks of the sedimentary section, the individual layers of rock. Of interest on the scale of outcrop observation are characteristics such as lithology (sandstone, evaporite, shale, etc.), thickness, areal extent, approximate age, relationship to overlying and underlying beds, and lateral variations (facies). Outcrops are usually more limited in areal extent than we would prefer, so many inferences are always required in lithostratigraphic field work. It is the soundness of these inferences that determines the quality of our geologic maps.

A **lithofacies** is a rock unit defined by its distinctive lithologic features, including composition, texture, bedding characteristics, and sedimentary structures. Lithofacies may be grouped into lithofacies associations or assemblages that are characteristic of particular depositional environments (Fig. 10-1), and these assemblages form the basis for defining lithofacies models. The assemblages commonly are cyclic. For example, a fluvial deposit may consist of a conglomerate lithofacies interbedded with a cross-bedded sandstone facies and a red mud-shale facies. This association may be repeated many times in a measured stratigraphic section and forms the basis for a "model" of fluvial deposition.

Some sequences of sedimentary rocks have distinctive features that permit them to be recognized as resulting from specific tectonic settings; such groupings have been termed tectofacies. A **tectofacies** is simply a lithofacies that is interpreted tectonically and refers to a group of strata of different tectonic aspect from laterally equivalent strata.

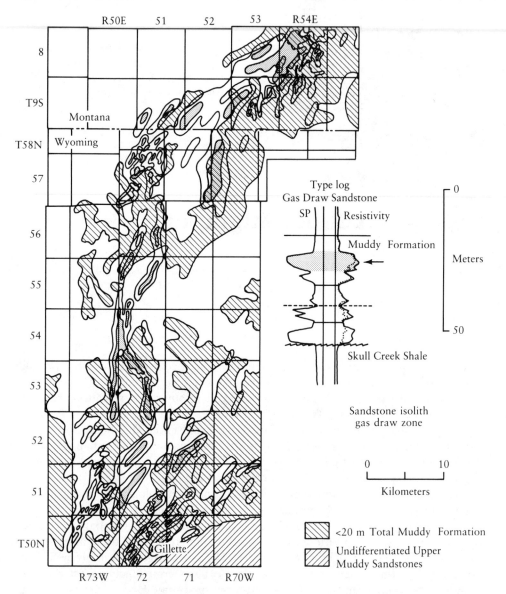

FIG. 10-1 • Isolith map of the oil-bearing Early Cretaceous Gas Draw Sandstone zone, Muddy Formation, Wyoming, showing trends of sandstone bodies deposited as beaches and offshore bars. Sand was supplied by a deltaic system to the northeast and carried southward by long-shore currents. (Reprinted by permission of the publisher from Stone, *Geologic Atlas of the Rocky Mountain Region*, Rocky Mountain Association of Geologists, 1972.)

Flysch

The term **flysch** was used originally in the 19th century by European geologists to describe rocks of Late Cretaceous to Oligocene age along the borders of the Alps. The sediments were deposited in "deep" water (below wave base) in front of northward-advancing nappes. They are the products of active orogenic activity and most commonly consist of an apparently unfossiliferous, thick sequence of thinly interbedded, graded sandstones and dark silty shales (Fig. 9-8) of great lateral continuity (at the time of deposition). The bedding is well marked, uniform, and rhythmic, and the thickness is measured in thousands of meters. Modern usage of the term flysch follows closely the original usage (Lajoie, 1970), except that it is not restricted to Alpine rocks. The relative amounts of shale and sandstone in flysch sequences are variable and depend on sediment supply, distance from shore (Walker, 1967), and whether the sea level was rising or falling during the time of formation of the beds (Mutti, 1985).

Flysch sequences are deposited in the bathyl depths of continental rises and on abyssal plains (Chapter 9) where sedimentation can be sustained at a fairly rapid rate by sediment-loaded currents (turbidity currents) moving down submarine canyons and forming extensive fan aprons at the base of continental slopes. Carbonate flysch is less common than silicate flysch. When present, it can consist of either coarse fragments of carbonate debris eroded from rising highlands or finer-grained material of shallow-water derivation containing benthonic fossils and possibly even oolites. Some flysch sequences contain volcanic materials such as pillowed lavas (commonly altered to greenstone and spilite) that erupted on the sea floor concurrently with sedimentation. Closely associated also are water-laid basic tuffs. Chert beds composed of crystallized radiolarian remains may also be present.

The average lithology of flysch sediments is about two-thirds shale, slate, or argillite (weakly metamorphosed mudstone) and one-third sandstone with a few percent of carbonate rocks and conglomerate. Lithostratigraphic correlations are rarely possible over distances greater than a few hundred meters because of poor exposures and the complex structural deformation characteristic of flysch terranes.

The type of transporting mechanism is suggested by the abundance of sedimentary structures characteristic of turbidity currents, such as graded beds, flutes, grooves, and several types of tool markings. None of these features individually is a foolproof criterion of turbidity current origin, but their great abundance is very suggestive. The absence of indigenous bottom-dwelling organisms and the presence of organisms such as planktonic graptolites and nectonic nautiloids argue for deposition below wave base and the maximum depth of light penetration (about 100 m). Shallow-water fossils that may be present (brachiopods, bryozoa, gastropods, corals, echinoderms) are all fragmented and abraded and occur in graded beds, indicating transport downslope by turbidity currents. The frequency of individual turbidite events, as determined from graded cycles, may be related to trends of increasing and decreasing tectonic uplift events (Klein, 1984). By using this concept it may be possible to evaluate rates of tectonic uplift.

The mineralogy of the sandstones is varied and depends on the types of rocks that were eroded to form the flysch sediment. The average flysch sandstone (graywacke) is rich in unstable lithic fragments and feldspars, perhaps one-third of each, and hence is rather deficient in quartz in comparison to the average sandstone. Furthermore, the feldspars very commonly are albitized, and secondary chlorite is present, giving the rocks a greenish color. The chlorite reflects a high grade of diagenesis related to deep burial and high heat flow in the depositional basin. These graywacke sandstones almost always contain appreciable muddy matrix, about 10–20%, some of which may be detrital but much of which was formed by diagenetic alteration of detrital unstable lithic fragments.

Molasse

A second lithofacies related to orogenic uplift is the **molasse** facies (Van Houten, 1974; Miall, 1981). The original usage of this term by European geologists in the 19th century referred to a thick clastic sequence of Tertiary age in the Swiss plain and Alpine foreland of southern Germany, the same region in which the term flysch was first used. The molasse facies, like that of flysch, consists mainly of sandstone and shale but represents postflysch sedimentation and appears very different in outcrop from flysch rocks. The environment of molasse sedimentation is a deltaic coastal plain and its inland extension to the mountain front, for example, the Catskill sequence (Devonian) in the central Appalachians (Fig. 10-2) (Woodrow and Sevon, 1985). In this proximal region, molasse, like the flysch that precedes it, may be several thousand meters thick. Thicker molasse belongs to the orogenic belt and is itself deformed; thinner molasse is a more distal portion of the same clastic wedge. It may cover part of the cratonic platform far beyond the orogenic belt.

Depositional environments represented in molasse rocks are varied, are both shallow-water marine and nonmarine, and include alluvial fans, rivers, swamps, beach, and tidal deposits. Lithologies are correspondingly variable. Rock types range from conglomerate to mud-shale and coal, fine sandstone and silty shale possibly being most common. As is typical in essentially alluvial sequences, fining-upward cycles are the rule, cycles being perhaps tens of meters thick. Sandstones tend to dominate in the lower part of each cycle and have sharp erosional contacts (channeling) with siltstones or shales below; shale-pebble conglomerates commonly occur at the base of the sandstone. Sandstone typically forms about 50% of the cycle. Large-scale cross-bedding is ubiquitous and is the dominant paleocurrent indicator, but flat-bedded, fine-grained sandstones with parting lineation are intercalated with the cross-stratified sandstones. Siltstones and mudstones generally show smaller-scale cross-bedding and commonly contain plant remains, root traces, and desiccation cracks.

Petrographically, molasse sandstones differ from the flysch graywacke in that

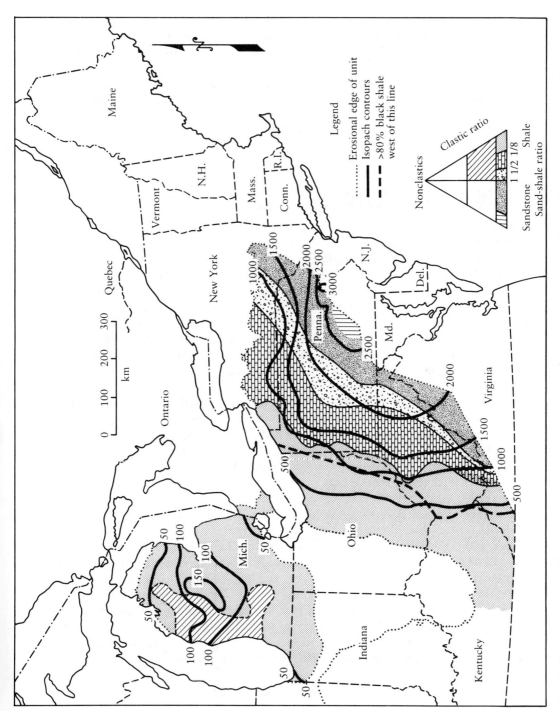

FIG. 10-2 ● Isopach and lithofacies map of Upper Devonian sediments of northeastern United States. Contour interval is in meters. Note the different contour interval between Appalachian and Michigan basin regions.

323

molasse normally lacks volcanic fragments and is richer in quartz, values in the 60–80% range being most common. The types of lithic fragments in molasse sandstones depend on local sources. Molasse sandstones distal from their highland source are progressively diluted by sands originating on the craton, and mineralogical and textural maturity increases correspondingly. The tectofacies of the molasse rocks is one of normal heat flow, so diagenetic changes are moderate and consist of compaction and decipherable chemical changes such as feldspar alteration to common clay minerals and precipitation of calcitic and hematitic cements.

Intracratonic Tectonic Arkose Lithofacies

Flysch and molasse have their origin in **orogenesis**, mountain building associated with tectonic activity along convergent plate margins. A third orogenic lithofacies is the **tectonic arkose** produced by intracratonic block faulting. This tectofacies can be the result of either compressional or tensional forces, and field examples of both types are well documented. Tensional forces associated with the early stages of rifting of North America from Europe and North Africa resulted in a series of horst-graben structures on both sides of the Atlantic Ocean during Triassic time (Chapter 11). These grabens were filled with arkosic coarse clastics, basalt flows, evaporites, and occasional sabkha deposits (Van Houten, 1977; von Rad et al., 1982). Intracratonic arkoses resulting from compressional forces, which lack interstratified basalt flows, are present in Pennsylvanian rocks of central and western Colorado (Fountain Formation and Minturn Formation) (Hubert, 1960; Boggs, 1966). These arkoses were a result of plate margin interactions 1600 km to the southeast (Kluth and Coney, 1981). Few areas of the world have had such a widespread development of this lithofacies during Phanerozoic time (Fig. 10-3). The epoch of block faulting lasted for perhaps 30–40 million years and resulted not only in extensive arkoses but also in a thick series of evaporites because of semiaridity produced downwind of the rising highlands.

Tectonic arkoses have many similarities to molasse deposits. They are very wedge-shaped in three dimensions, quite coarse-grained at the thicker side of the wedge, and grade basinward into finer-grained fluvial and shallow marine deposits (Fig. 10-4). The main distinctions between classic molasse and tectonic arkoses are in the style of tectonics that produces them and in the mineral composition of the sediment generated. Classic molasse deposits result from convergent plate margin tectonics and associated nappes and surficially shallow-dipping faults and tend to be rich in sedimentary and low- to medium-grade metamorphic lithic fragments. The classic tectonic arkose deposits result from intracratonic high-angle block faulting and are rich in granitoid debris, particularly potassic feldspars and quartzofeldspathic lithic fragments.

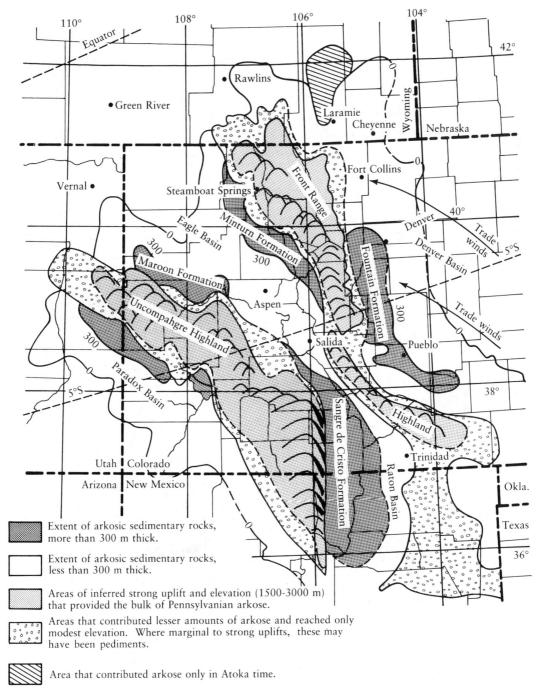

FIG. 10-3 • Restoration of the ancestral Rocky Mountains in Pennsylvanian time and thickness of associated arkosic sedimentary rocks. Also shown are the paleoequator, 5° South paleolatitude line, and paleowind direction at this latitude. The orographic rain shadow produced by the Front Range Highland is evident and resulted in the formation of evaporites in the Eagle Basin. (Reprinted by permission of the publisher from Mallory, *Geologic Atlas of the Rocky Mountain Region*, Rocky Mountain Association of Geologists, 1972.)

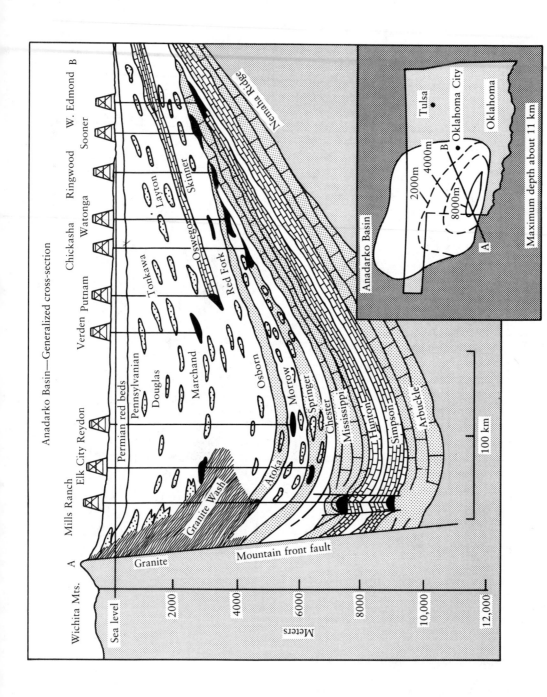

FIG. 10-4 • Cross-section through the Anadarko Basin, showing the stratigraphic position of intracratonic tectonic arkose ("granite wash") produced by Pennsylvanian uplift along a preexisting bounding aulacogen fault of possible Cambrian age. All other pre-Permian sedimentary rocks are shallow-water marine carbonates and quartz sandstones. (Reprinted by permission of the American Association of Petroleum Geologists from Fritz, *American Association of Petroleum Geologists Explorer*, 1985.)

Cratonic Lithofacies

The **cratonic lithofacies** forms a relatively thin but areally extensive stratigraphic section on the peripheries and interiors of Precambrian continental nuclei. Characteristic features of cratonic sequences include many disconformities and paraconformities and extensive development of depositional environments of high kinetic energy, such as desert dunes, beaches, peritidal deposits, and offshore bars. The source of detrital sediment is the continental nucleus, either crystalline rocks or derived sediments. The combination of low relief and a dominance of environments of high kinetic energy results in only thin sediment accumulations and repeated reworking of those grains that are available. As a result, the rock suite is dominated by texturally and mineralogically mature and supermature sandstones (Potter and Pryor, 1961) and shallow marine carbonate rocks (Fig. 10-5). The carbonates are both biohermal and biostromal and commonly are partly dolomitized.

In the United States the stereotypical cratonic sequence is the sediments deposited in one of the two major Cambro-Ordovician transgressive episodes. Each of these episodes starts with a pure quartz sandstone (sometimes with a thin basal arkose) and is overlain by a thicker succession of shales and carbonate rocks, the carbonates usually being more abundant than the shales. In the northern midcontinent the formational names used are St. Peter Sandstone and overlying Mohawkian-Cincinnatian Series of carbonates; in New York and Pennsylvania, Potsdam-Antietam Sandstone and Beekmantown Group carbonates; in Missouri, Lamotte–St. Peter sandstones and many named carbonate units; in Oklahoma, the Reagan Sandstone and Arbuckle Group of carbonates. Examples of facies analyses of cratonic sequences include those of the Cambrian rocks of northern and central Arizona (Hereford, 1977; Wanless, 1975; McKee, 1969), an analysis of lower Paleozoic transgression in South Africa (Hobday and Tankard, 1978), and those of late Precambrian sequences in Norway (Johnson, 1975).

FIG. 10-5 • Generalized cross-section of cratonic Lower Paleozoic rocks in midcontinental North America. Carbonate rocks are shallow-water, fossiliferous limestones and dolomites containing many small reefs; sandstones are composed almost entirely of fine- to medium-grained quartz grains cemented by quartz and calcite. (Reprinted by permission of the author from Potter and Pryor, *Geological Society of America Bulletin*, v. 72, 1961.)

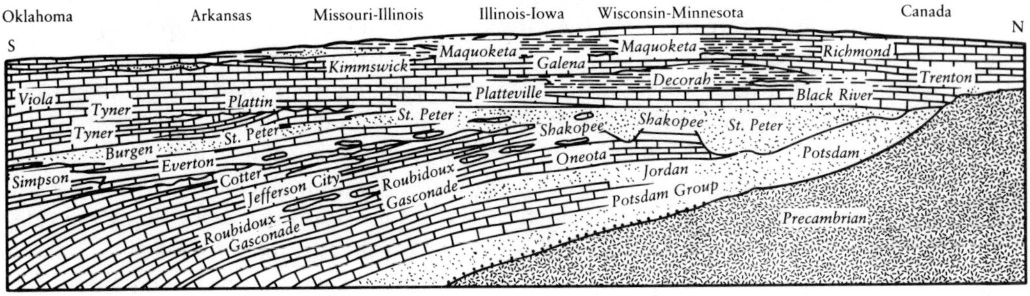

Black Shales

Black and dark gray shales are a recurrent facies that occurs sporadically throughout the geologic record. Commonly, these carbonaceous beds are sharply bounded by units of conspicuously lighter shade, and the sharp change in color suggests a geologically rapid change in the conditions that control rock color, the conditions that control the accumulation of organic matter. The organic carbon content of black shales exceeds about 1%; when it is conspicuously higher, the rock may yield petroleum on distillation and can be described as an oil shale (Chapter 13). Examples of this are some of the beds in the lacustrine Green River Formation (Eocene) in Wyoming, which contain 10–30% organic matter and reach a high of 80% in some laminae.

In outcrop, black shales are easily recognized because of their color. Other than color, their most characteristic lithologic feature is laminae of organic matter arranged parallel to bedding, which is responsible for the pronounced fissility of the rocks. Fissility is so well-developed that on weathering, the rocks are often described as "paper shales." The anoxia that permits accumulation of organic matter on the floor of the depositional basin means that water currents are sluggish to nonexistent, and therefore current-generated sedimentary structures are rare to absent. The anoxia also prevents the establishment of a diverse benthonic fauna. If any occur, they are low in diversity and high in numbers, indicating a high-stress environment. Occasional bedding surfaces may be crowded with large numbers of remains of a single species, which flourished at times when the degree of anoxia was slightly lower. Anoxia on the basin floor also prevents development of a burrowing infauna, so the thin laminae made possible by the slow settling and pronounced orientation of organic matter are not destroyed by bioturbation. Planktonic organic remains and nectonic fossils—such as ammonites or fish remains may be quite abundant in black shales and in an excellent state of preservation because of the lack of bottom-scavengers and burrowers.

Inorganic materials in black shales are those typical of low-energy environments and include clay, quartz silt, calcite, and dolomite, evaporite minerals occurring in some lacustrine settings. Pyrite cubes are sometimes prominent in black shales because the mineral is composed of reduced forms of iron and sulfur, the forms that would be stable in the oxygen-deficient environment where organic matter accumulates.

The only requirements for the formation of black shales are a supply of organic matter, absence of dissolved oxygen, and lack of strong currents. Such requirements can be met in a wide variety of environments, ranging from inland lacustrine (Bradley, 1931) to shallow shelf (Leggett, 1980) to bathyal-abyssal (Jenkyns, 1980). Recognition of this fact has long been obscured by the circumstance that the present is a very orogenic period in the Earth's history, so epicontinental and marginal seas are not prominent. The geologically uncommon accumulation site of the Black Sea (Degens and Ross, 1974) became the major model for the formation of black shales and anoxic environments. Research during the past ten years, however, has rapidly dispelled the idea that waters at least 500 m deep are the only environment in the marine realm where anoxia is likely to occur. Based on these studies of regional stratigraphy, local facies relationships, organic matter, and fossil content, an important consensus has been reached. Marine black shales on continental shelves tend to occur near the base of transgressive sequences (Hallam and Bradshaw, 1979; Jenkyns, 1980). Many of these bituminous horizons are extremely widespread, even extending across whole continents, with the associated facies of the shallow-water type. Such deposits are known from both Paleozoic and Mesozoic sequences. For example, many black shales occur in the lower part of Cretaceous transgressions (Waples, 1983).

The depositional model for major periods of formation of black shales is as follows. At times of increased rate of sea floor spreading, transgressions of the sea occur onto continental margins, increasing the area and volume of shallow epicontinental and marginal seas. The expansion of sea surface causes an increase in the proportion of solar energy absorbed at the Earth's surface (decrease in albedo or reflectivity). As a result, ice caps are greatly restricted in the polar regions; the quantity of cold, dense, oxygen-rich surface water available is reduced; and the supply of cold oxygenated bottom water in the world ocean is lessened. The reduced volume of cold seawater also increases the extent of salinity-density stratification at depth, so anoxia can become a more widespread phenomenon in pelagic areas than would otherwise be possible. On the expanded marine shelf area the water is warmer, so production of organic carbon is greater owing to increased volume of neritic habitats. Perhaps coincidentally, the Cretaceous and mid-Paleozoic were times of relative tectonic quiescence over much of North America, so topographic relief was subdued and a major part of the continent was inundated by shallow seas.

Force (1984) has noted that times of increased rate of sea floor spreading seem to correlate with times of stability in magnetic polarity (Fig. 10-6), a finding that implies that the Earth's interior processes and its climate are related and their status recorded by both magnetic polarity and anoxic event chronologies of the Earth. It is mind-boggling to think that processes at the core-mantle boundary may have an influence on climate as great as variations in the amount of solar radiation. Geologists are only beginning to understand the interrelationships among the Earth's interior, its surface processes, the atmosphere, the sun, and the other objects in the solar system.

One of the best documented examples of the interrelationship among tectonics, climate, and the widespread occurrence of black shales is provided by Devonian-Mississippian rocks in midcontinental North America (Ettensohn, 1985; Ettensohn and Elam, 1985; Ettensohn and Barron, 1981). The regional tectonic setting in which the black shales formed (Fig. 10-7) was established during Silurian and Early Devonian time by the collison of North America and the western edge of the European landmass (Caledonian and Acadian orogenies). The mountain range erected by the collision is estimated to have been at least 4 km in height. By late Early Devonian time (Emsian), about 390 m.y. ago, the mountain belt had extended from the northeastern United States to the southwest along the southeastern edge of the North

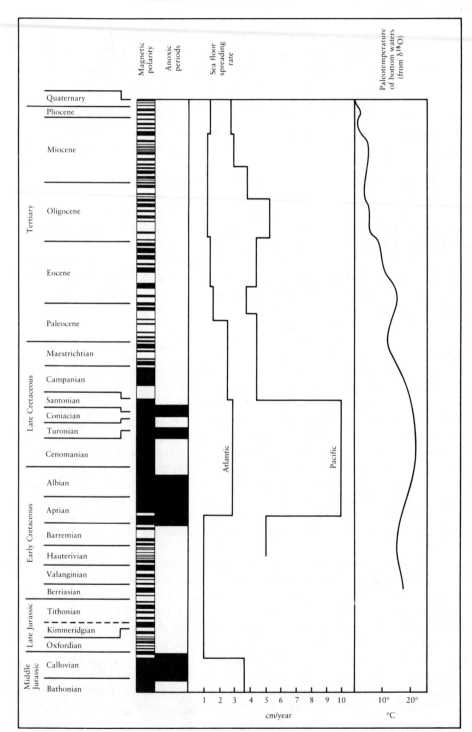

FIG. 10-6 ● Comparison of magnetic polarity time scale with periods of widespread anoxia, average half-rates of sea floor spreading, and bottom paleotemperatures from the north Pacific Ocean based on oxygen isotope studies. (Adapted from Force, 1984.)

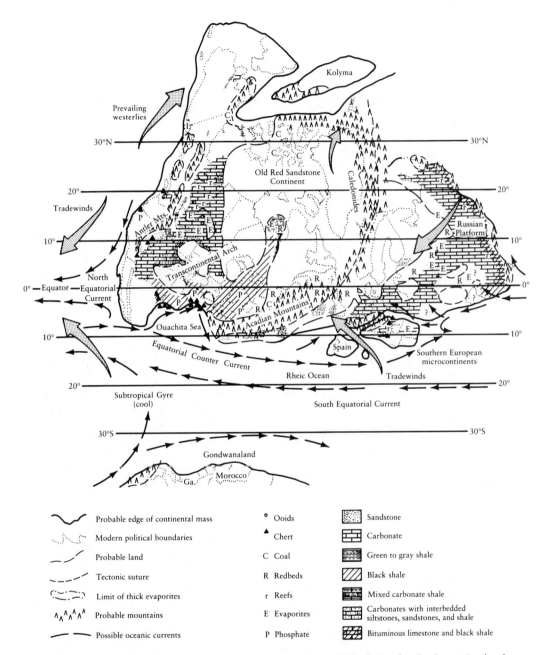

FIG. 10-7 • Generalized Late Devonian paleogeography and lithofacies for the Laurasian land mass. (Reprinted by permission of the author from Ettensohn, *Geological Society of America, Spec. Paper* 201, 1985.)

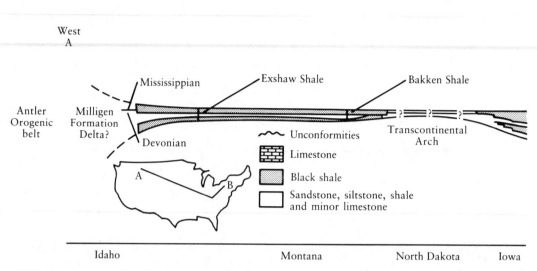

FIG. 10-8 • Schematic cross-section showing the distribution of black shales and related coarser clastic facies from western New York to Idaho. The large body of coarser clastics cyclically inter-

American landmass (Fig. 10-7). The mass load of the mountain belt produced a subsiding peripheral basin on the cratonic side of the belt because of regional isostatic adjustment to the load by the lithosphere. Hence a close relationship exists between the mountain belt and its adjacent peripheral basin into which water from the world ocean flowed to produce the epicontinental sea. The sea was bounded to the north by the Old Red Sandstone Continent and to the west by the Transcontinental Arch. The main, if not the only, consistent entrance to the Devonian-Mississippian cratonic sea was from the southwest, through the embayment occupied by the Ouachita Sea. Because increased spreading rates are accompanied by displacement of ocean water onto the continents, the waters of the epicontinental sea transgressed farther onto the craton as the Acadian Orogeny proceeded, so by latest Devonian time its deposits had topped the Transcontinental Arch (Fig. 10-8) into the northwestern United States.

The climatic regime in which the epicontinental sea was located was controlled both by its latitudinal position and by the mountains that bordered its eastern and southern sides (Fig. 10-7). Nearly all of the sea lay within 5° of the paleoequator, a belt characterized by warm, humid conditions and heavy rainfall, more than 200 cm/year in the modern equatorial zone. The northeasterly tradewinds from Europe and Asia were forced to rise as they encountered the Acadian Mountains, so they dropped their moisture on the eastern side of the mountains, creating the potential for a rain-shadow desert west of the mountains. Such a regional desert did not form, however, because of the pronounced convectional air flow from the surface upward that is characteristic near the equator. The sea surface provided the moisture for rainfall without the help of the northeast tradewinds.

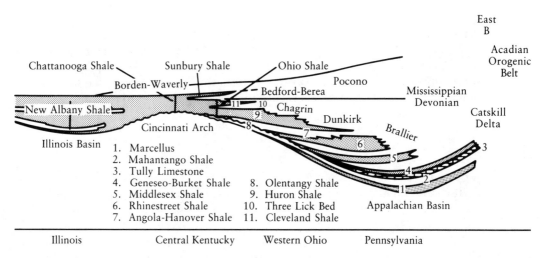

tonguing with the black shales in the Appalachian Basin is the Catskill Delta complex. (Reprinted by permission of the author from Ettensohn, *Geological Society of America, Spec. Paper* 201, 1985.)

How did anaerobic conditions develop in the Devonian-Mississippian epicontinental sea in North America? Development of anoxia is typically related to restrictions in the vertical mixing of surficial and deeper waters. These restrictions can arise from two mechanisms.

1. The water mass may be barred from free interchange with the world ocean by the presence of a sill that extends upward to a near-surface position. The best modern example of this is the Black Sea, whose outlet into the Mediterranean Sea is restricted by the Dardanelles sill that rises to about 40 m below the sea surface.

2. The alternative means of producing anoxia, the one that was the cause of the mid-Paleozoic black shales, is density stratification caused largely by climatic effects rather than by topographic or structural ones. In the equatorial belt, the surface waters are nearly always warmer and less dense than those lower in the water column, so vertical overturn and oxygenation of bottom waters cannot occur. Downward vertical diffusion of oxygen is very slow and ineffective as a replenishment mechanism. Total absence of oxygen may occur at depths less than 50 m. As a result, the abundant supply of organic matter produced by phytoplankton in the warm tropical surface waters settles unoxidized to the floor of the sea.

The thermal stratification may have been complemented by salinity stratification, which would have enhanced the stability of the oxic/anoxic contrast. Because the sea was equatorial, large amounts of rain and runoff would at least periodically have formed a lighter, less saline layer in parts of the sea. The presence of supposed brackish-to-marine pelagic algae such as *Tasmanites* and *Foerstia* throughout all or parts of the black shale supports the inference of salinity stratification (Ettensohn and Barron, 1981). Further supporting the effect of rain and runoff from the land surface is the presence of abundant terrestrial organic material in the black basinal shales, and a study of the carbon isotopes from the shales indicates that the proportion of nonmarine carbon in the organic matter increases toward the land area.

Bentonites

A **bentonite bed** is a layer of montmorillonitic clay and colloidal silica produced by devitrification and accompanying chemical alteration of a glassy igneous material, usually a tuff or volcanic ash. Its color ranges from white to light green and light blue when fresh, becoming light cream on exposure and gradually changing to yellow, red, or brown. The rock is greasy and soaplike to the touch and, because of the near-absence of detrital quartz, is not gritty. When wet, bentonites may swell to many times their original volume as the montmorillonitic clay absorbs water. For this reason, bentonite is commercially valuable as an increaser of viscosity in drilling muds. However, the ability to absorb large amounts of water means that bentonite beds form very unstable slopes in outcrop.

A bed of bentonite is the result either of a single eruption or of several eruptions within a very brief period, perhaps a few years. As a result, sensibly single bentonite beds are normally less than 50 cm thick, and even these may be composites of several eruptions. Because they form unstable slopes and are very thin, bentonites are easily overlooked during routine field work, even in areas where many are present. Sometimes a geologic unit will contain as many as 20 individual bentonite beds separated by nonbentonitic material, as in Cretaceous rocks of Wyoming and Montana (Slaughter and Early, 1965). Individual beds of bentonite may be laminated or massive. The massive variety frequently breaks along irregular curved surfaces that have been described as "picture puzzle" fractures. Because of repeated expansion and contraction of the montmorillonite during wetting and drying, bentonite outcrops frequently have a "popcorn" appearance (Fig. 10-9).

Recognizable beds of bentonite can be present to distances of several hundred kilometers from the volcanic vent. The beds decrease in thickness away from the vents (Elder, 1988) (Fig. 10-10), and if any of the original volcanic fragments are still visible in an ancient bentonite, they show a decrease in grain size with increasing distance from the vent. Also, they may be graded within the bed because larger and denser grains settle more rapidly through the atmosphere. The amount of time transgressed from the volcanic vent to the outer limit of the bentonite unit is geologically negligible, so bentonites are commonly referred to as "time surfaces," and as such, they are of great value in stratigraphic studies. In a basin where facies changes are numerous and rapid and where zonal fossils are not abundant, bentonites may provide the only means of stratigraphic correlation. In marine ash deposits the precision of correlation can be decreased by the burrowing and sediment-mixing activities of benthonic organisms.

FIG. 10-9 • Exposed surface of Miocene bentonite, Canterbury, New Zealand, showing a characteristic polygonal shrinkage crack pattern of angular clay fragments about 6 mm in diameter known as popcorn topography. (Reprinted by permission of the publisher from Ritchie et al., *New Zealand Journal of Geology and Geophysics*, v. 12, Science Information Publishing Centre, 1969.)

In general, the basal contact of bentonite with underlying beds is sharp, whereas there is a gradational contact with the overlying beds. The overlying contact may be an interlayering of beds or a gradual transition of the bentonite to shale, sandstone, or carbonate rock. In some cases, overlying beds of sand contain nodular masses of bentonite sharply separated from the sand (Grim and Güven, 1978).

Perhaps the best area in the world in which to apply bentonite stratigraphy is in the Western Interior Cretaceous seaway (Fig. 10-11), where more than 400 bentonites are scattered throughout the Albian to Maestrichtian section (45 m.y.). Most of them have extents in the hundreds of kilometers, and a few of the major ones extend for more than 1600 km outward from their source areas. The most extensive bentonite bed is recognizable from Alberta, Canada as far south as northern Texas. Major bentonites in the Upper Cretaceous rock section may reach thicknesses as great as 2 m, but the great majority are a few centimeters to less than 1 cm thick. Many thinner and less extensive ashes are probably also present and will be discovered during future stratigraphic studies (Kauffman, 1970, 1977).

Complementing the large number of bentonite beds in this section about 3000 m thick is a highly refined biostratigraphic zonal system that has been developed for the basin, a development made possible by the rapid evolutionary rates of the ammonites and bivalves that dominate the fossil assemblages (Fig. 10-12). Seventy biozones have been defined. Although conditions defined by the large number of numerically dateable bentonites and the many tens of recognizable zonal fossil assemblages seem like a field setting too good to actually exist, there are still problems in integrating the two sets of data. For example, the uncertainty on each bentonite date is 2–4 m.y. and thus covers the time ranges of two to eight assemblage zones. Further, most of the 400 bentonites have not yet been dated. It is obvious that

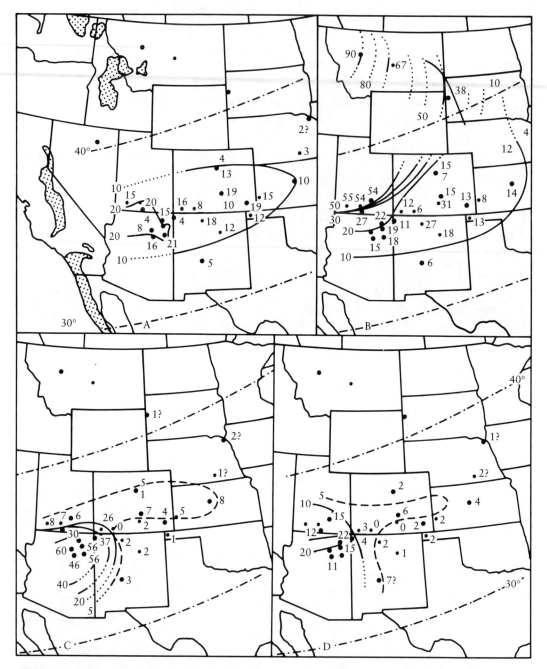

FIG. 10-10 • Isopach maps showing thickness (in centimeters) of four Cretaceous bentonite marker beds in the U.S. western interior. Solid isopach lines mark 10-cm contour intervals, dashed lines mark 5-cm contour, and dotted lines indicate questionable projections of isopach lines. Dot-dash lines are approximate paleogeographic orientation of latitude 30° and 40° N 90 million years ago. Bed A map shows major Cretaceous batholiths (stippled). (Reprinted by permission of the author from Elder, *Geology*, v. 16, 1988.)

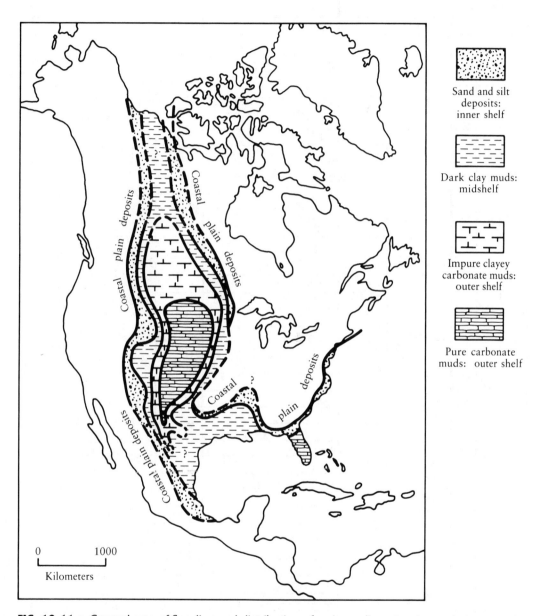

FIG. 10-11 • General area of flooding and distribution of major sedimentary types during maximum transgression (Latest Cenomanian–Early Turonian) of the Greenhorn Marine Cycle (Albian–Middle Turonian), Western Interior of North America. (Reprinted by permission of the editor from Kauffman, *in* Yochelson, E., (ed.), *Proceedings of the North American Paleontological Convention,* 1970.)

FIG. 10-12 Generalized radiometric time scale for the marine Cretaceous sequence in the Western Interior of North America, based on K-Ar dating of biotite and sanidine in ashes and

Stages: Western interior U.S.A. usage			Stratigraphy, Central Western interior U.S.A			Biostratigraphy: Main zonal indices	Generalized time scale (Ma)	Transgressive-regressive cycles
UPPER CRETACEOUS	Maastrichtian	M. U.		Hell Creek Formation		Triceratops sp.	64 / 65 / 65.5	R / T
		Lower		Fox Hills Sandstone		Discoscaphites nebrascensis	66	
						Hoploscaphites nicolletti	67 / 67.5	
						Sphenodiscus (Coahuilites)	67.75 / 68	
				South	North			
	Campanian	Upper	Pierre Shale	"Transition" Member	Upper unnamed shale Mbr.	Baculites clinolobatus		R
						Baculites grandis	68.25	
						Baculites baculus	68.5	
					Kara Bentonite M.	Baculites eliasi	68.75	
				"Tepee Butte" zone	Lower unnamed shale unit	Baculites jenseni		
						Baculites reesidei	69	
						Baculites cuneatus	70	
						Baculites compressus	71 / 71.5	
						Didymoceras cheyennense	71.75	
						Exiteloceras jenneyi	72	
						Didymoceras stevensoni	72.25	
						Didymoceras nebrascense	72.5	
				"Rusty zone"	Redbird Silty Mbr.	Baculites scotti	73	T
						Baculites gregoryensis	74	
						Baculites perplexus (late form)		
						Baculites gilberti	75	R
		Lower		Sharon Springs Member	Mitten Black Shale Mbr.	Baculites perplexus (early form)	76	
						Baculites sp. (smooth)	77	
						Baculites asperiformis	77.5	
					Sharon Springs Mbr.	Baculites mclearni	78	T
						Baculites obtusus	78.25	
				Apache SS Mbr.		Baculites sp. (weak flank ribs)	78.5	
				"Transition Member"	Gammon Ferruginous Mbr.	Baculites sp. (smooth)	79	
						Haresiceras natronense	80	
						Haresiceras placentiforme	81	R
	Santonian	Upper	Niobrara Formation			Haresiceras montanaense	82	
				Smoky Hill Member		Desmocaphites bassleri	82.5	T
						Desmocaphites erdmanni	83	R
		Middle				Clioscaphites choteavensis	83.5	
						Clioscaphites vermiformis	84	
						Clioscaphites saxitonianus	84.5	T
		L.				Scaphites depressus	85	R
	Coniacian	Up.				Scaphites ventricosus	86	T
		Mid.				Scaphites preventricosus Inoceramus deformis	86.5	R
		Lower		Fort Hays Limestone Member		Inoceramus erectus (late form)	86.75	
						Inoceramus erectus, s.s. Barroisiceras, Peroniceras	87	T

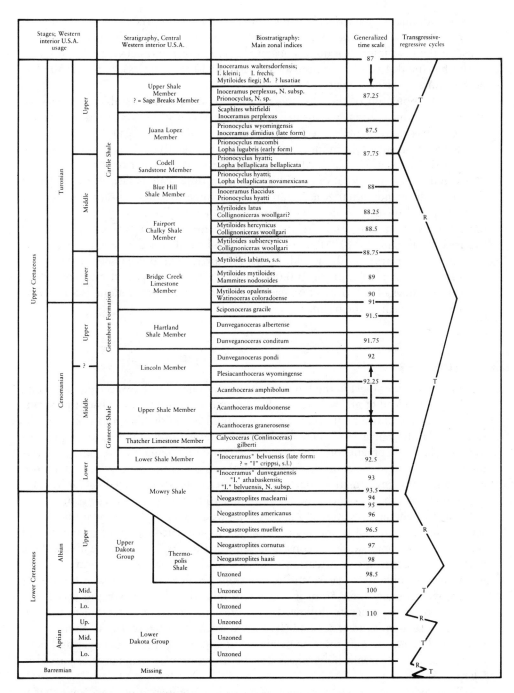

Stages; Western interior U.S.A. usage			Stratigraphy, Central Western interior U.S.A.		Biostratigraphy: Main zonal indices	Generalized time scale	Transgressive-regressive cycles
Upper Cretaceous	Turonian	Upper	Carlile Shale		Inoceramus waltersdorfensis; I. kleini; I. frechi; Mytiloides fiegi; M. ? lusatiae	87	
				Upper Shale Member ? = Sage Breaks Member	Inoceramus perplexus, N. subsp. Prionocyclus, N. sp.	87.25	T
					Scaphites whitfieldi Inoceramus perplexus		
				Juana Lopez Member	Prionocyclus wyomingensis Inoceramus dimidius (late form)	87.5	
					Prionocyclus macombi Lopha lugubris (early form)	87.75	
		Middle		Codell Sandstone Member	Prionocyclus hyatti; Lopha bellaplicata bellaplicata		
				Blue Hill Shale Member	Prionocyclus hyatti; Lopha bellaplicata novamexicana		
					Inoceramus flaccidus Prionocyclus hyatti	88	
				Fairport Chalky Shale Member	Mytiloides latus Collignoniceras woollgari?	88.25	R
					Mytiloides hercynicus Collignoniceras woollgari	88.5	
					Mytiloides sublhercynicus Collignoniceras woollgari	88.75	
		Lower	Greenhorn Formation	Bridge Creek Limestone Member	Mytiloides labiatus, s.s.		
					Mytiloides mytiloides Mammites nodosoides	89	
					Mytiloides opalensis Watinoceras coloradoense	90 / 91	
	Cenomanian	Upper		Hartland Shale Member	Sciponoceras gracile	91.5	
					Dunveganoceras albertense		
					Dunveganoceras conditum	91.75	
		?		Lincoln Member	Dunveganoceras pondi	92	
					Plesiacanthoceras wyomingense	92.25	T
		Middle	Graneros Shale	Upper Shale Member	Acanthoceras amphibolum		
					Acanthoceras muldoonense		
					Acanthoceras granerosense		
				Thatcher Limestone Member	Calycoceras (Conlinoceras) gilberti		
		Lower		Lower Shale Member	"Inoceramus" belvuensis (late form: ? = "I." crippsi, s.l.)	92.5	
Lower Cretaceous	Albian	Upper		Mowry Shale	"Inoceramus" dunveganensis "I." athabaskensis; "I." belvuensis, N. subsp.	93 / 93.5	
					Neogastroplites maclearni	94 / 95	
			Upper Dakota Group		Neogastroplites americanus	96	R
				Thermo-polis Shale	Neogastroplites muelleri	96.5	
					Neogastroplites cornutus	97	
					Neogastroplites haasi	98	
					Unzoned	98.5	
		Mid.			Unzoned	100	T
		Lo.			Unzoned	110	
	Aptian	Up.	Lower Dakota Group		Unzoned		R
		Mid.			Unzoned		T
		Lo.			Unzoned		
	Barremian		Missing				R / T

bentonites. (Reprinted by permission of the publisher from Kauffman, *Mountain Geologist*, v. 14, Rocky Mountain Association of Geologists, 1977.)

current radiometric data do not constitute a sufficient matrix of geologic time to determine how closely the ammonite zones approach ideal isochroneity. And because of the instrumental imprecisions associated with radiometric dating, it is unlikely that the degree of precision will be obtained that is necessary to test the hypothesis of zonal isochroneity.

But the potential of the time matrix extends far beyond its ability, or lack of ability, to establish zonal isochroneity. A principal object of constructing a radiometric matrix is to provide adequate data for a detailed time scale for these rocks that is based on sufficient numbers of dates that the relatively large error factors of individual dates can be averaged out to give the whole system regional reproducibility and thus a high level of confidence. Such a time scale is ideal for calculating the duration of individual zones, patterns of change in zonal duration related to environmental history, evolutionary rates of individual lineages, and various geological phenomena that can be related to the geochronology and biostratigraphy of the area (Kauffman, 1988). All marine events can be rather accurately dated in comparison to other areas of the world, be widely correlated, and have their durations measured with a resolution of 100,000 years or less. Events that occurred at or adjacent to times of ash deposition can be related to each other along isochronous surfaces over much of the basin.

Cyclic Sequences

The idea that cycles exist in the Earth's history is not new but has received increasing emphasis since Vail et al. (*in* Payton, 1977) proposed, on the basis of seismic stratigraphy, that the entire stratigraphic column could be subdivided into cycles of different durations. Since 1977 there has been a veritable flood of publications proposing a cyclic development for almost every geologic phenomenon. Stratigraphers, as interpreters of the sedimentary record, are in the midst of the controversy. In a modern stratigraphic investigation, one of the important questions asked is "Do regular patterns exist in the sequence, and how are they to be interpreted?" For example, in tectonically generated continental successions we frequently find repetitions of the series conglomerate-sandstone-shale. Because this sequence has three members and the chance that they will occur consistently in the same order is remote unless there is an outside control, our attention is drawn to the nature of this control. When many tens of millions of years and a thousand or more meters of section are involved, we generally infer an orogenic origin related to relatively rapid uplift that generates the conglomerate, followed by decreasing intensity of tectonism that results in subsequent sandstones and shales. On a smaller time scale and thinner section of sediments we may infer an origin related to meteorologic changes and periodically increased stream flow. These are the fining-upward sequences discussed in Chapter 7.

We may define a **cyclic sediment** as an ordered sequence of lithologies that is repeated in a predictable pattern, although the orderliness of the sequence is never perfect. The sequence may carry time information, but this is not a requirement for the lithologies to be cyclic. For example, a repeated alternation of fine-grained sandstone and shale that resulted from a migrating stream channel would be cyclic because of the predictability of the repetition, but the time required for each sandstone-shale pair to form might vary greatly, as might the relative thicknesses of each rock type. Lacustrine varves, on the other hand, carry rather precise temporal information, each couplet indicating one year of elapsed time. Clearly, in cyclic sequences, all recurrence intervals are possible, from "very exact" to "approximately periodic" to "apparently regular" to simply repetitive at irregular or unknown intervals. Commonly, we cannot be certain which category we are dealing with and may suggest an explanation either more precise or less precise than is warranted by the field or laboratory data we can record. Statistical analyses (time series, Markov chains) can be useful but cannot supply certainty.

The scale of cyclicity is limited on the "rapid end" by our ability to recognize small-scale variations in outcrop. The fining-upward sequences produced by waning fluvial flood waters or laterally migrating streams (Chapter 7) are perhaps the most rapidly cyclic sedimentary package, involving perhaps only a few weeks to form several repetitions. Next upscale are lacustrine or evaporitic varves, which change lithology every 6 months for as long as the lake or evaporitic condition exists. A bit less rapid are the sequences of graded beds that are deposited by recurrent turbidity currents (Chapter 9) that may occur on a time scale of years. Still slower are punctuated aggradational cycles (PACs) and pelagic limestone-marl cycles (see below), which require periods on the order of 10^4-10^5 years to develop. At a larger time scale are cyclothems, the cyclic sequences of transgressive and regressive deposits that are very common in Carboniferous sediments of the midcontinental United States and elsewhere and that require periods of a few hundred thousand years to form. At the largest recognizable scale in Phanerozoic rocks are the six unconformity-bounded, worldwide packages of sedimentary rock first recognized by Sloss (1963). These require periods of tens of millions of years to develop. Still longer cycles in sedimentologic and stratigraphic phenomena may exist, but because only the last 12% of the Earth's history (Phanerozoic time) is adequately dated, there is no way to evaluate the possibility. Benkö (1985) has suggested the presence of cycles related to the rotation of the Milky Way Galaxy or to the position of our solar system in it or related to possible periodic variations in the universal gravitational constant. There is a marked trend in stratigraphy at present to seek interrelationships between astronomical factors and stratigraphic variations (Barron et al., 1985; House, 1985; Fischer, 1986; Research on Cretaceous Cycles Group, 1986; Williams, 1986), and it seems likely that many remain to be discovered.

Punctuated Aggradational Cycles

The concept of **punctuated aggradational cycles** (PACs) was suggested by Goodwin and Anderson (1985) and proposes that the stratigraphic record accumulated episodically in response to an allogenic (outside control) mechanism. Examples of allogenic controls would be sea floor spreading and astronomical influences. This contrasts sharply with the generally accepted model of small-scale stratigraphic accumulation, which relies heavily on autocyclicity. **Autocyclicity** refers to stratigraphic variations that result from lateral migration of environments such as meandering channels or prograding tidal flats. Implicit in the autocyclic model is that changes in the stratigraphic succession are gradual rather than episodic. This is the basis for the concept of a **formation,** which is assumed to be the product of broad environmental bands that migrate to produce generally diachronous rock units. Small-scale cyclicity is usually explained by environmentally specific (autogenic) processes. Walther's Law, commonly applied at the scale of formations or members, is used to explain vertical facies succession.

Adherents of the PAC hypothesis believe that stratigraphic accumulation occurs episodically as thin (1–5 m thick) shallowing-upward cycles separated by sharply defined nondepositional surfaces (Fig. 10-13). From analysis of facies in these cycles it appears that these nondepositional surfaces are created by geologically instantaneous basinwide relative base-level rises (punctuation events) and that deposition

FIG. 10-13 • *Dicoelosia* Zone PAC at Catskill, New York. Arrows mark upper and lower boundaries at which major facies changes occur abruptly. (Reprinted by permission of the publisher from Goodwin and Anderson, *Journal of Geology,* v. 93, p. 526, The University of Chicago Press, 1985.)

occurs during the intervening periods of base-level stability. Thus the hypothesis predicts that the fundamental units of stratigraphic accumulation are thin, laterally extensive, time-stratigraphic asymmetric cycles bounded by isochronous surfaces. Walther's Law can be applied within a PAC but not across PAC boundaries, which separate environmentally disjunct facies. Stratigraphic analysis of the number of PACs between numerically dated horizons indicates that the interval of recurrence of punctuation events is less than 100,000 years. Because PACs are bounded by actual surfaces of environmental discontinuity, paleoenvironmental analysis is most appropriately conducted at the scale of PACs, rather than at the scale of the formation. Use of the formation as a fundamental unit is inappropriate because formations normally consist of many PACs and therefore contain many stratigraphically significant environmental discontinuities.

The allogenic cause of PACs is believed by Goodwin and Anderson (1985) to be the known variations in the Earth's orbital characteristics (Fischer, 1986; Herbert and Fischer, 1986). These variations are termed **Milankovitch cycles** after their discoverer. These cycles do not change the annual total energy received from the sun, but they do change its latitudinal and seasonal distribution. Three varieties of orbital variation are documented.

1. The Earth's axis precesses with two dominant periods averaging 19,000 and 23,000 years.
2. The Earth's rotational axis changes its inclination by up to 3° in an obliquity cycle of 41,000 years.
3. The Earth's orbit changes from almost circular to more elliptical in two eccentricity cycles, one with a period ranging from 95,000 to 136,000 years, the other with a period of 413,000 years.

When obliquity is high, the temperature gradient between equator and poles is decreased and seasonality is increased, with attendant changes in carbonate productivity, precipitation, erosion, and other factors affecting stratigraphic accumulations. Figure 10-14 is a generalized diagram showing how 20,000- and 100,000-year Milankovitch-type periodicities can be superimposed to produce sea level changes and the way in which these changes interact with various rates of continuous subsidence. The striking result of this analysis is that base level rises stepwise with rapid rates followed by periods of stability, exactly the pattern that would generate PACs.

Goodwin et al. (1986) illustrate the use of PACs as fundamental units of stratigraphy in a reinterpretation of the Helderberg Group (Devonian) in New York state (Fig. 10-15). Based on the assumption that PAC sequences are time-stratigraphic units of basinwide extent, they divided the lower Helderberg Group into 15 PAC units rather than into two formations as has been customary in stratigraphic analysis by standard methods. The interpretation of the Manlius Formation changes from a complex disordered mosaic of environments in time and space to a highly ordered "layer-cake" of time-stratigraphic units bounded by synchronous surfaces. Each

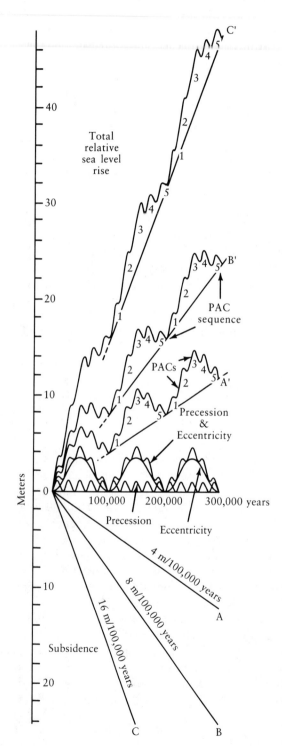

FIG. 10-14 • General model for episodic stratigraphic accumulation produced by eustatic responses to orbital perturbations superimposed on continuous subsidence. The precessional cycle (20,000 years) and eccentricity cycle (100,000 years) are represented by sine waves and are summed in a third curve representing the general form of eustatic response to the selected simple pattern of orbital fluctuations. This total eustatic response is then added to each of three subsidence curves (A, B, and C) to give three sample patterns of total relative sea level rise (lines A′, B′, and C′). The resulting patterns could produce PACs and PAC sequences; PAC sequences might be truncated by erosion in regions of low subsidence (line A′) or be more complete in areas of high subsidence (line C′). (Reprinted by permission of the publisher from Goodwin and Anderson, *Journal of Geology*, v. 93, p. 528, The University of Chicago Press, 1985.)

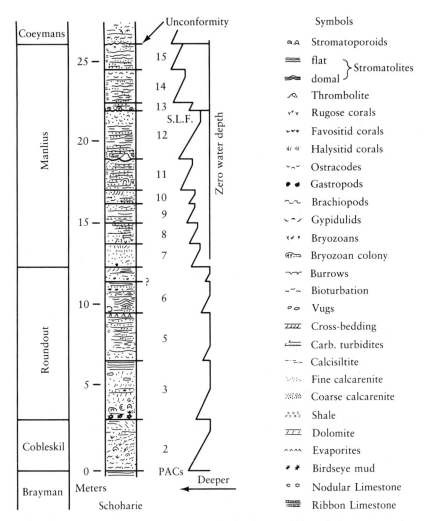

FIG. 10-15 ● Columnar section of the lower Helderberg Group on Interstate 88, Schoharie, New York, with traditional stratigraphic units on the left and numbered PACs on the right. Water depth curve shows location and relative magnitude of punctuation events (horizontal segments), amounts of gradual shallowing by aggradation (diagonal segments), and proportion of each PAC attributed to gradual subsidence (vertical segment). Sea level falls (S.L.F.) are recognized at the tops of PACs 12 and 15. Missing PACs 1 and 4 are recognized at other localities, while the questioned PAC boundary within PAC 6 is recognized only at Schoharie and South Wilbur. (Reprinted by permission of the publisher from Goodwin, et al., *Paleoceanography*, v. 1, Elsevier Science Publishers B.V., 1986.)

PAC is a genetic unit defined on patterns of facies change, not a unit of particular facies or lithology. As a result of the reinterpretation of the Helderberg Group according to the PAC hypothesis, the Rondout-Manlius formational boundary occurs at different PAC boundaries at different localities. The PAC hypothesis also leads to reinterpretation of regional Appalachian basin dynamics because it implies episodic deposition in response to repeated rapid base level fluctuations rather than gradual and continuous deposition. In addition, some formational contacts interpreted as environmental boundaries in non-PAC analysis are interpreted as small-scale unconformities by PAC methodology.

The concept of punctuated aggradational cycles is an interesting one that deserves careful testing in areas other than the shallow marine carbonate sequence in which it was developed. Van Tassell (1987) has made such a test in the Catskill Formation (Upper Devonian) deltaic section of the central Appalachians. He did not find the PACs to be separated by sharply defined surfaces of abrupt deepening, as was predicted by Goodwin and Anderson (1985), because rates of clastic sediment input apparently kept pace with subsidence during deposition of the PAC sequences. This suggests that the PAC hypothesis needs to be modified to include the effects of variations of sediment input in order to be applicable to clastic sequences.

The existence of discrete lithostratigraphic units called PACs is currently the subject of much debate in the stratigraphic/sedimentologic community; but even if they are real, their cause is uncertain. Algeo and Wilkinson (1988) examined data from more than 200 Phanerozoic mesoscale cycles and observed that the period of the cycles is strongly correlated with the thickness of the cycles and that all cycles with thicknesses between 1 m and 20 m will have periods in the Milankovitch range (21,000–413,000 years), irrespective of the actual causes of the cyclicity. No doubt the importance of PACs as fundamental stratigraphic units will be a subject of debate for some years to come.

Pelagic Limestone-Marl Rhythms

Many pelagic shales, marls, and carbonates show a rhythmic oscillation in carbonate content, resulting in carbonate-richer and carbonate-poorer beds (couplets). Spacing of couplets (Fig. 10-16) may be even, irregular, or patterned (bundled), and statistical analysis may reveal the presence of larger groupings termed superbundles. The limestone-marl cycles are known from pelagic deposits ranging in age from Oxfordian (Late Jurassic) to Late Tertiary. The age of the oldest known limestone-marl cycles corresponds to the widespread occurrence of the first pelagic calcareous microfossils, the coccolithophores, which appeared in earliest Jurassic time. Limestone-marl couplets visible in outcrop are normally less than 1 m thick and consist of a dominant limestone bed, generally containing more than 80% calcium carbonate, and a thinner marl unit containing subequal amounts of calcium carbonate and clay minerals. If the limestone contains less than 80% $CaCO_3$ it is unlikely to appear

FIG. 10-16 • Bridge Creek Member, Greenhorn Formation (Late Cretaceous), Pueblo, Colorado. Shale-limestone couplets of mixed productivity-dilution origin. Even spacing suggests obliquity cycle. Change in spacing probably reflects change in sedimentation rate. (Reprinted by permission of the publisher from Kauffman, *in* Pratt et al., *Society of Economic Paleontologists and Mineralogists Guidebook No. 4*, 1985.)

visibly different from the marl, but cyclicity may still be present and can be detected by laboratory analyses of insoluble residues.

As summarized by Einsele (*in* Einsele and Seilacher, 1982), there are several ways in which pelagic limestone-marl couplets may be formed.

1. **Periodic fluctuations of carbonate production.** In this model there is a steady influx of clay, but biogenic productivity varies. To produce the common type of cycle in which thick limestone beds alternate with thin marl layers, productivity must vary by a factor of about ten, and this must be reflected in the sea floor accumulation. Periodic changes in productivity are known to occur in response to waxing and waning of glacial ice, but this mechanism is difficult to apply to pre-Quaternary times, and localized upwelling is the only other way known to greatly change an already high level of biogenic carbonate generation.

2. **Periodic increase of detrital silicate supply.** This method of cycle generation assumes a steady production of biogenic carbonate but a fluctuating input of the detrital phase. This model has a high level of support among students of limestone-marl cycles, the cause of periodically increased clay being eustatic changes in sea level. However, it is not clear why such eustatic changes should occur repeatedly and at such brief intervals in the absence of glaciation.

3. **Periodic dissolution of carbonate.** Dissolution of biogenically generated calcium carbonate occurs both in the water column during settling and on the sea floor before burial. Thus we can postulate a steady production of biogenic carbonate at the sea surface and constant influx of clay but vary the depth of the carbonate compensation depth and the lysocline. This will give rise to limestone-marl cycles, as has been documented in Quaternary pelagic sediments affected by Pleistocene glacial episodes. The difficulty with this model is finding a cause of geologically rapid and repeated fluctuations in the depth of the lysocline in the absence of glaciers.

4. **Early diagenetic enrichment.** In this model for producing pelagic limestone-marl rhythms the sea floor muds are homogeneous in respect to the carbonate/clay ratio but inhomogeneous in the nature of the carbonate and/or the texture of the sediment. During early diagenesis, carbonate is selectively dissolved from the marl layers-to-be and migrates to texturally different limestone-to-be layers, where it is precipitated. The marls we see in outcrop have lost perhaps 50% of their original carbonate to the limestone beds, which, as a result, have lost their porosity and perhaps increased in thickness as well.

Because both the limestone and marl beds are normally rather thin, on the order of a few centimeters in thickness, they are easily homogenized after deposition by benthonic organisms. This is particularly true when depositional rates are slow and the sea floor is well oxygenated. This effect is clearly displayed in sections where thoroughly bioturbated carbonates are interbedded with well-laminated and undisturbed black shales. In fact, one of the best evidences for primary rather than diagenetic origin of limestone-marl rhythms is preservation of delicate sedimentary structures. Diagenetic migration of calcium carbonate among different levels in a stratigraphic sequence would destroy features such as lamination and small-scale bedding features such as cross-lamination or lenticular bedding.

In summary, several conditions must be satisfied for pelagic limestone-marl cycles to form. (1) Sediment accumulation must take place below the storm wave base to permit the accumulation of sediment as fine as silicate clay and porous and/or hollow carbonate shells. (2) The site of accumulation must be far removed from ready sources of coarse clastic debris, that is, turbidity currents. (3) There must be maintenance of a carbonate/clay ratio of about 3–4. Because carbonate productivity is limited (by nutrient availability in surface waters), the ratio implies a sedimentation rate of 0.5–3 cm/1000 years. Assuming a rate of 1 cm/1000 years, cycle-producing conditions must persist for 10^7 years to produce 100 m of stratigraphic section. The frequency of cyclic limestone-marl sequences in pelagic rocks indicates that the necessary conditions are often realized.

It is clear from the Quaternary stratigraphic record that climate has a marked effect on seawater temperatures and hence carbonate productivity in the oceans. But the last pre-Pleistocene glacial episode recorded on the continents occurred in Permo-Carboniferous time, so it seems we cannot appeal to glaciers to explain the numerous limestone-marl rhythms of Mesozoic and Tertiary age in pelagic rocks

(Laferriere et al., 1987). Yet recent evidence from the Deep Sea Drilling Project seems to indicate contemporaneity of many limestone-marl cycles based on microfossil ages and statistical analyses of cycle lengths and periodicities using techniques similar to those used for varve correlations (Chapter 9). As was the case for the punctuated aggradational cycles discussed earlier, Milankovitch periodicities have been suggested as the cause of pelagic limestone-marl cycles. The orbital periodicities would affect both carbonate productivity in the oceans and the generation of silicate mud on the land surface, the source of the mud in the deep ocean basin.

Cyclothems

Another important stratigraphic cyclicity is the **cyclothem**, which was named in the 1930s to describe Pennsylvanian and Early Permian cycles in the midcontinental United States but occurs also in other areas in either complete or truncated form. A typical cyclothem is 10–20 m in thickness and may be repeated many times in any one basin (Fig. 10-17). More than 100 repetitions have been mapped in Kansas, and some individual beds can be traced for more than 300 km. It is clear that each cyclothem records a cyclic advance and retreat of the sea over the craton, and a greater number of them are present in geographic areas where the shoreline passed most frequently (Heckel, 1986). Temporal analysis indicates relatively rapid transgression followed by slower regression, the same asymmetric periodicity interpreted from the punctuated aggradational cycles (PACs) considered earlier in this chapter.

Watney (*in* Watney et al., 1985) has reviewed the evidence accumulated over the past 60 years concerning the cause of the midcontinent cyclothems. He concluded that glacially controlled advances and retreats of the shallow sea over the continental margin were the proximate cause, local tectonism and subsidence playing significant but relatively minor roles. During the past 700 years the melting of glaciers has occurred at a rate of 5000 km^3 of ice per year giving an average rate of change in sea level of 1 cm/year, which is extremely fast on a geologic time scale. The maximum lowering of sea level attributed to glaciation during the Pleistocene is 100–150 m, which, at a rate of 1 cm/year, could be accomplished in only 10,000 years. Thus on the low slopes that are characteristic of the cratonic margins during Paleozoic time (2 cm/km) (Chapter 7) the shoreline could migrate rapidly seaward and landward for hundreds of kilometers in only 10^3 years as the Permo-Carboniferous glaciers waxed and waned. A very large number of cyclothems might be generated during the roughly 40 m.y. of deposition represented by the cyclothems, depending on the number of advances and retreats of the southern hemisphere glaciers.

Attribution of cyclothems to repeated glaciations invokes once again Milankovitch periodicities as the ultimate cause. Because of uncertainties concerning the quantitative relationship between the various periodicities and the amount of cooling needed for ice sheets to grow, as well as the effect of the positions of the continental plates on growth of the sheets, we can do no more at present than give a qualitative assessment. The idea seems reasonable.

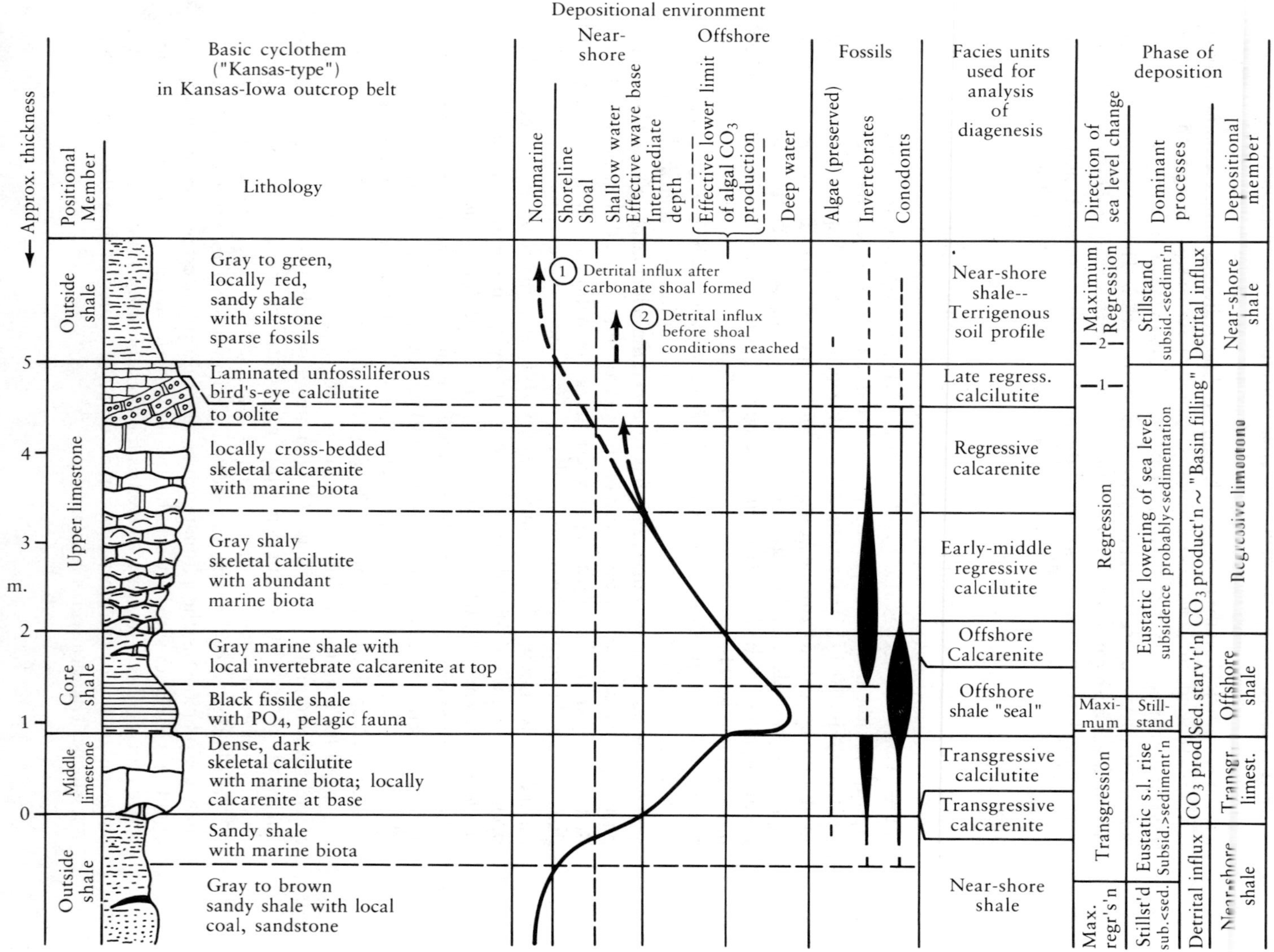

FIG. 10-17 ● Basic sequence of eustatic Midcontinent Pennsylvanian "Kansas cyclothem" showing depositional interpretation, gross distribution of fossil groups, and lithic facies units used for analysis of carbonate diagenesis. (Reprinted by permission of the publisher from Heckel (ed.) *Kansas Geological Survey Bulletin*, v. 169, 1964.)

Tectonoeustatic Cycles

Milankovitch periodicities are of geologically short duration, the longest being less than a half million years. Of much longer duration are **tectonoeustatic cycles**, periodicities whose cause lies in the change in volume of the deep ocean basin when magma from the mantle is transferred to the ocean floor. Displacement of seawater upward during times of active plate spreading, ridge construction, mantle plume development, and doming of oceanic crust over active hot spots cause global eustatic rise and synchronous epicontinental transgression. The sea level may be raised several hundred meters by this mechanism. Sea level falls of long duration (tens of million of years) are explained by subsidence of these topographically high areas on the sea floor by loss of heat. Tectonoeustacy cannot explain most eustatic cycles in sedimentary rocks because they are of too short a duration based on biostratigraphic zonation and numerical dating of bentonites.

In a now-classic paper, Sloss (1963) recognized six interregional unconformities in North American cratonic rocks, between which lie six major rock-stratigraphic units (Fig. 10-18). At the cratonic margins the bounding unconformities tend to

FIG. 10-18 • Time-stratigraphic relationships of sequences in North American craton. Black areas represent nondepositional hiatuses; white and stippled areas represent deposition. (Reprinted by permission of the author from Sloss, *Geological Society of America Bulletin*, v. 74, 1963.)

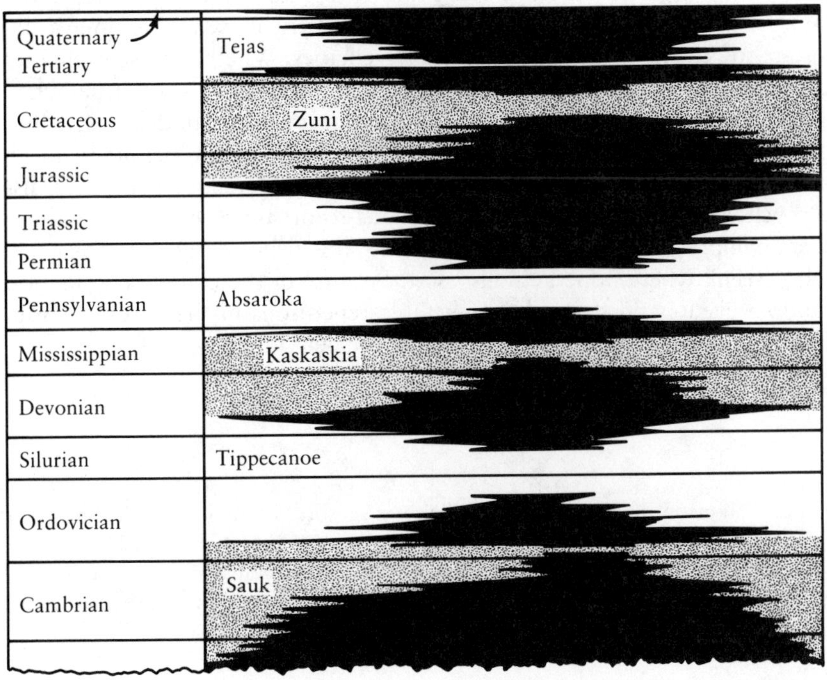

disappear in continuous rock successions, and the cratonic sequences are replaced by others controlled by events in the marginal basins. The six sequences represent the largest natural groupings known in sedimentary rocks and are recognizable from the Cordillera in central Nevada on the west to the Appalachian basin on the east, a distance of about 2700 km. The concept of major rock groupings of the type suggested by Sloss (1963) was enlarged upon in 1977 by P. R. Vail and co-workers (Fig. 4-9), who identified three orders of cycles that may be traceable over major areas of ocean basins as well as continents. Sloss's six lithostratigraphic groups would correspond with either the first- or second-order cycles of Vail, but it is difficult to be certain of such correlations because of the different methodologies and areas in which data were obtained. Sloss's data were obtained from cratonic rocks, and he used both outcrop and subsurface data. Vail and associates relied almost entirely on subsurface seismic data obtained from basins marginal to the craton and from deeper offshore areas. It remains for future stratigraphers to sort out the suggested cyclicities in stratigraphic sections and correlations between tectonoeustatic events as reflected on cratons and within the ocean basins.

Summary

In the late 1980s the field of stratigraphy was in the throes of an exciting development, the first serious attempt to relate features seen in the stratigraphic record to variations in astronomical parameters. The smallest of these now being considered is the Milankovitch suite: precession, obliquity, and eccentricity, whose time scales are appropriate to "explain" most stratigraphic cyclicities. Whether these cyclicities are in fact caused by Milankovitch periodicities is still uncertain. As is normal when a new development is being promoted, there are extremists on both sides; either everything is cyclic or else nearly all cyclicity is in the eye of the beholder. In an important but generally neglected publication, Zeller (*in* Merriam, 1964) has discussed the psychological (genetic) need for humans to organize their surroundings to make them comprehensible and has shown how this primal urge may cause cyclicities to be "seen" where none actually exist. Because of the complete gradation between random events and perfectly predictable repetitions in the natural world, many cyclicities may be, like beauty, in the eye of the beholder.

References

Algeo, T. J. and Wilkinson, B. H., 1988. Periodicity of mesoscale Phanerozoic sedimentary cycles and the role of Milankovitch orbital modulations. J. Geol., v. 96, pp. 313–322.

Barron, E. J., Arthur, M. A., and Kauffman, E. G., 1985. Cretaceous rhythmic bedding sequences: A plausible link between orbital variations and climate. Earth Planet. Sci. Lett., v. 72, pp. 327–340.

Bayer, U. and Seilacher, A. (eds.), 1985. Sedimentary and Evolutionary Cycles. New York, Springer-Verlag, 465 pp.

Benkö, F., 1985. Geological and Cosmogonic Cycles. Budapest, Hungarian Academy of Sciences, 401 pp.

Boggs, S., Jr., 1966. Petrology of Minturn Formation, east-central Eagle County, Colorado. Amer. Assoc. Petroleum Geol. Bull., v. 50, pp. 1399–1422.

Bradley, W. H., 1931. Origin and Microfossils of the Oil Shale of the Green River Formation of Colorado and Utah. Washington, D.C., U.S. Geological Survey, Prof. Paper 168, 58 pp.

Degens, E. T. and Ross, D. A. (eds.), 1974. The Black Sea—Geology, Chemistry, and Biology. Tulsa, American Association of Petroleum Geologists, Mem. 20, 633 pp.

de Graciansky, P. C., Deroo, G., Herbin, J. P., Jacquin, T., Magniez, F., Montadert, L., Müller, C., Ponsot, C., Schaaf, A., and Sigal, J., 1986. Ocean-wide stagnation episodes in the Late Cretaceous. Geol. Rundschau, v. 75, pp. 17–41.

De Visser, J. P., Ebbing, J. H. J., Gudjonsson, L., Hilgen, F. J., Jorissen, F. J., Verhallen, P. J. J. M., and Zevenboom, D., 1989. The origin of rhythmic bedding in the Pliocene Tribi Formation of Sicily, southern Italy. Paleogeog., Paleoclim., Paleoecol., v. 69, pp. 45–66.

Einsele, G. and Seilacher, A. (eds.), 1982. Cyclic and Event Stratification. New York, Springer-Verlag, 536 pp.

Elder, W. P., 1988. Geometry of Upper Cretaceous bentonite beds: Implications about volcanic source areas and paleowind patterns, western interior, United States. Geology, v. 16, pp. 835–838.

Ettensohn, F. R., 1985. Controls on Development of Catskill Delta Complex Basin-Facies. Boulder, Geological Society of America, Spec. Paper 201, pp. 65–77.

Ettensohn, F. R. and Barron, L. S., 1981. Depositional model for the Devonian-Mississippian black shales of North America: A paleoclimatic-paleogeographic approach. *In* T. G. Roberts (ed.), Field Trip Guidebook No. 3, vol. II. Boulder, Geological Society of America, pp. 344–361.

Ettensohn, F. R. and Elam, T. D., 1985. Defining the nature and location of a Late Devonian–Early Mississippian pycnocline in eastern Kentucky. Geol. Soc. Amer. Bull., v. 96, pp. 1313–1321.

Fischer, A. G., 1986. Climatic rhythms recorded in strata. Ann. Rev. Earth Planet. Sci., v. 14, pp. 351–376.

Force, E. R., 1984. A relation among geomagnetic reversals, seafloor spreading rate, paleoclimate, and black shales. EOS, v. 65, pp. 18–19.

Fritz, M., 1985. Anadarko eases into middle age. Tulsa, Amer. Assoc. of Petroleum Geol. Explorer, Feb., pp. 24–27.

Goodwin, P. W. and Anderson, E. J., 1985. Punctuated aggradational cycles: A general hypothesis of episodic stratigraphic accumulation. J. Geol., v. 93, pp. 515–533.

Goodwin, P. W., Anderson, E. J., and Goodman, W. M., 1986. Punctuated aggradational cycles: Implications for stratigraphic analysis. Paleoceanography, v. 1, pp. 417–429.

Grim, R. E. and Güven, N., 1978. Bentonites. New York, Elsevier, 256 pp.

Hallam, A. and Bradshaw, M. J., 1979. Bituminous shales and oolitic ironstones as indicators of transgressions and regressions. J. Geol. Soc. London, v. 136, pp. 157–164.

Heckel, P. H., 1983. Diagenetic model for carbonate rocks in midcontinent Pennsylvanian eustatic cyclothems. J. Sed. Petrology, v. 53, pp. 733–759.

Heckel, P. H., 1986. Sea-level curve for Pennsylvanian eustatic marine transgressive-regressive depositional cycles along midcontinent outcrop belt, North America. Geology, v. 14, pp. 330–334.

Herbert, T. D. and Fischer, A. G., 1986. Milankovitch climatic origin of mid-Cretaceous black shale rhythms in central Italy. Nature, v. 321, pp. 739–743.

Hereford, R., 1977. Deposition of the Tapeats Sandstone (Cambrian) in central Arizona. Geol. Soc. Amer. Bull., v. 88, pp. 199–211.

Hobday, D. K. and Tankard, A. J., 1978. Transgressive-barrier and shallow-shelf interpretation of the lower Paleozoic Peninsula Formation, South Africa. Geol. Soc. Amer. Bull., v. 89, pp. 1733–1744.

House, M. R., 1985. A new approach to an absolute timescale from measurements of orbital cycles and sedimentary microrhythms. Nature, v. 315, pp. 721–725.

Hubert, J. F., 1960. Petrology of the Fountain and Lyons Formations, Front Range, Colorado. Colo. School Mines Quart., v. 55, no. 1, 242 pp.

Jenkyns, H. C., 1980. Cretaceous anoxic events: From continents to oceans. J. Geol. Soc. London, v. 137, pp. 171–188.

Johnson, H. D., 1975. Tide- and wave-dominated inshore and shoreline sequences from the late Precambrian, Finnmark, north Norway. Sedimentology, v. 22, pp. 45–73.

Kauffman, E. G., 1970. Population systematics, radiometrics and zonation—A new biostratigraphy. In E. Yochelson (ed.), Proceedings of the North American Paleontological Convention, 1969. Lawrence, Kansas, Allen Press, pp. 612–666.

Kauffman, E. G., 1977. Geological and biological overview: Western Interior Cretaceous Basin. Mountain Geol., v. 14, pp. 75–99.

Kauffman, E. G., 1988. Concepts and methods of high-resolution event stratigraphy. Ann. Rev. Earth Planet. Sci., v. 16, pp. 605–654.

Klein, G. deV., 1984. Relative rates of tectonic uplift as determined from episodic turbidite deposition in marine basins. Geology, v. 12, pp. 48–50.

Kluth, C. F. and Coney, P. J., 1981. Plate tectonics of the Ancestral Rocky Mountains. Geology, v. 9, pp. 10–15.

Laferriere, A. P., Hattin, D. E., and Archer, A. W., 1987. Effects of climate, tectonics, and sea-level changes on rhythmic bedding patterns in the Niobrara Formation (Upper Cretaceous), U.S. Western Interior. Geology, v. 15, pp. 233–236.

Lajoie, J., 1970. Flysch Sedimentology in North America. Toronto, Geological Association of Canada, Spec. Paper No. 7, 272 pp.

Leggett, J. K., 1980. British Lower Paleozoic black shales and their paleo-oceanographic significance. J. Geol. Soc. London, v. 137, pp. 139–156.

Long, D. G. F. and Young, G. M., 1978. Dispersion of cross-stratification as a potential tool in the interpretation of Proterozoic arenites. J. Sed. Petrology, v. 48, pp. 857–862.

Mallory, W. W. (ed.), 1972. Geologic Atlas of the Rocky Mountain Region. Denver, Rocky Mountain Association of Geologists, 331 pp.

McKee, E. D., 1969. Stratified rocks of the Grand Canyon. In The Colorado River Region and John Wesley Powell. Washington, D.C., U.S. Geological Society, Prof. Paper 669-B, pp. 23–58.

Merriam, D. F. (ed.), 1964. Symposium on Cyclic Sedimentation, Vols. 1, 2. Lawrence, Kansas, Kansas Geological Survey, Bull. 169, 636 pp.

Miall, A. D. (ed.), 1981. Sedimentation and Tectonics in Alluvial Basins. Toronto, Geological Association of Canada, Spec. Paper No. 23, 272 pp.

Mount, J. F. and Ward, P., 1986. Origin of limestone/marl alternations in the Upper Maastrichtian of Zumaya, Spain. J. Sed. Petrology, v. 56, pp. 228–236.

Mutti, E., 1985. Turbidite systems and their relations to depositional sequences. *In* G. G. Zuffa (ed.), Provenance of Arenites. NATO Advanced Scientific Institute. Dordrecht, Holland, Reidel Publishing Company, pp. 65–93.

Payton, C. E. (ed.), 1977. Seismic Stratigraphy—Applications to Hydrocarbon Exploration. Tulsa, American Association of Petroleum Geologists, Mem. 26, 516 pp.

Pratt, L. M., Kauffman, E. G., and Zelt, F. B. (eds.), 1985. Fine-Grained Deposits and Biofacies of the Cretaceous Western Interior Seaway: Evidence of Cyclic Sedimentary Processes. Tulsa, Society of Economic Paleontologists and Mineralogists, Field Trip Guidebook No. 4, 249 pp.

Potter, P. E. and Pryor, W. A., 1961. Dispersal centers of Paleozoic and later clastics of the Upper Mississippi Valley and adjacent areas. Geol. Soc. Amer. Bull., v. 72, pp. 1195–1250.

Research on Cretaceous Cycles Group, 1986. Rhythmic bedding in Upper Cretaceous pelagic carbonate sequences: Varying sedimentary response to climatic forcing. Geology, v. 14, pp. 153–156.

Ricken, W., 1986. Diagenetic Bedding. New York, Springer-Verlag, 210 pp.

Ritchie, J. A., Gregg, D. R., and Ewart, A., 1969. Bentonites of Canterbury. New Zealand J. Geol. and Geophys., v. 12, pp. 583–608.

Ross, C. A. and Ross, J. R. P., 1985. Late Paleozoic depositional sequences are synchronous and worldwide. Geology, v. 13, pp. 194–197.

Schopf, T. J. M., 1983. Paleozoic black shales in relation to continental margin upwelling. *In* J. Thiede and E. Suess (eds.), Coastal Upwelling—Its Sediment Record, Part B: Sedimentary Records of Ancient Coastal Upwelling. New York, Plenum Publishing, pp. 579–596.

Slaughter, M. and Early, J. W., 1965. Mineralogy and Geological Significance of the Mowry Bentonites, Wyoming. Boulder, Geological Society of America, Spec. Paper No. 83, 116 pp.

Sloss, L. L., 1963. Sequences in the cratonic interior of North America. Geol. Soc. Amer. Bull., v. 74, pp. 93–114.

Stone, W. D., 1972. Stratigraphy and exploration of the Lower Cretaceous Muddy Formation, northern Powder River Basin, Wyoming and Montana. Mountain Geol., v. 9, pp. 355–378.

Tillman, R. W. and Siemers, C. T. (eds.), 1984. Siliciclastic Shelf Sediments. Tulsa, Society of Economic Paleontologists and Mineralogists, Spec. Pub. No. 34, 268 pp.

Van Houten, F. B., 1974. Northern Alpine molasse and similar Cenozoic sequences of southern Europe. *In* R. H. Dott, Jr., and R. H. Shaver (eds.), Modern and Ancient Geosynclinal Sedimentation. Tulsa, Society of Economic Paleontologists and Mineralogists, Spec. Pub. No. 19, pp. 260–273.

Van Houten, F. B., 1977. Triassic-Liassic deposits of Morocco and eastern North America: Comparison. Amer. Assoc. Petroleum Geol. Bull., v. 61, pp. 79–99.

Van Tassell, J., 1987. Upper Devonian Catskill delta margin cyclic sedimentation: Brallier, Scherr, and Foreknobs Formations of Virginia and West Virginia. Geol. Soc. Amer. Bull., v. 99, pp. 414–426.

von Rad, U., Hinz, K., Sarnthein, M., and Seibold, E., 1982. Geology of the Northwest African Continental Margin. New York, Springer-Verlag, 703 pp.

Walker, R. G., 1967. Turbidite sedimentary structures and their relationship to proximal and distal sedimentary environments. J. Sed. Petrology, v. 37, pp. 25–43.

Wanless, H. R., 1975. Carbonate tidal flats of the Grand Canyon Cambrian. *In* R. N. Ginsburg (ed.), Tidal Deposits. New York, Springer-Verlag, pp. 269–277.

Waples, D. W., 1983. Reappraisal of anoxia and organic richness, with emphasis on Cretaceous of North Atlantic. Amer. Assoc. Petroleum Geol. Bull., v. 67, pp. 963–978.

Watney, W. L., Kaesler, R. L., and Newell, K. D. (eds.), 1985. Recent Interpretations of Late Paleozoic Cyclothems. Tulsa, Society of Economic Paleontologists and Mineralogists, Mid-Continent Sec., Proceedings of the Third Annual Meeting, Guidebook, 273 pp.

Weissert, H., 1981. The environment of deposition of black shales in the Early Cretaceous: An ongoing controversy. *In* The Deep Sea Drilling Project: A Decade of Progress. Tulsa, Society of Economic Paleontologists and Mineralogists, Spec. Pub. No. 32, pp. 547–560.

Williams, G. E., 1986. The solar cycle in Precambrian time. Sci. Amer., v. 255, pp. 88–96.63.

Woodrow, D. L. and Sevon, W. D. (eds.), 1985. The Catskill Delta. Boulder, Colo., Geological Society of America, Spec. Paper 201, 246 pp.

The stratigraphic system analysis network consists of several concept levels of abstraction, interconnected through the associations of components in terms of their functional transformation.

UNESCO

11
Physical Framework for Stratigraphic Analysis

The Earth's surface is irregular with respect to any arbitrary plane such as sea level, the irregularities ranging in scale from pits between sand grains on a beach to "holes" the size of an ocean basin. Irregularities on the smaller end of the scale are generally transitory, disappearing with the next wave or pawprint; those of larger scale are more permanent. The origin of the smaller features is the province of sedimentologists, soil scientists, or geomorphologists. The larger features are studied by tectonocists or geophysicists, and both scales of features are of fundamental concern to stratigraphers. Included in this group of large-scale features are the sediment sources—mountain belts, high-standing plateaus, and horsts—and the sediment traps—ovoid intracontinental basins, elongate fault-bounded grabens, and various types of low areas on continental margins and within the ocean basins, such as aulacogens, trenches, volcanic arc-related oceanic areas, and abyssal plains. These large-scale features are the subject of this chapter. Why are they located where they are? How long-lasting are they? How does tectonics control stratigraphic accumulations?

Crustal Geotectonics

About 250 distinct sedimentary basins of various geologic ages have been recognized on the Earth, ranging in size from 10^4 km^2 to more than 10^7 km^2 (American Association of Petroleum Geologists, 1984), the largest ones being larger than the United

States. All of them sit in the crust, some directly on the lower basaltic part, but most on the sialic portion. Sediment thicknesses in individual basins range up to perhaps 25 km, and sediment volumes range to more than 10^6 km^3. How are these basins formed, and how do the sediment sources originate?

The bulk of the 250 sedimentary basins that have been recognized seem to have resulted from interactions among the various plates that have wandered over the face of the globe through geologic time. At present there exist perhaps 12 major plates, some of which are contained entirely within an ocean basin but most of which contain both oceanic and continental crust. The sizes, shapes, and boundaries of the plates are determined by factors not now known and perhaps unknowable. For example, why did the Eurasian-African plates separate from the North American–South American plates at the site of the Mid-Atlantic Ridge? Why did the separation not occur along some other line? Presumably, the line of separation marks a boundary between convection cells in the mantle, but why were (are) these cells located there? (Why do the bubbles in a pot of hot water on a stove form where they do?) Field studies in ancient terranes reveal plate sutures indicating earlier plate boundaries within present continental masses. So it is clear that the locations of ancient plate boundaries are not predictable from theory and can appear anywhere on the Earth's surface. No doubt, as criteria for recognition improve, additional ancient plate margins will be discovered. The meaning of this for stratigraphers is that ancient depocenters must be sought by "old-fashioned" field work. To recognize them requires an insightful understanding of tectonics (orogenesis, epeirogenesis) and the causes of thick sediment accumulations.

Ancient basins can occur anywhere. Of great importance for stratigraphers is the fact that the thickness, volume, and character of the basin fill can be used to identify its origin. Some basins contain thousands of meters of thinly interbedded dark sandstones and shales with intercalated bedded cherts. Others are filled mostly by carbonate rocks and evaporites. Still others are rich in volcaniclastic rocks and flows. What do such lithologic assemblages tell us about their origin?

The evolution of a sedimentary basin is determined by several interdependent factors of importance to stratigraphers (Ingersoll, 1988).

1. The size and shape of the basin are determined by the evolving configuration of the bounding basement rocks that form the floor and flanks of the basin. The pattern and timing of changes in the gross shape of the basin largely control regional tilts of strata and other structural features that strongly influence source rock exposures, topography, and stream drainage within the basin. In general, basin configurations directly reflect tectonic setting.
2. The nature of the stratigraphic fill in the basin is the product of the depositional systems that were active during basin evolution. These systems will vary through time and space as a result of the interaction between rates of subsidence, rates of sedimentation, and climate during basin evolution. At one extreme, initial structural subsidence to form a deep topographic depression is followed by fill-

ing that causes further isostatic subsidence under the sediment load. At the other extreme, sedimentation keeps pace with subsidence, and an empty hole never develops. Climate varies in response to paleolatitude (continental drift) and the orientation of mountainous areas in relation to dominant wind patterns.

3. Initial stratigraphic relationships may be modified by subsequent tectonic or diagenetic events that are not directly related to those responsible for creating the basin of deposition. For example, faulting may juxtapose originally non-adjacent sedimentary facies; the original extent and thickness of evaporitic units may be greatly lessened by subsurface dissolution; limestones may have their depositional thickness reduced 20–30% by pressure solution.

For convenience in discussion, as well as for fundamental geotectonic reasons, we can divide sedimentary basins into two main categories: (1) those clearly resulting from interactions at plate boundaries, such as trenches, forearc basins, intra-arc basins, retroarc basins, marginal rift basins, and aulacogens, and (2) those with no obvious relation to interactions at plate margins, such as abyssal plains and intra-continental basins. In actual fact these latter two types may be related to activity at plate margins as well, but the relationship is less evident. At the present state of our understanding of geotectonics, many different groupings of tectonic factors and basin characteristics are possible.

Most of the time when stratigraphers consider the complimentary topics of uplift and basin formation, the subject of plate interactions dominates the discussion. This is appropriate because it is orogeny that generates most uplifts and basins, and the proximate cause of orogeny is plate movement. As we observed in Chapter 10, the plate tectonics concept combined with Milankovitch cycles also explains the repeated transgressions and regressions of the sea that are seen in the rock record. However, the usual emphasis in tectonostratigraphic studies on interactions at plate margins does not mean that **epeirogenic movements** of the crust are unimportant. These movements, suggested in 1890 by G. K. Gilbert, are defined as those that are primarily flexural and vertical and that affect large areas of cratonic crust. Gilbert suggested the concept based on his studies of shoreline uplift surrounding Lake Bonneville in Utah, but other observations can also be made in support of pronounced epeirogeny. For example, both the central Appalachians and the Ouachita Mountains were uplifted during late Paleozoic time and are still elevated areas 300 million years later. If they had not been continually uplifted, they would have been flattened by erosion in about 10 m.y. Both areas were divergent margins during Mesozoic and Cenozoic time, so compression along plate margins cannot have been the cause of the apparent continual elevation. Friedman (1987, 1988) has suggested that the central Appalachians have been epeirogenically uplifted as much as 7 km, and mathematical modeling of sedimentary basins in undisturbed areas suggests erosion of more than 15 km of strata. Such drastic unroofings have very important implications for paleogeography and stratigraphy, most of which are probably not now appreciated.

Plate-Associated Basins

The crust of the Earth is fragmented into very large segments (**plates**) that move in various directions over the surface. These plates do not all move at the same rate in the same direction, so intense interactions are expected at plate margins, and it is these interactions (convergences and divergences) that form most large basins of deposition. The most prominent are formed along convergent plate margins (Fig. 11-1): trenches, forearc basins, intra-arc basins, and retroarc basins. Each of these can be characterized by the amount and type of sediment it contains (Dickinson and Seely, 1979; Dickinson and Suczek, 1979; Valloni and Maynard, 1981; Carey and Sigurdsson, 1984).

Trench Sediment

Sediment deposited in a trench at a convergent plate margin is chaotic in both texture and mineral composition and commonly is partly metamorphosed. Because of structural complexity, it is difficult if not impossible to determine the total thickness of the accumulation but estimates on the order of 10^3 m are common. In terms of stratigraphic coherence, trench sediment is a disaster area composed of a tectonically mixed assortment of very large fragments of older sedimentary and crystalline rocks set in a clayey and micaceous matrix (Fergusson, 1985) (Fig. 11-2). Some of the fragments may be kilometers in length, and in areas of inadequate outcrop these are easily mistaken for normal stratigraphic units. This intensely disturbed rock mass is called a **mélange**, a French word meaning "mixture."

The stratigraphy of a mélange cannot be established on presumptions of lateral

FIG. 11-1 • Generalized sketch of an arc-trench system along a convergent continental margin, showing spatial relationships and nomenclature of plate tectonics and related sedimentary basins. Marine sediment accumulations in basins are stippled. (Reprinted by permission of the publisher from Blatt, *Sedimentary Petrology*, W. H. Freeman & Co., 1982.)

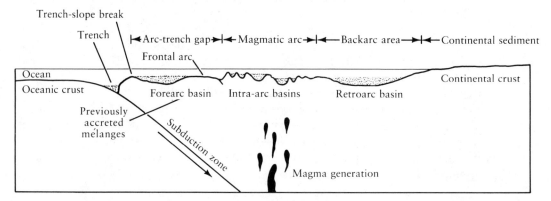

FIG. 11-2 • Partially rounded exotics in mélange, Franciscan in age. (Reprinted by permission of the publisher from Hsü, *Journal of Sedimentary Petrology*, v. 44, Society of Economic Paleontologists and Mineralogists, 1974.)

stratal continuity and superposition. Assignment of an age of deposition to a mélange on the basis of fossil occurrences alone can be incorrect. Different parts of a mélange may have different ages, and this can be made apparent by mapping of changes in lithology, mineral composition, or ages of the blocks in the mélange. One must adopt the approach used to interpret the stratigraphy of the clasts in a conglomerate or a breccia by attempting to (1) recognize lithologically distinct clasts, (2) date or assign an age to the clasts on the basis of fossils or other criteria, and (3) relate the clasts sequentially on the basis of probable ages and origins. But first and foremost, the rock mass must be identified as a mélange deposit, and because the contained blocks can be so much longer than the length of an outcrop, this is not always easy (Fig. 11-3). Hsü (1968) discusses in detail the problems involved in the recognition, characterization, and stratigraphic evaluation of mélanges.

Two different sets of processes could lead to stratal disruption and mixing of sediment and rock on a crustal slab moving downward into a trench at a convergent plate margin (Hsü, 1974; Underwood, 1984). In a mélange the fragmentation and mixing are tectonic. However, the disintegration of a rock-stratigraphic unit into large blocks and mixing of such blocks with muds could also result from sedimentary processes such as submarine sliding or a combination of erosion and sliding. This latter type of chaotic mixture is called an **olistostrome**. Mélanges are tectonic

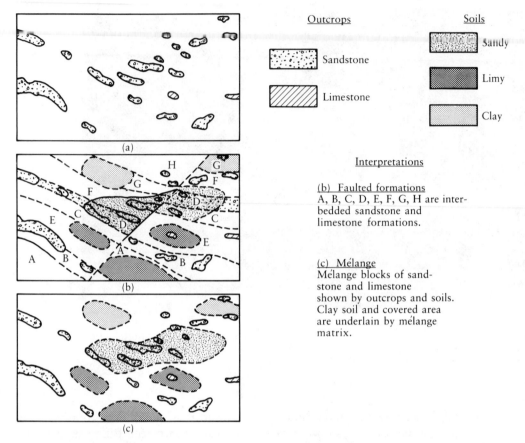

Outcrops

Sandstone

Limestone

Soils

Sandy

Limy

Clay

Interpretations

(b) Faulted formations
A, B, C, D, E, F, G, H are interbedded sandstone and
limestone formations.

(c) Mélange
Mélange blocks of sand-
stone and limestone
shown by outcrops and soils.
Clay soil and covered area
are underlain by mélange
matrix.

FIG. 11-3 • Two different interpretative geologic maps. (a) Outcrop map. (b) Interpretative map, based upon a presumption of stratal continuity. (c) Interpretative map, based upon a recognition of a mélange. (Reprinted by permission of the author from Hsü, *Geologic Society of America Bulletin*, v. 79, 1968.)

units bounded by shear surfaces. In contrast, olistostromes are stratigraphic units, generally separated from underlying and overlying formations by depositional contacts. Regional mapping is perhaps the best tool for distinction. In a mélange terrane, one can observe different degrees of severity of fragmentation and mixing, grading from internally intact allochthonous slabs to broken formations and to pervasively sheared and intimately mixed mélanges. In an olistostrome terrane, one finds gradations from boulder beds to graded turbidites. Mélanges are massive and they are developed on a regional scale in most instances. Olistostromes, as sedimentary units, are commonly more limited in size. Mélanges are deformed under an overburden, so the pelitic materials flow and other rocks fracture. Partially broken blocks are commonly bounded by shear fractures. Olistostrome blocks are commonly sedimentary boulders, already rounded prior to sedimentary transport. Although these various criteria are easy to write and read, they commonly are less easy to use in the field setting, where intense shearing after olistostrome emplacement may make the unit indistinguishable from a tectonic mélange.

Mélanges have been described in North America from Newfoundland, New York, California, and Alaska (Raymond, 1984) and are known from many other parts of the world as well, including all continents except Antarctica. The study by Hsü (1969) of the Franciscan mélange near San Francisco can serve as an example of the methodology and results of a stratigraphic investigation in a mélange terrane.

Field mapping since 1895 had determined the distribution of lithologies in the unit (Jurassic-Cretaceous) to be extremely irregular (Fig. 11-4) and to consist of interwoven outcrops of chert, graywacke, and greenstone. Hsü's field work was devoted to (1) observation of deformational styles, stratal orientations, and lithologic associations of the Franciscan rocks and (2) collection of sandstone samples for laboratory studies. Sandstone was sampled on a grid with centers spaced no more than 400 m apart, and at least two samples were taken at each location. Approximately 600 samples were examined, largely by determination of the orthoclase content of stained rock slabs.

On the basis of feldspar content in sandstones within the mélanges and the lithologies of mélange blocks, Hsü differentiated three mélange units within the Franciscan.

1. The Presidio Mélange is characterized by an abundance of blocks of serpentinite, serpentinized peridotite, and graywacke devoid of orthoclase. Small outcrops of chert and volcanic rocks also occur in the mélange.
2. The Mount Davidson Mélange consists largely of radiolarian chert, graywacke containing from zero to 10% orthoclase, greenstone, and pillow lava.
3. The Sweeney Ridge Mélange is characterized by blocks of thin-bedded, foraminiferal, cherty limestone accompanied by diabasic volcanic blocks.

Also present in the Franciscan unit is a 25-km allochthonous slab of feldspathic San Bruno Sandstone of probable Late Cretaceous age. In this slab the original stratal continuity of the interbedded graywacke and shale is largely preserved; the slab is interpreted to be perched on top of a Franciscan mélange and to have been emplaced *en bloc* by gravity-sliding deformation. It is the youngest unit of the Franciscan assemblage.

On the basis of the few fossils present in some of the block types and the regional geologic setting, Hsü concluded that the tectonic inclusions of the Franciscan complex were derived from two stratigraphic units: a pre-Tithonian unit of ophiolite, chert, and nonfeldspathic graywacke and a Tithonian to Late Cretaceous unit of pelagic limestone, submarine volcanic rocks, and graywacke with varied amounts of K-feldspar. The rocks of the lower unit were tectonized during a Late Jurassic deformation. The lower strata of the upper unit were fragmented and mixed with the lower unit mélange during a Late Cretaceous deformation, the San Bruno slab riding piggyback on underlying mélanges during the deformation. The deformation re-

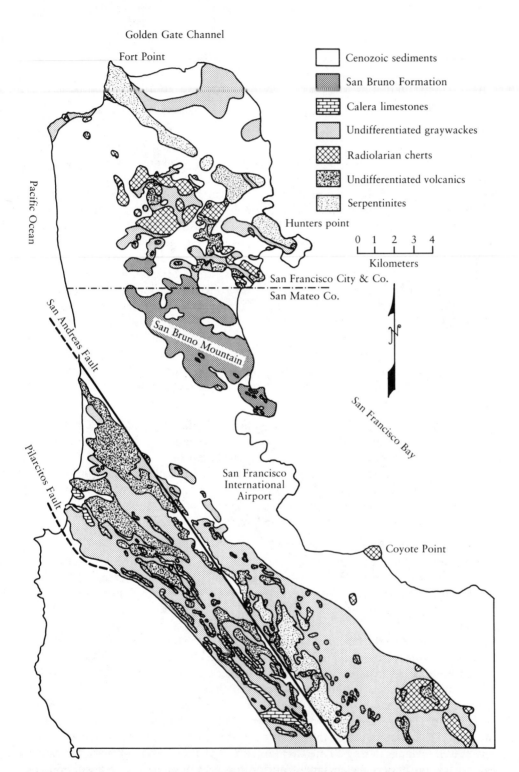

FIG. 11-4 ● Outcrop map of type Franciscan, San Francisco Peninsula, California. (Reprinted by permission of the American Association of Petroleum Geologists from *American Association of Petroleum Geologists Bulletin*, v. 53, 1969.)

sulted in the mélanges being subhorizontal tabular sheets. Subsequent Pleistocene folding caused synforms in mélanges northeast of the San Andreas Fault.

Forearc Basins

The **forearc region** along a convergent plate margin lies landward of the trench and seaward of the volcanoes and is an area of great structural complexity. Determination of the thickness and nature of the crust underlying the forearc is difficult in modern arc-trench systems because the crust is typically buried beneath the thick sediments of the forearc basins, but both oceanic and lithospheric crust are probably present, either as continuous strips or in isolated pockets. Modern forearc terranes ring the Pacific Ocean Basin.

Forearc basins can become major repositories for sediment accumulation. The arc massif provides a nearby source of sediments; the subduction complex serves as a dam to pond sediment in the forearc basin. As subduction proceeds in the adjacent trench, the morphology of forearc regions may adopt varied configurations and can be descriptively classified as shelved, sloped, terraced, and ridged. Distinctions among the various types of forearc basins depend on the structural evolution of the subduction complex and the history of accompanying sedimentation. An idealized stratigraphic sequence in a forearc basin might consist of deep-water, arc-derived smectitic shales, ash falls, and lesser amounts of fine-sand turbidites lying on abyssal plain sediments at the base of the sequence and grading upward to more abundant and coarser turbidite, shelf, or deltaic sands derived, in part, from the uplifted roots of the arc. On the inner edge of the basin these would interfinger with lava flows, lahars, agglomerates, and fans. For most forearc basins the general trend in paleobathymetry is toward shallower water depths for higher stratigraphic zones, a large-scale fining-upward sequence.

Variations on this idealized stratigraphy will be present where significant nonvolcanic source areas occur, as along the Peruvian coast and in Cook Inlet, Alaska. Where river systems empty into forearc basins, the basins fill quickly and are composed of large volumes of shallow-marine to nonmarine facies deposited during the subsidence phase(s) of basin evolution. In such cases, volcanic contributions to the basin can be volumetrically minor. In equatorial regions, thick sections of carbonate reefal sediments may occur, as in some modern forearc regions of southeast Asia.

It is important for a stratigrapher to recognize that in arc-trench systems the processes of sedimentation, metamorphism, plutonism, and diastrophism are inseparably linked as related manifestations of the same geodynamic system. Radiometric data on the ages of igneous rocks in the arc massif and mineralogic evidence on the conditions of metamorphism within the subduction complex may be as important as paleontologic data on the ages of sedimentary beds in the forearc basin for understanding the overall evolution of the forearc region.

Descriptions of forearc basins and their stratigraphic development have been given by Dickinson and Seely (1979), Ward and Stanley (1982), and Dalziel (1984). The study of the Haslam Formation by Ward and Stanley can serve as an example of forearc basin stratigraphy and structural development.

TABLE 11-1 • Stratigraphy of the Nanaimo Group

Age	Formation
Maestrichtian	Spray shale, turbidites 320–580 m
Upper Campanian	Geoffrey conglomerate 130–480 m
	Northumberland turbidites 160–260 m
	De Courcy sandstone, conglomerate 320–480 m
Middle Campanian	Cedar District shale, turbidites 230–650 m
	Protection sandstone 100–320 m
Lower Campanian	Pender siltstone, shale 100–230 m
	Extension conglomerate 30–480 m
	Haslam shale 60–480 m
Santonian	Comox sandstone, conglomerate 50–650 m

(Reprinted by permission of the publisher from Ward and Stanley, *Journal of Sedimentary Petrology*, v. 52, Society of Economic Paleontologists and Mineralogists, 1982.)

The Haslam Formation of Late Cretaceous age (Late Santonian-Early Campanian) is 70–500 m in thickness and is exposed largely on Vancouver Island, British Columbia, Canada. It is one of the lower formational units in the Nanaimo Group (Table 11-1), a sequence of clastic rocks deposited in a forearc basin that formed to the west of a Late Cretaceous island arc (Fig. 11-5). The Haslam Formation is the oldest unit that clearly contains debris derived from the continental mainland; the underlying Comox Formation is composed entirely of sediment derived from the insular belt in which the forearc basin was formed. The sediment that formed the Haslam resulted from the major mid-Cretaceous thrusting (see Fig. 11-5) that raised rocks from older subduction-arc complexes so they could be re-eroded.

In its stratotype area near Nanaimo on southeastern Vancouver Island the Haslam consists of massive, poorly bedded, dark gray siltstones and shales containing many concretions. The formation is easily differentiated from the underlying Comox, which is composed of a basal conglomerate overlain by coarse-grained sandstones, and from the overlying Extension Formation, which is composed of pebble and cobble conglomerates and interbedded sands. Contacts between the Extension, Haslam, and Comox formations are sharp and clearly seen.

South of the stratotype area, lateral equivalents of the three formations are litho-

FIG. 11-5 ● Cordilleran arc-trench system in Late Cretaceous time. Solid triangles denote the edge of the overriding block above the subduction zone. Stipples denote undeformed sediments. BI = Baranof Island; GVS = Great Valley sequence; NAN = Nanaimo Group; SUS = Sustut assemblage. The exotic Insular Belt extends from the U.S./Canadian border northward to Baranof Island. (Reprinted by permission of the publisher from Dickinson, *Canadian Journal of Earth Science*, v. 13, National Research Council of Canada, 1976.)

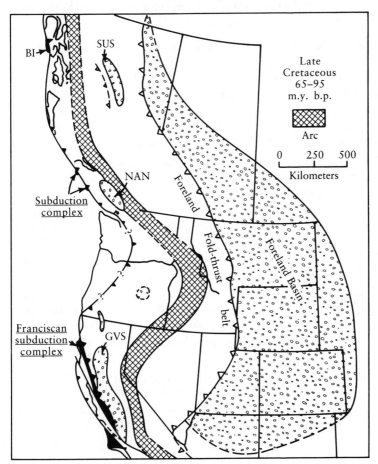

logically much more heterogeneous. Haslam Formation equivalents are composed of a variety of submarine-fan facies, including channelized conglomerates and sandstones and a wide variety of turbidites. Two members were defined on the basis of these southern area lithologies: the lower unit, named the Haslam Creek Member, is composed of massive siltstone and shale; the upper unit, named the Cowichan Member, is composed of interbedded sandstones and shales or siltstones (turbidites) with subordinate, thinly bedded sandstones and conglomerates. Most exposures of the Comox, Haslam, and Extension formations are on island shorelines and river banks, so facies relationships among the three units and within the Haslam can only be inferred by comparison of rock types within megafossil ranges (*Inoceramus* species) in each section and from vertical sequences of stratification. The fossil evidence indicates that the base of the Haslam is time-transgressive from west to east, but the Comox Formation, the basal unit of the Nanaimo Group, is unfossiliferous. As a result, the temporal significance of its base is unknown.

The Cowichan Member of the Haslam Formation on southern Vancouver Island, the San Juan Islands, and the Gulf Islands consists of gray shale and sandstone with sparse conglomerate that can be grouped into five facies.

1. **Conglomerate and sandstone facies.** This facies consists of coarse-grained sandstone, with or without scattered pebbles and cobbles, and conglomerate beds 1 m or more thick with very irregular basal contacts. The beds in this facies occur as channel fills cut into sandstone or shale beds and are usually internally massive. If nonmarine, these sandstone and conglomerate beds would be classed as fluvial channels; in these marine rocks they are interpreted as inner fan (coarse grain size) channel deposits in the turbidite sequence.

2. **Interbedded discontinuous sandstone and shale facies.** This facies occurs adjacent to or above the previous facies and consists of interbedded, thin (1–10 cm thick) lens-shaped or irregular and discontinuous sandstone and shale beds with a sandstone/shale ratio greater than 5/1. The sandstones are not graded but are cross-bedded and exhibit sharp contacts with underlying shales. This facies is interpreted as a submarine-fan channel margin facies.

3. **Interbedded sandstone and shale facies.** This facies consists of interbedded medium- to fine-grained sandstone and shale with sandstone beds 10–100 cm thick and a sandstone/shale ratio greater than 1/1. Sandstone beds have sharp bases with sole marks, are normally graded, and begin with Bouma division A or B. This facies commonly is intercalated with facies 1 and underlain by facies 2 (see Fig. 11-6) and is interpreted as an interchannel turbidite, that is, an unchanneled proximal turbidite deposit.

4. **Interbedded shale and sandstone facies.** This facies is similar to facies 3 except that the sandstone/shale ratio is less than 1/1 and the sandstones are finer-grained, are less than 10 cm thick, and lack division A of the Bouma sequence. This facies is interpreted to be simply a finer-grained version of facies 3, an interchannel turbidite deposit.

FIG. 11-6 • Interbedded sandstone and shale facies of the Cowichan Member. (Reprinted by permission of the publisher from Ward and Stanley, *Journal of Sedimentary Petrology*, v. 52, Society of Economic Paleontologists and Mineralogists, 1982.)

5. **Shale facies.** This facies consists of massive and horizontally laminated shale and mudstone, commonly visibly silty. This is the only facies that contains abundant fossils, and it is the dominant lithology of the Haslam Creek Member.

The five facies occur in the Haslam Formation in regional and vertical associations that can be easily related to classic submarine-fan facies sequences described elsewhere. Regionally, shale-rich facies dominate outcrops to the west on southern Vancouver Island; conglomerate and sandstone facies dominate exposures to the east in the San Juan Islands. In addition, the Haslam stratigraphic sections on each island coarsen upward with shale-rich facies more common in the lower part of the formation. The increase of the sandstone/shale ratio, the thickness of sandstone beds, and sandstone grain size from east to west and from bottom to top is consistent with (1) upper- and mid-fan deposition to the east, (2) lower fan-basin deposition to the west, and (3) progradation of the fan complex from east to west during Haslam time.

Paleocurrent data from Haslam Formation rocks are consistent with the facies data in reflecting east-to-west sediment transport. Measurements of directional structures such as flute clasts, ripple lamination, and small-scale cross-bedding indicate grain movement to the west and northwest, with mean azimuths in various outcrop areas averaging 279° and with standard deviations at each outcrop of 30–50°. Such small standard deviations are characteristic of nonmeandering fluid flow.

As was noted previously, the Comox Formation, which underlies the Haslam, consists of sediment derived entirely from within the Insular Belt. The sandstones consist of about 40% volcanic rock fragments (almost all basaltic), almost 20% feldspar, and about 35% quartz, mostly monocrystalline. In addition, heavy minerals average 7.5%, a value so extraordinarily high (most sandstones contain less than 1%) that derivation from great distances is precluded. The sediment clearly is indigenous to the insular volcanic belt. Paleocurrent data from the Comox indicate very variable directions of provenance, consistent with an "internal" derivation from scattered, mostly volcanic outcrops.

In contrast to the underlying Comox Formation the Haslam contains 25–30% chert, only 11% volcanic fragments, mostly silicic rather than basaltic, and nearly 10% of sedimentary and very low-grade metasedimentary rock fragments (argillite, shale, slate, phyllite). Clearly, an appreciable proportion of the sediment was not derived from an oceanic basinal source but has a continental origin.

Intra-Arc and Backarc Basins

On the basis of the system of basinal nomenclature that uses the location of the oceanic trench as the starting point, **backarc basins** lie landward of (in back of) the volcanic arc (Fig. 11-1). In contrast to basinal types located farther seaward the crust underlying these basins is of continental type (granitoid) rather than of mixed (transitional) or oceanic (basaltic) type. But the relationship between backarc basins and the arc-trench system is revealed by the general parallelism maintained by the trench or subduction complex, volcanic chain or batholith belt, and the backarc basins. Landward of these basins lie basins more closely related to cratonic sequences than to arc-trench sequences. On a regional scale it appears that perhaps half a continental block can be affected directly by plate interactions along a convergent plate margin system.

The most characteristic strata of backarc basins are deep-water clastics containing significant proportions of volcanogenic debris from the orogenic flank, but sediment contributions from the craton become more prominent as distance from the orogen increases. The relative importance of orogenic and cratonic sources can be indicated by features such as depth of deposition (deeper toward the arc), composition of the sediment (less quartzose toward the arc), or sedimentary structures (more sole markings toward the arc). Basins whose sedimentary accumulation reflects largely continental origins are commonly termed **foreland basins** rather than backarc basins; that is, the reference point for their name is their location relative to the craton rather than relative to the convergent margin orogen.

The Jurassic and Cretaceous crystalline rocks and sedimentary units on the isolated island of South Georgia (Fig. 11-7) are an example of intra-arc and backarc basement and basinal fill. Four major units record the development of the arc-basin system. These are (1) extended continental basement rocks of the basin floor (Drygalski Fjord Complex), (2) mafic rocks of the basin floor (Larsen Harbour Formation), (3) sedimentary fill of the basin (Ducloz Head, Cooper Bay, Sandebugten, and Cumberland Bay Formations), and (4) calc-alkaline volcanic, sedimentary (Annenkov Island Formation), and intrusive rocks of the magmatic arc. The island is split by the Cooper Bay dislocation zone, a ductile shear zone 1.5 km wide that cuts across the southwestern corner of the island. Rocks of the basin floor and island arc lie to the southwest of this dislocation, and all the basin fill rocks lie to the northeast.

The Drygalski Fjord Complex consists of multiply deformed siliceous metasedimentary rocks cut by a wide variety of igneous rocks including calc-alkaline granitic plutons of a subduction-related magmatic arc. A large volume of tholeiitic basalt similar to modern midocean ridge basalts was intruded into this sialic crust and

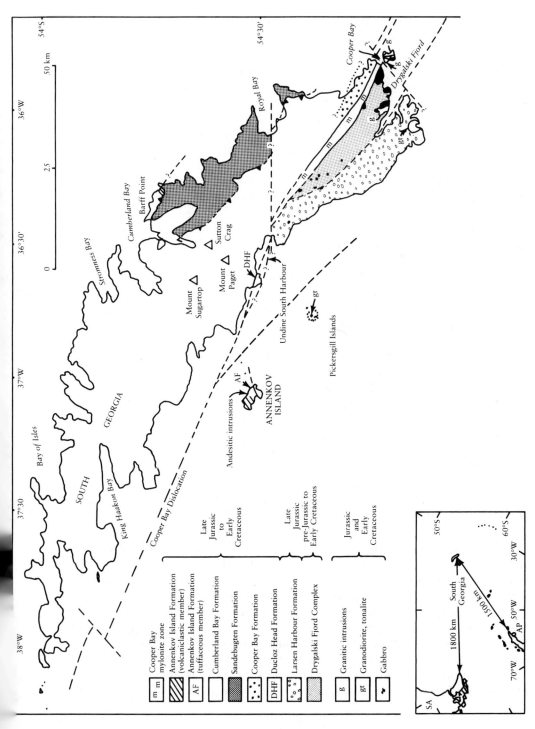

FIG. 11-7 ● Geologic map of South Georgia. The inset shows the location of the island in relation to South America (SA) and the Antarctic Peninsula (AP). The mylonite trend marks the site of the Cooper Bay dislocation zone. (From P. W. G. Tanner, "Geologic Evolution in South Georgia," *in* C. Craddock, editor, *Antarctic Geoscience* (Madison: The University of Wisconsin Press; © 1982 by The Board of Regents of the University of Wisconsin System), p. 168.)

represents the early stage of the formation of the backarc basin, in the middle- to late Jurassic. Sheeted dikes like those associated with modern oceanic ridges are common.

The Larsen Harbour Formation is everywhere bounded by faults and consists of at least 2 km of probable ophiolitic rocks consisting of stratiform metabasalts, pillowed lavas, glassy breccias, mafic tuffs, and basaltic dike complexes.

Because of metamorphism, intense tectonic activity, and limited exposures, the stratigraphic relationships among the various basin fill units are uncertain. Fossils indicate that all the units are of Late Jurassic to Early Cretaceous age, but whether they have different ages or are simply lateral facies equivalents of the same age cannot be reliably determined. The Ducloz Head Formation consists of two unfossiliferous detrital units. One is fault-bounded and equivalent to the (volcanogenic) Annenkov Island Formation of the volcanic-arc terrane (described below). The other member consists of massive lithic (metamorphic rock fragments) sandstones and sandstone breccias, interbedded with thin-bedded fine sandstones, siltstones, and black shales and massive, felsitic volcaniclastic breccias. Beds are structureless or crudely graded, up to 10 m thick, and are poorly sorted. The metamorphic rock fragments in the sandstones indicate derivation from a nearby area of continental basement. The association of the sediments with basic pillow lavas is interpreted to mean deposition in an active submarine rift.

The Cooper Bay Formation and Sandebugten Formation are turbidite flysch shales and sandstones containing about 40% quartz, 40% lithic fragments of various types of silicic igneous and metamorphic rocks, and 20% feldspar. Paleocurrent indicators point to a northerly or northeasterly provenance for the detritus.

The Cumberland Bay Formation is the most widely exposed and studied unit (Tanner and Macdonald, 1982; Macdonald and Tanner, 1983), although precise stratigraphic and sedimentologic analysis is hampered by the presence of high-grade diagenesis/low-grade metamorphism in some areas. In these areas, shales have been changed to slates, and andesitic detritus has altered to prehnite, indicating temperatures of 150°–250°C, high partial pressure of water, and/or nonhydrostatic stresses in various parts of the outcrop belt during deformation. Neither the base nor the top of the formation is visible because of deformation, but the thickness of the unit exceeds 8 km.

Three facies associations have been recognized in the Cumberland Bay unit based on the shale/sandstone ratio (Macdonald and Tanner, 1983). The shale facies (shale/sandstone > 2) is typically formed of structureless shale with parallel-laminated siltstone, occasional scours filled with cross-laminated fine sandstone, and fine-grained sandstones with starved ripples. Sandstone beds are generally a few centimeters thick and are graded, but sole structures are rare. Most beds are extremely persistent laterally despite their thinness. Rocks of this facies have the features of lower fan and basin plain deposits of a submarine fan.

The transitional facies (shale and sandstone subequal) has a much greater maximum bed thickness (30 cm). Sandstone turbidite beds are graded from medium to

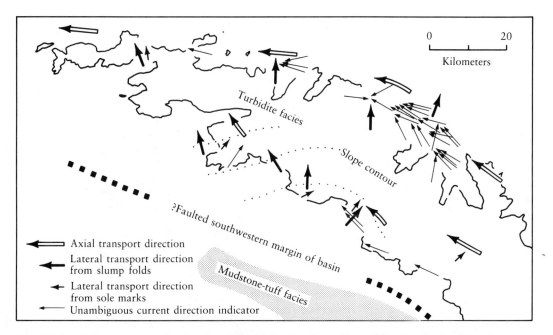

FIG. 11-8 • Paleocurrent and paleoslope model for the deposition of the Cumberland Bay Formation. (Reprinted by permission of the publisher from Macdonald and Tanner, *Journal of Sedimentary Petrology*, v. 53, Society of Economic Paleontologists and Mineralogists, 1983.)

fine sandstone, contain all Bouma structures (divisions A–E), are usually 10–20 cm thick, and frequently have sole markings. Intercalated shales have similar thicknesses. Lateral continuity is again extremely good. The rocks of this facies are typical of lobe-fringe deposits of a submarine fan.

The sandstone facies predominates along the southwest coast of South Georgia. The contour separating areas with more than 50% sandstone facies in the local succession from those with less than 50% defines a series of lobes projecting from the southwest toward the northeast. Individual beds are typically 20–80 cm thick, are usually graded from very coarse sand and granules at the base to fine sand toward the top, contain some but not all of the Bouma subdivisions, and have a wide variety of sole structures. Amalgamation with underlying beds is common, particularly in the southwest. Less than 1% of all beds show any significant lateral thickness change, even over distances of 500–1000 m. The sandstone facies resembles classical proximal turbidites.

Compositionally, the Cumberland Bay Formation is a sequence of andesitic volcaniclastic sandstones containing about 75% volcanic detritus interbedded with shales of undetermined composition and a few randomly occurring tuff beds 5–20 cm thick and composed of prehnite after glass. Paleocurrent evidence (Fig. 11-8)

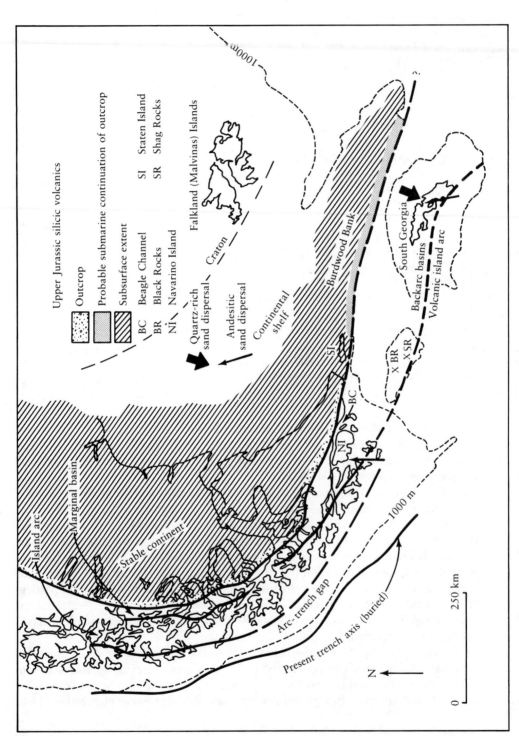

FIG. 11-9 ● Diagram showing the tectonic setting of South Georgia Island in Early Cretaceous time. The map is based on present geography and does not consider tectonic shortening across strike or possible oroclinal bending. Arrows indicate the general dispersal of sandstone types. (The Tierra del Feugo arrow represents 446 cross-lamina readings from 13 localities.) (Reprinted by permission of the author from Dalziel et al., *Geological Society of America Bulletin*, v. 86, 1975.)

shows that although most transport was to the northwest along the basin axis, the components were derived from many points along the volcanic arc to the southwest.

The Annenkov Island Formation is a classic intra-arc stratigraphic sequence, consisting of 2–3 km of thinly interbedded andesitic crystal-lithic turbidite tuffs and mudstones overlain by at least 1 km of poorly bedded volcaniclastic breccia with subordinate sandstone (Storey and Macdonald, 1984). The tuff beds are generally 1–5 cm thick, almost all have graded bedding, and many contain mudstone rip-up clasts. The sandstone beds that overlie the tuffs range up to 2 m in thickness and consist of very coarse sand to granule-sized fragments. The soles of the beds have sharp, undulating bases and are either amalgamated or separated by thin beds of structureless mudstone. All sandstone beds contain mudstone clasts up to 30 cm in diameter.

The sandstones are overlain by andesitic volcanic breccias in beds 1–5 m thick. Bed bases are highly irregular and locally cut out up to 1.5 m of the underlying beds. Clasts vary from a few centimeters to 1.5 m in size and are set in a matrix of andesitic sand and granules. In some areas the breccias contain large fragments of laminated mudstone and tuff.

Figure 11-9 illustrates an integrated paleogeographic map based on tectofacies on South Georgia Island and on the tectonically related southern extremity of South America. The volcanic arc occurs today as the spine of South America along its western edge, and its Cretaceous location in the South Atlantic Ocean is given by the Annenkov Formation, an intra-arc stratigraphic unit that consists almost entirely of volcanogenic detritus deposited by turbidity currents in deep water. The five back-arc basinal fill formations are located to the northeast of the arc, and these units grade farther northeastward onto the continental shelf of Gondwana. To the south of South Georgia the volcanic arc extends through the Antarctic Peninsula.

Marginal Rift Basins

On a global scale the major Phanerozoic rift systems have a strongly developed north-south orientation, examples being the mid-Atlantic rift, the East African–Middle Eastern rift system, the North African–Rhinegraben system (Fig. 11-10), the Reelfoot rift of Paleozoic age in which the modern Mississippi River system flows, and the very extensive system of rifts of Triassic age along the eastern border of North America (Manspeizer, 1988). The same tendency is present on a smaller scale as well, exemplified by the Rio Grande Rift in New Mexico. Exceptions to the north-south orientation are commonly tears normal to a continental margin (aulacogens). The explanation of the dominant orientation pattern is east-west tensional stress in the crust due to events such as continental splitting or upwelling of magma.

Probably the best understood rift basin development is the type associated with a newly rifted continental plate, such as those developed as North and South

FIG. 11-10 ● Aerial view of the Rhinegraben of Tertiary age, West Germany, between approximately 48°N and 50°N latitude and at about 8°E longitude. (Reprinted by permission of the publisher from Illies and Mueller, *Graben Problems*, E. Schweizerbart's-che Publishers, 1970.)

America split from Europe and Africa (Fig. 11-11). During Triassic time, tensional stresses in the crust caused intensive fracturing in a zone centering on the present-day mid-Atlantic Ridge and extending for several hundred kilometers on either side of it. Those rifts most distant from the major tear, the mid-Atlantic Ridge, were completely isolated from the sea and accumulated a thick sequence of nonmarine arkosic clastics and lacustrine deposits in developing grabens and horsts. Intercalated with these sediments are mafic volcanic flows extruded from lower crustal sources.

Closer to the major tear, in the zone where the present Atlantic Ocean was to enter, tearing was irregularly distributed in time and space. This resulted in the development of a sublinear belt of unconnected or poorly connected grabens with erratic access to seawater, and in suitably dry climates, thick sequences of evaporites and salt diapirs developed. Because of different geologic ages, rates of graben sinking, and access to the sea, a wide variety of types of stratigraphic sequences is found in circum-Atlantic grabens. Major deltas commonly prograde down the length of failed rift arms because they are open to the ocean at only one end; examples include the Amazon, Mississippi, and Niger (Benue Trough). Grabens parallel to the continental margins may develop carbonate sequences many kilometers thick, as occurred along much of the length of the Atlantic coastline of the United States from New Jersey to Florida, where subsidence of the graben floors was slow enough that reef growth could keep pace with basin sinking. In areas of rapid sinking of graben floors, extensive deep-water turbidites formed.

Also found in the stratigraphic sequences of some rift basins are metalliferous black shale deposits, reflecting a development of noncirculating bottom waters and resultant anoxia in the lower parts of some graben waters. The economically valuable metals commonly occur as sulfides deposited by hydrothermal solutions circulating through the rifts, black muds, and possibly the evaporites as well. Stratabound metal sulfide ore deposits associated with evaporites and black shales are well known from many areas of the world, although not all are clearly related to rifting tectonics (Morganti, 1981).

Baltimore Canyon Trough

As an example of one of the large number of graben troughs formed during the Triassic and Early Jurassic Periods in eastern North America (Fig. 11-12), we can examine the Baltimore Canyon Trough. This structural basin is better understood than most of its contemporaries on the continental shelf because of its presumed potential as a petroleum reservoir, which has resulted in extensive seismic exploration and drilling within it. No oil or gas in commercial quantities has yet been found, although a few "shows" have been logged.

The trough extends for about 500 km beneath the Outer Continental Shelf between Virginia and New Jersey. It varies in width from 50 km off Virginia to 150 km off New Jersey (Fig. 11-13), and in depth from 10 km off Virginia to more than 18 km off New Jersey, with an abrupt increase in width and depth north of Delaware Bay.

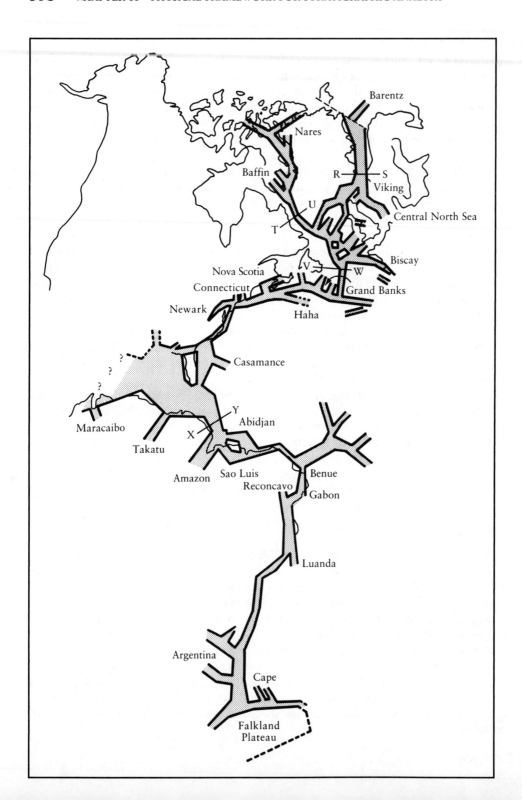

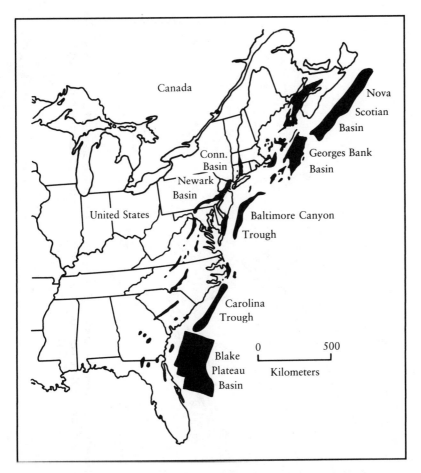

FIG. 11-12 • Locations of major rift grabens produced in early Mesozoic time along the eastern coast of the United States and Canada.

◀ **FIG. 11-11** • Sketch map showing the distribution of major grabens around the Atlantic Ocean that formed in association with Mesozoic continental rupture. Grabens between lines V-W and X-Y developed 210–170 m.y. ago; those south of X-Y, 145–125 m.y. ago; north of T-U, 80 m.y. ago; north of R-S, 60 m.y. ago. Within the area of the North Atlantic bounded by V-W, T-U, R-S, and central Europe there were seven episodes of graben formation between the Permian and Late Tertiary, although only a Late Cretaceous–Early Paleocene event gave rise to ocean floor. Troughs ("failed rifts") normal to continental margins are termed aulacogens, e.g., Benue, Amazon. (Reprinted by permission of the publisher from Burke, *Tectonophysics*, v. 36, Elsevier Science Publishers B. V., 1976.)

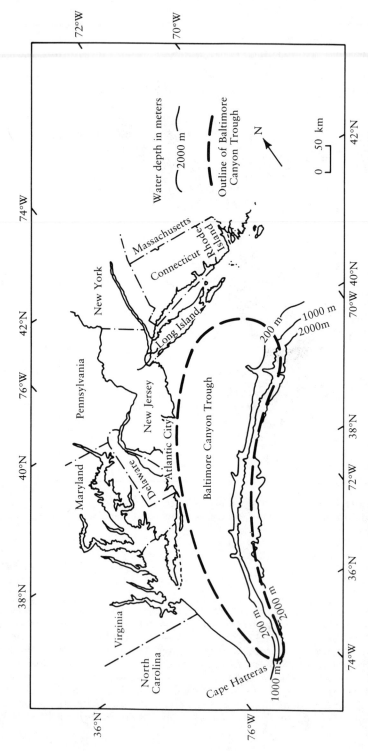

FIG. 11-13 • Extent of the Baltimore Canyon Trough on the continental shelf of the northeastern United States. The trough has been traversed by numerous seismic lines, and several tens of holes have been drilled in search of oil and gas accumulations. (Adapted from Scholle, 1977.)

Pre–Upper Jurassic stratigraphy of the Baltimore Canyon Trough (Fig. 11-14) is inferred mostly from seismic data, because few wells have penetrated these older rocks. The trough is floored by Triassic(?) nonmarine lacustrine and swamp deposits, which may represent either sediments deposited during the early rift stage or prerifting continental sedimentation, and these are overlain by Lower Jurassic salt deposits of uncertain thickness. One deep well has penetrated this salt. Presumably, the evaporite formed during the early stage of continental breakup at a point when the rate of evaporation still exceeded the rate of inflow of oceanic waters. As was noted previously, evaporites are common near the base of Atlantic coast Mesozoic grabens. In the Baltimore Canyon sequence, as in most of the Atlantic margin grabens, diapirs of salt protrude upward hundreds to thousands of meters into overlying sediments.

FIG. 11-14 • Schematic cross-section normal to axis of Baltimore Canyon Trough, starting 100 km east of Atlantic City, New Jersey, based on both geophysical measurements and drilling. (Reprinted by permission of the American Association of Petroleum Geologists from Libby-French, *American Association of Petroleum Geologists Bulletin*, v. 68, 1984.)

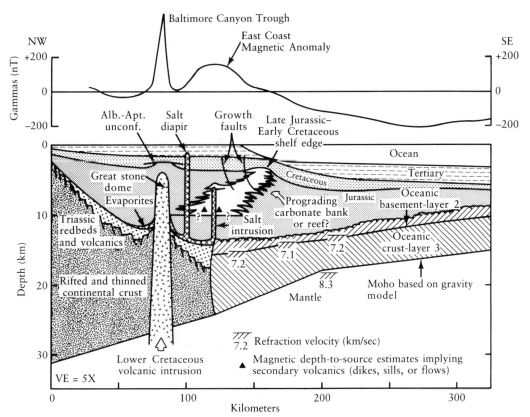

With the exception of the single penetration of diapiric Lower Jurassic(?) salt, the oldest strata penetrated so far in the Baltimore Canyon Trough are of Late Jurassic age. Most of the Upper Jurassic section east of the paleoshelf consists of gray to black micaceous shale and siltstone with occasional thin sandstone interbeds. Both laminated and bioturbated beds occur. The SP log pattern is a predominant shale line with short sandstone and siltstone deflections (Fig. 11-15), which is characteristic of prodelta shales.

Also within the Upper Jurassic stratigraphic section but located only at the shelf

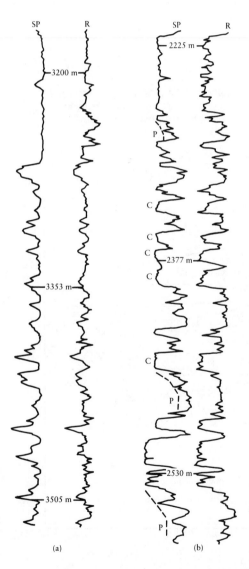

FIG. 11-15 • Spontaneous potential (SP) and resistivity (R) curves of (A) prodelta sediments of Late Jurassic age, mostly shales, and (B) delta front deposits of Early Cretaceous age, rich in sandstones. Channel (C) and progradational units (P) are indicated. (Reprinted by permission of the American Association of Petroleum Geologists from Libby-French, *American Association of Petroleum Geologists Bulletin,* v. 68, 1984.)

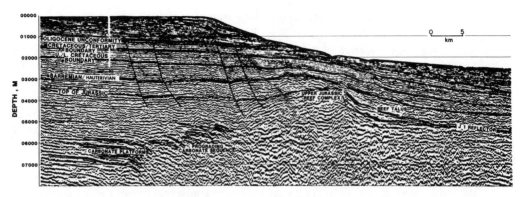

FIG. 11-16 • Seismic cross-section 75 km in length approximately normal to the axis of the Baltimore Canyon Trough, from 125 km east of Atlantic City, New Jersey onto the continental slope. Stratigraphic and sedimentologic interpretations based on intensities and orientations of seismic reflections and on drill data. This depth section proves useful in understanding the evolution of the Middle to Late Jurassic depositional environment as it changed from stacked carbonate platforms—indicating a balance between sedimentation and subsidence—to prograding carbonate ramps, where sedimentation exceeded subsidence, and finally to development of a major barrier reef. (Reprinted by permission of the American Association of Petroleum Geologists from Gamboa et al., *American Association of Petroleum Geologists Bulletin*, v. 69, 1985.)

edge, as defined by the changing slope of strong seismic reflectors and the lithology of drill cuttings and cores, is a carbonate buildup. As shown in Figure 11-16, the seismic character of the buildup shows a horizontal platform at 6–7 km depth to the west, changing eastward and upward (at 5–6 km depth) into a clinoform pattern of strong reflectors dipping at low angles to the east, and this pattern changes further eastward and upward (at 4 km depth) into a reef complex complete with talus apron on the side facing the open ocean. The reef shows no progradation during its existence, indicating that the location of the shelf-slope break must have stabilized. The death of the reef is indicated by a pronounced unconformity of earliest Cretaceous age (microfossils in cuttings) that truncates it. Landward of the limestone buildup, gray shale, coal, and thin limestone beds interfinger with the carbonate. Judging by the dimensions of the reef, it must have acted as a major barrier to the transport of terrigenous sediments to the deep sea in Late Jurassic time.

Lower Cretaceous sediments in the Baltimore Canyon Trough (Fig. 11-14) are dominated by thick-bedded sandstones that, according to cuttings, are dominantly fine- to medium-grained and poorly sorted; minor amounts of coal also occur. In the lower part of the Early Cretaceous sequence, dark gray shales are prominent. The sequences of alternating thick-bedded sandstone, siltstone, and shale characterize delta-front deposits, and the abundance of sandstone is interpreted to indicate major deltaic progradation. The SP curve of several of the sandstones is funnel-shaped and serrated, indicating the coarsening- and thickening-upward pattern typical of delta-front deposits. The blocky SP patterns of the sandstones may represent channel sands. Areally, the sandstone channels are most common in northern well sites, providing evidence of a source of clastic debris from that direction in Early Cretaceous time. Examination of cores from deep wells reveals the sand units to be bioturbated and to contain ripple laminations and slump and load-cast structures with bioturbation decreasing upward in favor of sedimentary structures reflecting higher kinetic energies. On the basis of log character and lithologic data, the Lower Cretaceous rocks are interpreted as lower delta plain and channel to predominantly delta-front environment. Water depths probably were less than 100 m during deposition of the sequence.

The remainder of the Lower Cretaceous and the Upper Cretaceous sequence in the Baltimore Canyon Trough reflects a series of transgressions of the sea and deltaic progradations, as evidenced by lithologic variations, sedimentary structures, and the shapes of log curves. These are described in detail, accompanied by photographs of cores, by Libby-French (1984). In general, the sea level fluctuated during the Late Cretaceous, and depositional environments ranged between inner and outer shelf. Dark shales are the dominant lithology, with sandstones subordinate.

The Cenozoic section represents continued marine sedimentation. Lithologic interpretation of the various marine environments is difficult, however, owing to the lack of characteristic environmental features, and it is necessary to rely on paleontologic data. Numerous unconformities and sea level fluctuations have been recognized on the basis of foraminiferal data.

Aulacogens

Aulacogens are elongate, fault-bounded sedimentary basins that extend, as gradually narrowing wedges or pie slices in plan view, from the margins of cratons toward their interiors. Aulacogens differ from previously discussed marginal rift basins mainly in their orientation with respect to the edge of the craton. Plate tectonic interpretations view aulacogens as aborted oceans, that is, as the failed arms of branching rift systems (triple junctions) whose other members continued to evolve into full-fledged ocean basins. Examples of aulacogen basins include the Reelfoot Rift, the structural flume of Late Paleozoic age in which the Mississippi River flows; the Amazon rift, which serves as the sluiceway for the largest river system on the Earth; and the Benue Trough of Cretaceous age, in which the Niger River is located. These three examples make it apparent that, in areas with adequate rainfall, aulacogens can be very important features in the structural, stratigraphic, sedimentologic, and topographic development of a continent or plate. Because aulacogens always have one end extending into the heart of a plate and the other end open to the ocean, it is very common to find large deltas at the oceanic end.

The sedimentary sequences of aulacogens are similar in general character to facies equivalents in platformal sequences of the cratons adjacent on both sides but are much thicker. Sediment fills in the grabens are mainly alluvial fans and shallow-marine shelf strata inland, grading into deeper-water facies deposited on the continental shelf and slope. Sandstone mineralogies generally reflect mature cratonic sources, although immature coarse clastics of partly nonmarine origin may occur because of tectonic activity along marginal fault scarps. In stratigraphic sequences associated with the early phases of aulacogen development, lavas associated with rift development may be prominent.

Benue Trough, West Africa

The Benue Trough is the failed arm of a triple junction associated with the opening of the South Atlantic during Early Cretaceous time. Current theory attributes the initial stages in the evolution of the trough to a rising mantle plume in the region of the present Niger delta (Fig. 11-17). This rise caused doming and rifting, which resulted in the development of a triple junction. Subsequent stratigraphic development within and at the margin of the trough was controlled by repeated mantle upwelling followed by subcrustal contraction, supplemented by isostatic subsidence under sedimentary loading. Interacting with these basinal tectonic features were worldwide eustatic changes in sea level during Cretaceous and Tertiary time. The net result of these interactions is a sequence of Cretaceous and younger sediments in the trough more than 5 km thick for at least 500 km along the trough axis. The location of the Benue depression presumably determined the course of the ancestral Niger River and the position of its delta. The delta is a continental embankment containing sediment about 10 km thick built into deep water at the mouth of the aulacogen.

The oldest rocks in the southern part of the Benue Trough are basaltic pyroclastics ejected into a shallow-marine environment during Aptian and Early Albian time, about 115–110 m.y. ago. Overlying the volcanics is a series of fine sandstones and mudrocks 3000 m thick, the Asu River Group (Fig. 11-18), which is composed mostly of sparsely fossiliferous, carbonaceous, and pyritic shale containing some aragonitic foraminifera. From a stratigraphic viewpoint the shale represents the initial encroachment of the new ocean into the trough, and the high content of organic matter and scarcity of fossils suggest a brackish water environment, as might occur in an estuary.

In the upper part of the Benue Trough, the time-equivalent rocks are coarse sandy to conglomeratic, ferruginous feldspathic sandstones and shales that contain scattered pebbles and cobbles of granite and gneiss. This unit, the Bima Sandstone, is of very variable thickness because of deposition over irregular basement topography and because of proximity of depositional sites to repeated block faulting in the trough. The average thickness of the unit is about 3000 m, about three-quarters of which is sandstone. Closest to the Asu River Group, the shaley beds are more abundant and have been interpreted as mud flats, and these beds grade eastward (uptrough) into deltaic deposits. Farthest uptrough, continental coarse sands dominate. The sandstones are poorly sorted, cross-bedded, and typically have the concave bases of fluvial channels. Their aggregate characteristics are those of conglomeratic alluvial fan (fanglomerate) deposits produced by basement uplift close to a cratonic margin.

The Bima Sandstone is unconformably overlain by the Yolde Formation of Cenomanian age, about 95 Ma old. The base of the formation is marked by the first

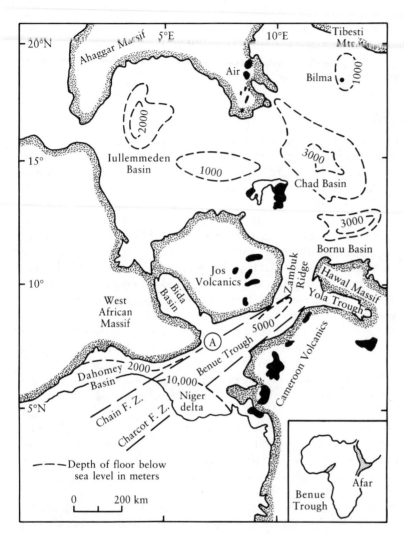

FIG. 11-17 • Tectonic framework of the Benue Trough showing nearby West African sedimentary basins. Shield areas are shown with stipples. The circled area north of the delta is the Anambra Basin (A). (Reprinted by permission of the author and the publisher from Petters, *Journal of Geology*, v. 86, p. 312, The University of Chicago Press, 1978.)

appearance of marine shales and the top by the onset of limestone deposition; its thickness varies areally between 100 m and 300 m. Various species of oysters are the dominant fossils in the Yolde. Further offshore the Eze-Aku Group (Fig. 11-18) was deposited, consisting of calcareous black and dark gray shales and siltstones, thin, sandy or shelly limestones, and calcareous sandstones. The Yolde and Eze-Aku are essentially diachronous paralic deposits representing the basal coastal, shoreline, and shallow sublittoral facies of a major onlap phase of sedimentation of early Late

Cretaceous age. This progressive shallow-marine transgression was oscillatory, the position of the strandline being governed by the rate of sediment supply.

The Eze-Aku Group and Yolde Formation grade upward into the Awgu Shale and the Pindiga Formation. As was the case with the stratigraphic units lower in the section, fully marine deposits dominate toward the western end of the Benue Trough, and transitional and marginal nonmarine deposits are more frequent eastward. The Pindiga consists of marine shales and limestones near its base and is richly fossiliferous; the upper part consists of blue-black shales that are mostly unfossiliferous but may have a monospecific arenaceous microfauna of marsh origin. Apparently, salt marshes were common in the eastern part of the trough during Turonian-Coniacian time. The total thickness of the Pindiga is 1300 m. The Awgu Shale consists of black shales, sandstones, and limestones up to 700 m thick, and in the eastern part of the outcrop at least 30 coal seams have been found.

The Turonian-Coniacian depositional cycle ended with compressional folding of the Benue Trough, which produced more than 100 anticlines and synclines in Santonian time, about 85 m.y. ago. The folding was accompanied by extensive magmatic activity. Extrusive and intrusive igneous rocks increase in amount and diversity from northeast to southwest, with maximum reported thicknesses of 1300 m of mafic rocks. The magmatism was accompanied by large-scale lead-zinc mineralization.

Following the Santonian deformation and magmatism the depositional axis of the trough was displaced westward, and the Anambra Basin subsided (Fig. 11-17). Sediments derived from the Santonian uplifts via the ancestral Niger River filled the basin, which until then had been a shallow shelf area. The basal Nkporo Shale of

FIG. 11-18 • Restored stratigraphic section along the Benue Trough and Niger delta miogeocline (after showing depositional sequences and facies relationships. During major regressions the shoreline lay farther southward supplying turbidites to the Cretaceous continental rise. (Reprinted by permission of the author and the publisher from Petters, *Journal of Geology*, v. 86, p. 314, The University of Chicago Press, 1978.)

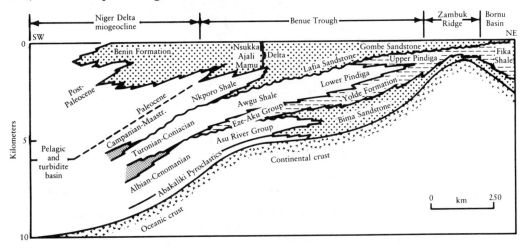

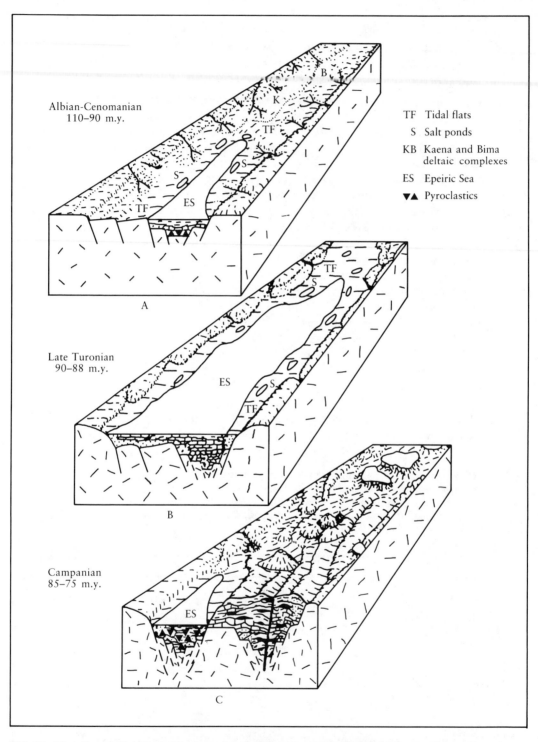

Albian-Cenomanian
110–90 m.y.

TF   Tidal flats

S   Salt ponds

KB   Kaena and Bima
       deltaic complexes

ES   Epeiric Sea

▼▲   Pyroclastics

A

Late Turonian
90–88 m.y.

B

Campanian
85–75 m.y.

C

FIG. 11-19 ● Sequential block diagrams showing paleogeographic model for the Benue Trough. (Reprinted by permission of the author and the publisher from Petters, *Journal of Geology*, v. 86, p. 315, The University of Chicago Press, 1978.)

the Anambra Basin oversteps the outcrops of the Awgu Shale and Eze-Aku Group (Fig. 11-18). The 2000 m of Campanian-Maestrichtian (85–65 m.y.) fill of the Anambra Basin is divided into (1) an Upper Campanian transgressive cycle represented by the shallow-marine Nkporo Shale and its paralic equivalents and (2) a major Maestrichtian-Danian (75–65 m.y.) deltaic offlap complex represented by the coal-bearing Mamu Formation, fluvial Ajali Sandstone, and coal-bearing Nsukka Formation. The Nkpopo Shale contains a shallow-water benthonic foraminiferal assemblage, while the Enugu Shale Formation contains a limited marsh arenaceous microfauna. The equivalents of these deposits in the northeast (uptrough) are the deltaic Gombe Sandstone and Lafia Sandstone. A generalized summary of paleogeographic events is shown in Figure 11-19.

In the south the Maestrichtian-Danian deltaic outbuilding ended with a brief early Late Paleocene (63–61 m.y.) marine transgression. Rejuvenation of the structures generated during Santonian time is evidenced by the thickening of Paleocene sandstone bodies toward the uplift, and general thinning of these strata over the nose of the uplift. By Eocene time (58–37 m.y.) the Anambra Basin was filled, and the Niger delta prograded southward across the shallow Anambra shelf, so the Cretaceous continental slope of the Benue Trough started to migrate farther into the Gulf of Guinea, thus producing the 200-km overlap in the computed fit of Africa and South America on predrift paleogeographic reconstructions. A summary of transgressive and regressive events in the Benue Trough and surrounding basins is shown in Figure 11-20.

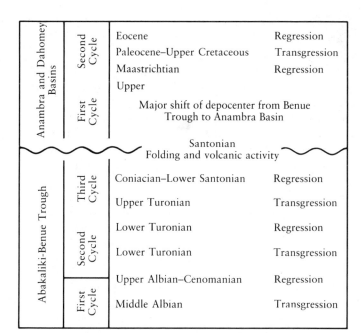

Anambra and Dahomey Basins	Second Cycle	Eocene	Regression
		Paleocene–Upper Cretaceous	Transgression
		Maastrichtian	Regression
	First Cycle	Upper	
		Major shift of depocenter from Benue Trough to Anambra Basin	

Santonian
Folding and volcanic activity

Abakaliki-Benue Trough	Third Cycle	Coniacian–Lower Santonian	Regression
		Upper Turonian	Transgression
	Second Cycle	Lower Turonian	Regression
		Lower Turonian	Transgression
	First Cycle	Upper Albian–Cenomanian	Regression
		Middle Albian	Transgression

FIG. 11-20 • Cretaceous and tertiary transgressions and regressions recognized in the Benue Trough and adjacent areas of Nigeria. (Reprinted by permission of the author and the publisher from Whiteman, *Nigeria: Petroleum Geology Resources and Potential*, Graham and Trotman, 1982.)

Intracontinental Basins

Intracontinental basins are those that have no apparent relationship to interactions at either modern or ancient plate margins. Examples in midcontinental North America include major depocenters such as the Williston Basin (350,000 km²), the Michigan Basin (250,000 km²), and the Illinois Basin (250,000 km²). Each of these large basins contains about 5000 m of sedimentary rocks. Numerous intracontinental basins of smaller size and volume are also known. The causes of the structural depressions that outline these basins are uncertain, but suggestions include thinning of the base of the crust by mantle convection and migration of "hot spots" in the mantle.

The types of sediments deposited in intracontinental basins are normally either quartzose medium- to fine-grained sandstones, siltstones, and mud-shales or else limestones, dolomites, and/or evaporites. Normally absent are volcanic rocks, tuffs, bedded cherts, volcanic rock fragments, calcic feldspar grains, and pieces of coarse-grained crystalline rocks such as granites and gneisses. Depositional environments are continental to shallow-marine; maximum water depths probably will not exceed 200 m. The abundant sedimentary structures reflect the depositional environments: cross-bedding, ripple marks, parting lineations, mud cracks, and ichnofossils in the clastic deposits; mounds, reefs, stromatactis, and birdseyes in the carbonates. Structures such as flutes, grooves, and graded bedding are uncommon in the clastic units. Fossils in the carbonate units are generally benthonic shallow-water forms such as crinoids, trilobites, pelecypods, or hermatypic corals; planktonic forms are rare. In general, only the tectonic setting distinguishes an intracratonic basin from a continental platform along a divergent plate margin. The types of sedimentary rocks, their internal characteristics, their depositional environments, and their facies relationships tend to be very similar in the two tectofacies.

As an example of a stratigraphic-sedimentologic description and analysis of an intracratonic basinal unit, we can examine the Tar Springs Sandstone, an Upper Mississippian unit in the southern part of the Illinois Basin (Wescott, 1982). The Tar Springs is part of a transitional sequence of alternating sandy and calcareous strata separating a dominantly limestone section below from a dominantly clastic section of Pennsylvanian age above. The unit is exposed only in the southeastern corner of Illinois and in adjacent Kentucky but extends in the subsurface over much of southern Illinois.

In outcrop it is clear that the unit is of paralic and near-shore marine origin, because it contains several coal beds as well as shallow-water marine fossils in some of the sandstones. Shales and sandy carbonate rocks also are present. Outcrop and subsurface log data concerning the Tar Springs unit reveal the following features.

1. Typical thicknesses are 10–20 m but exceed 30 m in some areas. The unit overlies limestone, conformably in some areas; but in others, channels more than 10 m in depth were carved in the limestone prior to deposition of Tar Creek sand. The sandstone is overlain by another limestone unit in its southern part,

but to the north the limestone pinches out, and the Tar Creek is succeeded by a sandstone similar in character to the Tar Creek.

2. Isopach data based on subsurface logs, coupled with cross-bedding directions from surface outcrops, reveal the geometry of the Tar Creek Sandstone to be linear to slightly "shoestring" in plan view with a dominant direction of sediment transport from northeast to southwest (Fig. 11-21). Toward the southwest,

FIG. 11-21 • Direction of cross-bedding and thickness of sandstone in Tar Springs Sandstone in southern Illinois. The isopach pattern shows sand bodies that tend to parallel the strongly preferred southwesterly direction of cross-bedding. (Reprinted by permission of the American Association of Petroleum Geologists from Potter et al., *American Association of Petroleum Geologists Bulletin*, v. 42, 1958.)

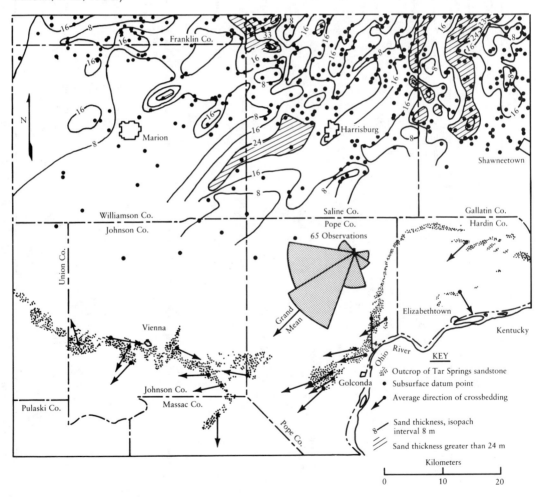

TABLE 11-2 ● Summary of facies and sedimentary characteristics of the Tar Springs Sandstone in southern Illinois

	Facies A	Facies B	Facies C	Lower Facies D	Middle Facies D	Upper Facies D
Lithology	Light brown to white, fine-grained quartz arenites with shale clasts and plant debris at base of sandstones.	Gray to brown, very fine- to medium-grained quartz arenites.	Gray to light-brown, very fine- to fine-grained quartz arenites, may be interbedded with mudstones. Fossils are rare.	Light brown to white fine-grained quartz arenites interbedded with gray mudstones. Some mudstones are fossiliferous.	Light brown to white, fine-grained quartz arenites interbedded with black shales.	Light brown to white, fine-grained quartz arenites interbedded with thick mudstone units. All units are fossiliferous.
Sandstone texture	Moderately to well-sorted; fining-upward sequences.	Moderately to well-sorted; coarsening-upward sequences.	Well- to very well-sorted; variable grain size trends.	Moderately sorted; coarsening-upward sequences.	Moderately to well-sorted; coarsening-upward sequences.	Well-sorted; coarsening-upward sequences.
Sedimentary structures	Trough and planar cross-beds, ripple marks, horizontal laminations; unimodal paleocurrent directions.	Low angle cross-beds, horizontal laminations; hummocky cross-beds, parting lineations, ripples and ripple lamination; bimodal paleocurrent directions.	Flaser, wavy, and lenticular beds, small herringbone cross-beds, reactivation surfaces, scour channels, horizontal laminations, parting lineations, ripples, abundant vertical burrows; polymodal paleocurrent orientations.	*Mudstones*: parallel laminations, lenticular and flaser beds, small-scale cross-beds, contorted beds; bimodal paleocurrent directions. *Sandstones*: horizontal beds, small-scale cross-beds, wavy and flaser beds.	*Mudstones*: parallel laminations; lenticular and wavy beds in silty units. *Sandstones*: trough and planar cross-beds, ripple marks, horizontal laminations, shale partings.	*Mudstones*: parallel laminations. *Sandstones*: shale partings, ripples, horizontal laminations, small-scale cross-beds, flasers, clay pebbles, dewatering structures, horizontal and vertical burrows
Geometry	Generally thick, elongate sand bodies.	Strike-oriented, relatively thick sheet sandstones.	Relatively thin, sheet sandstones.	Thin sheet sandstones interbedded with mudstones.	Thin lenticular or sheet sandstones interbedded with mudstones.	Thin sheet sandstones interbedded with thick mudstones.
Facies associations	Commonly overlies lower Facies D; may be overlain by Upper Facies D.	Overlies lower Facies D.	Overlies lower Facies D.	May be overlain by Facies A, B, or C.	Interbedded with Facies A.	Overlies Facies A.
Environments of deposition	Fluvial and distributary channels.	Littoral marine; lower and upper shoreface and foreshore.	Tidal flat; barrier bar, subtidal mud flat, intertidal mixed and sand flats.	Prodelta, distal bar, and lower shoreface transition.	Interdistributary bay fill and crevasse splay.	Delta destructional bars.

(Reprinted by permission of the publisher from Wescott, *Journal of Sedimentary Petrology*, v. 52, Society of Economic Paleontologists and Mineralogists, 1982.)

cross-bedding directions become more widely dispersed and a dominant flow direction is more difficult to define.

3. The Tar Springs Sandstone in outcrop can be divided into four major facies (Table 11-2) that are characterized by unique combinations of lithologies, sedimentary structures, and fossils (Wescott, 1982).

Facies A occurs as elongate sand bodies up to 20 m thick, composed of fine-grained and well-sorted quartz arenite. The beds occur in multiply stacked fining-upward sequences with scoured bases generally floored by clay pebbles and plant debris. Trough and planar crossbeds are the dominant sedimentary structure and show unimodal southwesterly paleocurrents. Facies A appears to be a classic channel deposit in streams of low to moderate sinuosity.

Facies B sand bodies are 3- to 6-m thick, slightly coarser-grained and less well-sorted quartz arenites that occur in stacked coarsening-upward sequences or else lack a vertical grain size trend. The beds have sheetlike or broadly lenticular geometries.

This facies exhibits a distinct vertical sequence of sedimentary structures. The basal sandstone beds are horizontally laminated with individual bedding surfaces showing current lineations. Interbedded with the horizontal laminae are ripple cross-laminated sandstones with bimodal dip orientations. Reactivation surfaces truncate some of the foresets. Hummocky cross-beds are locally interbedded with these thin-bedded units. The laminated sandstones are overlain by trough-cross-bedded sandstones. The complete coarsening-upward sequence, where present, is capped by horizontal beds and low-angle cross-beds. The sandstones of facies B are interpreted as shoreface and foreshore deposits, part of a delta-front sheet sandstone laterally adjacent to the distributary channel complex. The vertical sequence of sedimentary structures in this facies is similar to that formed in the exposed high-energy near-shore area of some modern coastlines.

Facies C is texturally and compositionally like Facies A but has a different suite of sedimentary structures. It has a sheet geometry with sandstones 1–3 m thick, commonly interbedded with mudstone units 1–2 m thick. Many types of small-scale sedimentary structures occur in this facies, including flaser, wavy, and lenticular bedding, small-scale herringbone cross-beds, reactivation surfaces, small channel scours, ripple marks, horizontal laminae with current lineations, small-scale load structures, and some mud cracks. The mudstone units are laminated and may have silty, lenticular beds near the top. Macerated plant fragments are abundant in some of the fissile shales. Paleocurrent indicators in this facies include small-scale cross-beds, herringbone cross-beds, ripple marks, and current lineations that have a bimodal north-south orientation. Locally, the sandstone units are dissected by cross-bedded channel-shaped sandstones.

The sandstones are generally bioturbated, and a diverse assemblage of trace fossils is recognizable in this facies. Individual traces include *Rhizocorallium*, *Teichichnus*, *Chondrites*, *Thalassinoides*, *Monocraterion*, *Skolithos*, *Conostichus*, and other unidentifiable ichnofossils. Most of the trace fossils are vertical forms and are

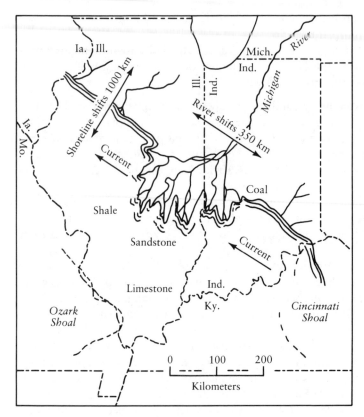

FIG. 11-22 • Typical paleogeography during deposition of Late Mississippian cyclic succession. During Tar Springs deposition the shoreline was about 200 km south of the location shown. (Reprinted by permission of the American Association of Petroleum Geologists from Swann, *American Association of Petroleum Geologists Bulletin*, v. 48, 1964.)

characteristic of the *Skolithos* ichnofacies. The physical structures of Facies C are diagnostic of tidal flat deposition, for example, the flaser bedding and bimodal herringbone cross-beds. The *Skolithos* ichnofacies indicates deposition in a sandy littoral environment in which the substrate is frequently reworked by waves and/or tidal currents. Both the lithofacies and biofacies of Facies C are very similar to those of the modern Wadden Sea in Holland.

Facies D is composed of interbedded sandstone and shale, and its characteristics vary slightly, depending on its relative stratigraphic position within the formation. The lower mudstones of Facies D are the basal units of the Tar Springs and rest on the limestones of the underlying formation. They are about 1 m thick, contain a shallow-water benthonic fauna, and have parallel lamination. They grade upward into siltstones and fine-grained sandstones with lenticular and flaser bedding and small-scale cross-beds with bidirectional dips; they total 2 m in thickness. This subfacies represents prodelta distal bar sediments.

In the middle mudstone facies the sedimentary structures are the same, but the shales are gray to black and contain abundant plant debris. Siltstones and fine sandstones are intensely bioturbated. This facies is commonly interbedded with sandstones of Facies A and is interpreted to be interdistributary bay and upper delta plain overbank deposits.

In the upper mudstone facies the shales are up to 3 m thick, interbedded with thinner sandstones 0.5–2.5 m thick. The sandstones are fine-grained and contain shale partings, ripple marks, horizontal laminae, small-scale crossbeds, flaser beds, clay pebbles, and dewatering structures, abundant horizontal and vertical burrows, and molds of brachiopods, bivalves, crinoids, and bryozoans. The sandstones are interpreted as deltaic sand bars.

The evidence from sedimentary structures and biofacies data suggest that the Tar Springs Sandstone is a progradational sequence deposited in fluvial-deltaic, shoreface-foreshore, and tidal flat environments along the margin of a Late Mississippian epeiric sea. The distribution of depositional environments indicates a variation in wave and current energy at the land-sea interface, changing from wave-dominated in some areas (Facies A) to tide-dominated in others (Facies D). The resulting distribution of lithologies and facies is complex, as is typical for deltaic sequences.

On a regional scale, subsurface data reveal that the delta containing the Tar Springs sediments varied in location through Late Mississippian time (Fig. 11-22) at the terminus of the "Michigan River," which had its headwaters in the crystalline rocks of the Canadian Shield far to the northeast. About 15 major cycles of advance and retreat of the shoreline have been documented (Swann, 1964), with diachronous lithologic units and facies boundaries migrating accordingly.

Summary

Plate interactions are the cause of most of the major basins in which sediment accumulates. At convergent margins between a continental and an oceanic plate, thick piles of clastic rocks form in trenches, forearc basins, intra-arc basins, and retroarc basins, each of which can be recognized in the ancient record by the type of sediment it contains.

Extensive rift systems typically form within a continent parallel to divergent plate margins prior to the major separation into two continental masses. Tensional, fault-bounded basins normal to the continental margin are termed aulacogens and represent rifts that failed to develop into ocean-forming tears. Major sediment transport from the continental interior typically occurs down the grabens formed by these failed rifts. Modern examples include the Mississippi River in the United States, the Amazon River in South America, and the Niger River in west Africa.

Intracontinental basins such as the Michigan Basin and Illinois Basin in north-central United States have no apparent relation to events at plate margins but subsided slowly during much of Paleozoic time. Several thousand meters of sediments accumulated in each basin.

Uplift and erosion of sedimentary rocks results from both orogenic movements near convergent plate margins and epeirogenic uplift of broad areas of the continental interior. The epeirogenic uplifts may continue for much longer periods of time than the orogenies at a plate margin. It is not clear whether orogeny or epeirogeny yields more sediment to sedimentary basins.

References

American Association of Petroleum Geologists, 1984. Sedimentary Provinces of the World (map, 1:31,000,000). Tulsa, American Association of Petroleum Geologists.

Blatt, H., 1982. Sedimentary Petrology. New York, W. H. Freeman, 564 pp.

Burke, K., 1975. Atlantic evaporites formed by evaporation of water spilled from Pacific, Tethyan, and southern oceans. Geology, v. 3, pp. 613–616.

Burke, K., 1976. Development of graben associated with the initial ruptures of the Atlantic Ocean. Tectonophysics, v. 36, pp. 93–112.

Carey, S. and Sigurdsson, H., 1984. A model of volcanogenic sedimentation in marginal basins. *In* B. P. Kokelaar and M. F. Howells (eds.), Marginal Basin Geology. Oxford, Blackwell Scientific Publications, pp. 37–58.

Dalziel, I. W. D., 1984. Tectonic Evolution of a Forearc Terrane, Southern Scotia Ridge, Antarctica. Boulder, Geological Society of America, Spec. Paper 200, 32 pp.

Dalziel, I. W. D., Dott, R. H., Jr., Winn, R. D., Jr., and Bruhn, R. L., 1975. Tectonic relations of South Georgia Island to the southernmost Andes. Geol. Soc. Amer. Bull., v. 86, pp. 1034–1040.

Dickinson, W. R., 1976. Sedimentary basins developed during evolution of Mesozoic-Cenozoic arc-trench system in western North America. Canadian J. Earth Sci., v. 13, pp. 1268–1287.

Dickinson, W. R. and Seely, D. R., 1979. Structure and stratigraphy of forearc regions. Amer. Assoc. Petroleum Geol. Bull., v. 63, pp. 2–31.

Dickinson, W. R. and Suczek, C. A., 1979. Plate tectonics and sandstone compositions. Amer. Assoc. Petroleum Geol. Bull., v. 63, pp. 2164–2182.

Emery, K. O. and Uchupi, E., 1984. The Geology of the Atlantic Ocean. New York, Springer-Verlag, 1050 pp.

Fergusson, C. L., 1985. Trench-floor sedimentary sequences in a Palaeozoic subduction complex, eastern Australia. Sedimentary Geol., v. 42, pp. 181–200.

Friedman, G. M., 1987. Deep-burial diagenesis: Its implications for vertical movements of the crust, uplift of the lithosphere and isostatic unroofing—A review. Sedimentary Geol., v. 50, pp. 67–94.

Friedman, G. M., 1988. Comment on "On orogeny and epeirogeny in the study of Phanerozoic and Archean rocks." Geosci. Canada, v. 15, pp. 230–231.

Gamboa, L. A., Truchan, M., and Stoffa, P. L., 1985. Middle and Upper Jurassic depositional environments at outer shelf and slope of Baltimore Canyon Trough. Amer. Assoc. Petroleum Geol. Bull., v. 69, pp. 610–621.

Hsü, K. J., 1968. Principles of melanges and their bearing on the Franciscan-Knoxville problem. Geol. Soc. Amer. Bull., v. 79, pp. 1063–1074.

Hsü, K. J., 1969. Melanges of the San Francisco Peninsula—Geologic reinterpretation of the type Franciscan. Amer. Assoc. Petroleum Geol. Bull., v. 53, pp. 1348–1367.

Hsü, K. J., 1974. Melanges and their distinction from olistostromes. *In* R. H. Dott, Jr. and R. H. Shaver (eds.), Modern and Ancient Geosynclinal Sedimentation. Tulsa, Society of Economic Paleontologists and Mineralogists, Spec. Pub. No. 19, pp. 321–334.

Illies, J. H. and Mueller, S. (eds.), 1970. Graben Problems. Stuttgart, West Germany, E. Schweizerbart'sche Verlagsbuchhandlung, 316 pp.

Ingersoll, R. V., 1988. Tectonics of sedimentary basins. Geol. Soc. Amer. Bull., v. 100, pp. 1704–1719.

Libby-French, J., 1984. Stratigraphic framework and petroleum potential of northeastern Baltimore Canyon Trough, mid-Atlantic outer continental shelf. Amer. Assoc. Petroleum Geol. Bull., v. 68, pp. 50–73.

Lorenz, J. C., 1987. Triassic-Jurassic Rift-Basin Sedimentology: History and Methods. New York, Van Nostrand Reinhold, 315 pp.

Macdonald, D. I. M. and Tanner, P. W. G., 1983. Sediment dispersal patterns in part of a deformed Mesozoic back-arc basin on South Georgia, South Atlantic. J. Sed. Petrology, v. 53, pp. 83–104.

Manspeizer, W. (ed.), 1988. Triassic–Jurassic Rifting. 2 vols., New York, Elsevier, 998 pp.

Miller, M. M., 1989. Intra-arc sedimentation and tectonism: Late Paleozoic evolution of the eastern Klamath terrane, California. Geol. Soc. Amer. Bull., v. 101, pp. 170–187.

Morganti, J. M., 1981. Ore deposit models. 4: Sedimentary-type stratiform ore deposits: Some new models and a classification. Geosci. Canada, v. 8, pp. 65–74.

Petters, S. W., 1978. Stratigraphic evolution of the Benue Trough and its implications for the Upper Cretaceous paleogeography of West Africa. J. Geol., v. 86, pp. 311–322.

Potter, P. E., Nosow, E., Smith, N. M., Swann, D. H., and Walker, F. H., 1958. Chester cross-bedding and sandstone trends in Illinois Basin. Amer. Assoc. Petroleum Geol. Bull., v. 42, pp. 1013–1046.

Raymond, L. A. (ed.), 1984. Mélanges: Their Nature, Origin, and Significance. Boulder, Geological Society of America, Spec. Paper 198, 170 pp.

Scholle, P. A., 1974. Diagenesis of Upper Cretaceous chalks from England, Northern Ireland, and the North Sea. *In* K. J. Hsü and H. C. Jenkyns (eds.), Pelagic Sediments: On Land and Under the Sea. International Association of Sedimentologists, Spec. Pub. No. 1. Oxford, Blackwell Scientific Publications, pp. 177–210.

Scholle, P. A. (ed.), 1977. Geological Studies on the COST No. B-2 Well, U.S. Mid-Atlantic Outer Continental Shelf Area. Washington, D.C., U.S. Geological Survey Circ. 750, 71 pp.

Sheridan, R. E., 1976. Sedimentary basins of the Atlantic margin of North America. Tectonophysics, v. 36, pp. 113–132.

Storey, B. C. and Macdonald, D. I. M., 1984. Processes of formation and filling of a Mesozoic back-arc basin on the island of South Georgia. *In* B. P. Kokelaar and M. F. Howells (eds.), Marginal Basin Geology. Oxford, Blackwell Scientific Publications, pp. 207–218.

Swann, D. H., 1964. Late Mississippian rhythmic sediments of Mississippi Valley. Amer. Assoc. Petroleum Geol. Bull., v. 48, pp. 637–658.

Tanner, P. W. G., 1982. Geologic evolution of South Georgia. *In* C. Craddock (ed.), Antarctic Geoscience. Madison, University of Wisconsin Press, pp. 167–176.

Tanner, P. W. G. and Macdonald, D. I. M., 1982. Models for the deposition and simple shear deformation of a turbidite sequence in the South Georgia portion of the southern Andes back-arc basin. J. Geol. Soc. London, v. 139, pp. 739–754.

Underwood, M. B., 1984. A sedimentologic perspective on stratal disruption within sandstone-rich melange terranes. J. Geol., v. 92, pp. 369–385.

Valloni, R. and Maynard, J. B., 1981. Detrital modes of recent deep-sea sands and their relation to tectonic setting. A first approximation. Sedimentology, v. 28, pp. 75–83.

Ward, P. D., 1978. Revisions to the stratigraphy and biochronology of the Upper Cretaceous Nanaimo Group, British Columbia and Washington State. Canadian J. Earth Sci., v. 15, pp. 405–423.

Ward, P. and Stanley, K. O., 1982. The Haslam Formation: A Late Santonian–Early Campanian forearc basin deposit in the insular belt of southwestern British Columbia and adjacent Washington. J. Sed. Petrology, v. 52, pp. 975–990.

Wescott, W. A., 1982. Depositional setting and history of the Tar Springs Sandstone (Upper Mississippian), southern Illinois. J. Sed. Petrology, v. 52, pp. 353–366.

Whiteman, A., 1982. Nigeria: Its Petroleum Geology, Resources and Potential. Vols. 1, 2. London, Graham and Trotman, 394 pp.

Scientists, on the whole, are amiable and well-meaning creatures.

Peter Medawar

12
Economic Stratigraphy I: Primary Deposits

The vast majority of geologists who specialize in stratigraphy are employed by the petroleum industry in the search for oil and gas. A much smaller number work for state and federal geological surveys or teach at one of the 700 or so colleges and universities that offer degrees in geology. Clearly, the economic application of stratigraphic principles and data is the pastime of most stratigraphers. However, relatively few published stratigraphic papers in major journals deal with economic stratigraphy, in part because of restrictions on publishing applied by many oil companies and in part because few academic stratigraphers have chosen to study such things as oil shales, coal facies, and the sedimentary copper deposits that supply more than 25% of the world's copper production. Nevertheless, each year every American is indirectly responsible for quarrying of 4150 kg of stone, 3900 kg of sand and gravel, 360 kg of cement, 220 kg of clays, 200 kg of salt, 140 kg of phosphate rock, and 490 kg of other nonmetals. Annual per capita use of metals and metallic products is 550 kg of iron and steel, 25 kg of aluminum, 10 kg of copper, 6 kg of lead, 6 kg of manganese, 5 kg of zinc, and 9 kg of other metals (Bates and Jackson, 1982). Each of us also consumes about 1000 liters of gasoline per year and similarly vast quantities of natural gas. Nearly all of these substances are obtained almost entirely from sedimentary rocks. In one or two chapters we cannot hope to do justice to the stratigraphic relationships of the wide variety of rocks and sediments from which these essential metallic and nonmetallic substances are obtained, but we hope a brief survey will be useful as an introduction to a fascinating and relatively unstudied area of stratigraphy.

Types of Occurrences

Mineral deposits in sedimentary rocks occur in many types of stratigraphic settings, ranging from deposits formed in continental weathering horizons to detrital sequences deposited in at least bathyal depths in the deep sea. These deposits can be grouped in several ways on the basis of factors such as the environment of formation, the nature of the ore, or the use to which the ore is put when extracted. We will use a system based on the first two of these criteria.

1. **Primary Deposits.** Primary sedimentary ore deposits may occur in specific geotectonic settings but are essentially controlled by the physical and chemical conditions of sedimentation. Climate and paleogeography may therefore be particularly important controls. In this category are building stone, sand and gravel, placer deposits, clay deposits such as bentonite and kaolinite, bauxite and ferruginous laterite, diatomite, evaporites, phosphate rock, bedded iron ore, manganese, coal, and oil shale. The stratigraphy of these deposits is the topic of this chapter.

2. **Deposits Formed by Secondary Mineralization.** These deposits are more closely related to basin evolution and fluid migration, although some may be primary. They may be more closely controlled by geotectonic processes, particularly because heat and rates and routes of fluid migration are important. Thus both geotectonic and environmental factors may be significant, particularly the existence of adequate porosity and permeability in the prospective host rocks. In this category of deposits we can include groundwater (although water might also be grouped in the first category), uranium, copper, lead and zinc, sulfur, petroleum and natural gas, and tar sands. The stratigraphy of these deposits will be the topic of Chapter 13.

Building Stone

Sedimentary rocks used for construction include all the major types: conglomerates, sandstones, mudrocks, limestones, dolostones, and gypsum, the physical and chemical properties desired depending on the use to which the material is to be put. The commonest use is as either building blocks or facing. Important physical properties for these uses include degree of lithification, joint presence and spacing, bedding and lamination, mineral composition, color, porosity/permeability, and, perhaps of most importance because of the high cost of transporting high-density materials, accessibility to the construction site. The need in construction materials for hardness and high compressive strength requires that the stone be well-indurated, be resistant to weathering, and have a low porosity.

In terms of the stratigraphy of building stones it is clear from their variety that

FIG. 12-1 • Quarrying Indiana limestone. The straight cuts are made by channeling machines (beyond the derrick). Long blocks, freed at the bottom, are "turned down" and split into smaller blocks for removal to the finishing plant. (Reprinted by permission of the Indiana Geological Survey.)

the rocks can be of any geologic age, formed in most depositional environments, and of most mineral compositions. However, the stratigraphic unit to be quarried must be laterally extensive and without sensible facies change so that a consistent quality is maintained in the sawed rock slabs. It also is desirable that the bedding surfaces be planar and regularly spaced and that the general appearance of the rock be aesthetically pleasing. Sedimentary rocks that have these desirable properties have attained widespread prominence as building stone. Examples include the well-known Indiana or Bedford Limestone (Mississippian, marine) (Fig. 12-1), Ohio Sandstone (Devonian, marine), and Italian travertine (Quaternary, nonmarine). Many other sandstones and carbonate rocks are commercially important in more restricted geographic areas but, because of their limited outcrop area or distance from large population centers, have not achieved national recognition. Any geotectonic regime may generate suitable building stones.

Sand and Gravel

The sand and gravel industry is the largest nonfuel mineral industry by volume in the United States. These materials are essential in modern construction, particularly in paving and building. About 50% of the sand and gravel used goes into concrete. Ideally, a commercial deposit contains 60% gravel and 40% sand, providing ample coarse material to crush for road base or bituminous aggregate and sand in the correct sizes and proportions for use in concrete.

Sand and gravel differ from other economically useful primary sedimentary deposits in being unconsolidated, highly variable mixtures of many constituents. In addition, they do not have to be used in the same physical state in which they are found. They can be artificially upgraded by screening, washing, and combining of size grades, so several deposits having different gross properties may all be commercially valuable. The ideal sand and gravel deposit is one that consists of clean (no clay, mica, or organic matter), hard (quartz preferred), sound (few cracks or partially healed fractures within the grains), inert particles that are present in quantity in a wide range of grain sizes (poorly sorted). The only durable fragment types that are undesirable are opal, chalcedony, and volcanic rocks in the rhyolite-andesite compositional range. These materials react chemically with the sodium and potassium that is present in some types of portland cement (alkali-aggregate reactivity) and cause expansion, cracking, and deterioration of concrete. As was true for building stone, the high cost of transporting sand and gravel makes it necessary that the deposit be located near the place where the material is to be used.

Most commercial sand (0.07 mm, 3.75ϕ, to 4.76 mm, -2.25ϕ) and gravel (up to 89 mm, -6.5ϕ) deposits are of Quaternary age because older deposits tend to be at least partially lithified. Streams flowing at high volume and velocity are the major environment of transportation and deposition of commercial gravel deposits. Wind shifts sand but is not competent to move pebbles. Glacial ice has essentially unlimited competency but has no sorting ability and so deposits till composed of particles ranging in size from clay to boulders. Waves can produce and sort sediments of gravel size, but such modern deposits are limited in extent to a narrow strip along the shoreline. Only fluvial waters move and concentrate large volumes of sediment in the sand-gravel size range. In addition, fast-moving streams act as grinding and washing mills, with the result that soft and structurally weak rock fragments tend to be ground up and removed, whereas hard, sound ones tend to be concentrated. Some amount of desirable rounding also is produced on the grains, particularly those coarser than 1–2 mm.

Within a fluvial complex, commercial sand and gravel deposits are most likely to be found in braided streams in the faster-moving waters of the channel center in nonbraided streams. In general, the closer to a highland area the better because gravels are not usually transported great distances across a piedmont area. Coincidence of large stream deposits and major urban areas is well illustrated in California, which produces more sand and gravel than any other state. The bulk of the production is from streams that drain the Coast Ranges and the Sierra Nevada. Gradual shifting of channels from Late Tertiary time to the present, coupled with uplift of the mountains, has produced unusually thick and extensive deposits that are poorly sorted but contain only about 1% silt and clay. The coarser particles are mostly fragments of igneous and metamorphic rocks, diorite and mafic gneiss being the most common types.

Placer Deposits

Placer deposits are mechanical concentrates of grains that are chemically inert and mechanically durable and have high specific gravities. Minerals that fit these criteria and that commonly occur as placers include gold, platinum, cassiterite (tin), magnetite, chromite, ilmenite, rutile, native copper, gemstones, zircon, monazite (rare earths), phosphate, and cinnabar (mercury).

The process of mechanical concentration depends on a few basic principles involving specific gravity, grain size, and grain shape as affected by the velocity of a moving fluid (Slingerland and Smith, 1986). In a fluid a grain with a greater specific gravity sinks more rapidly than one with lesser specific gravity (size and shape being equal), and the difference in specific gravity is accentuated in water compared to air. For example, the ratio of gold (specific gravity = 19) to quartz in water is $(19 - 1)/(2.6 - 1) = 11.25$; in the air the ratio is $19/2.6 = 7.3$. Thus gold is separated from quartz in water about 50% more efficiently than gold settling in air, leading to a richer placer. This is most noticeable in outcrop when the placered mineral is dark in color—for example, magnetite or ilmenite contrasted to platinum metal or zircon. After deposition the placer can be further enriched by the eddies generated along the bottom by turbulent stream flow, processes described by the fluid mechanics concepts of entrainment equivalence and dispersive equivalence. The eddies raise the lighter mineral grains off the bottom, so they can be easily carried farther downstream by the bulk of the moving water.

The effects of grain size and grain shape on mineral segregation are normally secondary to the effect of specific gravity because most of the placerable minerals are released from their parent rock in a narrow size range (fine sand to silt, although gold nuggets as large as cobbles are known) and with grain shapes that do not differ greatly from the shape of most nonplacerable minerals.

Stream channels are the most common site for placer deposits, particularly the middle reaches. If stream gradients are too steep, as occurs in mountain areas, placer minerals and gravel alike are swept along with little opportunity for settling. But where the stream debouches into lowland areas with gentler gradients, the conditions are ideal for placering to occur. Areal concentrations of heavy minerals in fluvial systems can be grouped into four categories, which correspond broadly to spatial scales at which hydraulic sorting processes operate (Smith and Minter, 1980).

1. **Large scale.** Concentrations occur on a regional scale in response to optimal combinations of average grain size of available heavy minerals and long-term

characteristics of the fluvial environment. Broad bands of placers parallel to depositional strike in alluvial fan deposits are an example. The lateral extent of these placers typically ranges from hundreds of meters to kilometers.

2. **Intermediate scale.** Concentrations arise from processes associated with major topographic features within channel systems; for example, the upstream part of point bars or midchannel bars that form at wide reaches of the channel, at short channel segments or channel junctions, and at meander bends. As a meander increases in amplitude and shifts downstream, these local concentrations become "paystreaks" in the floodplain sediments (Figs. 12-2, 12-3). Distribution scales of such deposits are on the order of meters to tens of meters both laterally and longitudinally.

3. **Small scale.** These concentrations and sorting processes occur at the bed-form level. Typical examples include heavy mineral-rich laminae in cross-beds and in horizontal (planar) strata. Linear scales commonly range from centimeters to tens of centimeters.

4. **Very small scale.** Concentrations result from processes operating at the millimeter scale. Examples include small clusters of grains within laminae or voids between larger grains. Such concentrations arise from very short-term fluid/sediment interactions set up by a rough sediment bed and stream turbulence.

Beach placers are formed along coastlines by the effects of waves and longshore currents. The longshore currents shift the sediment, the lighter minerals (quartz,

FIG. 12-2 • Gravel deposition and formation of paystreaks in a rapidly flowing meandering stream, in which meanders migrate laterally and downstream. Stream arrows represent point of cutting. 1, Original position; 2, intermediate position; and 3, present position of stream. Deposits formed at *a*, *b*, and *c*, or inside of meanders of stream 1, become extended downstream and laterally in direction of heavy arrow growth to *a'*, *b'*, *c'* and on the present stream, and buried pay streaks result. (Reprinted by permission of John Wiley & Sons, Inc., from Jensen and Bateman, *Economic Mineral Deposits*, copyright © 1979.)

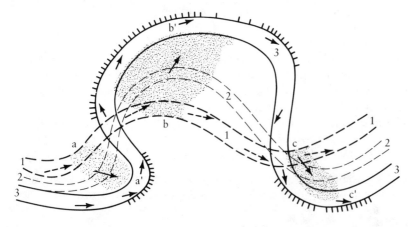

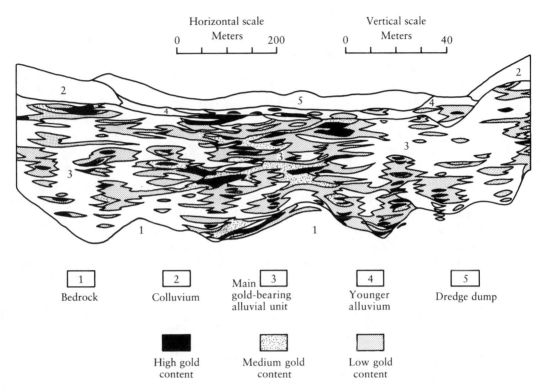

Horizontal scale
Meters
0   200

Vertical scale
Meters
0   40

| 1 | 2 | Main 3 | 4 | 5 |
Bedrock   Colluvium   gold-bearing   Younger   Dredge dump
                      alluvial unit   alluvium

High gold   Medium gold   Low gold
content   content   content

FIG. 12-3 • Stratigraphic cross-section showing placer development in a cross section of the Ku-ranakh River, southern Yakutia, U.S.S.R. The degree of placering (richness of the gold deposit) varies irregularly in both lateral and vertical dimensions as a result of irregular variation in current strengths during sediment deposition and winnowing as the deposits filled the valley. (Reprinted by permission of the publisher from Kartashov, *Economic Geology*, v. 66, Economic Geology Publishing Co., 1971.)

feldspar) being moved faster and farther than the heavy ones, resulting in placering of the heavy minerals. Waves carry sediment onto the beaches; the backwash and undertow carry out the lighter and finer material. Macdonald (1983, pp. 113–160) provides a detailed discussion of the relationship between the occurrence of placers and depositional environment.

Provenances of Placer Minerals

The igneous, metamorphic, and hydrothermal parent rocks of placer minerals are highly varied in character, as we would anticipate from the large number of minerals found as placer deposits. Igneous parent rocks range from high-silica granitoids to ultramafic rocks and kimberlites; metamorphics range from contact aureoles and skarns to high-grade schists and gneisses. However, many placer minerals are associated with particular types of parent rocks, either because of geochemical affinities among elements, variations in temperature pressure conditions in which particular types of rocks form, or chemical complexing between the heavy element in the placer mineral-to-be and sulfur or a halogen element. The relationship between parent rock and placer mineral is shown in Table 12-1.

There are few geotectonic controls on placer formation; it occurs in rivers and along coastlines on both stable and unstable margins. However, the high degree of concentration needed for the rarer minerals requires enriched source areas and re-peated reworking during transport. For example, the Kinta Valley tin deposits in Malaya exist only because of the previous concentration of tin in veins and skarns around granites, themselves the product of subduction. Intermittent but frequent reworking may occur during periods of progressively lower sea level related to gla-ciation but is more general in areas of tectonic instability. The circum-Pacific area has been the site of plate motion for at least 100 m.y., causing continual emergence of older auriferous belts and continual reworking in derivative clastic sediments. The placers in California, British Columbia, the Yukon, Alaska, Siberia, and New Zea-land fit this model nicely.

It is implicit from these descriptions that most placer deposits are relatively young. While there are older examples, they generally are small, and geotectonic understanding is unlikely to be of help in locating the ore. The generalization of the youthfulness of major placer ores is valid for rocks younger than 1.8 billion years

TABLE 12-1 • Provenances of the common placer minerals

Provenance	Placer Mineral (Specific Gravity)
Ultramafic and mafic terrains including pyroxenites and norites	The platinoids (14–21)
Granitoid terrains and related pegmatites and greisens	Cassiterite (7.0), monazite (5.0), zircon (4.6), rutile (4.2), gold (19)
Plateau basalts	Magnetite (5.2), ilmenite (4.7)
Syenitic rocks and related pegmatites	Zircon, rare earth minerals including uranium- and thorium-bearing minerals
Contact metamorphic aureoles-skarns	Scheelite (6.1), rutile, occasionally corundum (4.0)
Kimberlites	Diamonds (3.5)
High-grade metamorphic terrain	Gold, rutile, zircon, gemstones (19, 3.0–4.6)
Serpentine belts	Platinoids, chromite (5.1), magnetite
Carbonatites including associated rare basic igneous rocks	Rutile, ilmenite, magnetite, rare-earth minerals, uranium, niobium, thorium, and zirconium minerals

(Reprinted by permission of the publisher from Macdonald, *Alluvial Mining*, Chapman and Hall Ltd., 1983.)

but is not true for Lower Proterozoic strata, which in most areas are fluvial to shallow-marine sequences that lie unconformably on Archean basement.

Witwatersrand Placer Gold

As an example of economically important placer deposits, we can examine the famous hydrothermally altered gold placers developed in the 2.5-billion-year-old Witwatersrand system rocks in the Republic of South Africa (Minter, *in* Armstrong, 1981; Pretorius, 1981). More than half of the world's annual production of gold is obtained from these placers, and because of this, the stratigraphy and sedimentology of the area have been intensely studied by South African and other geologists. The structure of the Witwatersrand system is that of an elongated basin approximately 500 km by 200 km and extending over about 39,000 km^2 (Fig. 12-4); near the center the beds are domed (Vredefort Dome). The major gold placers crop out intermittently around the western and northern rim of the basin and dip inward at up to 60°; drilling reveals that the dips flatten at depth. Six **goldfields** (concentrations of placer deposits) have been discovered, and the possibility exists that still more fields may be present under a cover of younger Proterozoic and Phanerozoic rocks, because the basin is open to the east and south and the original depositional limits are unknown.

The Witwatersrand goldfield is a group of braided fluvial fans developed at the debouchment of a few major rivers flowing from a fault-elevated highland to the northwest toward a shallow intermontane intracratonic lake that covered the base of the fan. Maximum gold mineralization occurs midfan; the parent rocks in the highlands were granitic plutons and hydrothermal gold veins.

The auriferous units occur as regressive deposits on unconformities at the base of sedimentary units, as transgressive deposits on angular unconformities, or as terminal deposits on disconformities at the top of sedimentary units. Lithologically, the ore beds ("reefs") are usually more mature sediments than those above or below them and range from coarse pebble-supported channel fill conglomerates in a proximal setting (Fig. 12-5) to small-pebble anastomosing sand bodies in a distal setting. In proximal areas, hydrodynamic analyses have shown that the sand-sized gold was not deposited contemporaneously with the quartz/chert cobbles and pebbles, but was introduced during later pulses of sedimentation. Heavy-mineral-bearing sand was later washed over and into the openwork gravels or was incorporated into the matrix during disintegration of sediments from an earlier pulse as a result of the movement of coarse clastic material over the surface separating the two pulses. The best concentrations of gold occur in pebble-supported bars that were reworked and winnowed on several occasions. In distal areas of sedimentation the gold concentrates are more likely to be true placers formed penecontemporaneously with deposition of the coarser-grained quartz particles.

In summary, the Witwatersrand gold placers were formed in front of a fault-bounded highland composed of an igneous-metamorphic Archean complex in which gold-bearing granites and silicic hydrothermal gold veins occurred. Deposition of the gold-bearing detrital sediment occurred predominantly in coarse-grained detrital sediments on the midsection of braided alluvial fans in the piedmont area. Gold particles entered the openwork pebbly beds largely by infiltration from immediately overlying sand accumulations; distal sand units contain standard placers formed penecontemporaneously with the deposition of coarser quartz grains. Transgressions and regressions of waters located to the southeast of the fans periodically reworked and enriched the gold concentrations, resulting in the richest gold deposits known.

FIG. 12-4 • Geometry of the Witwatersrand Basin as revealed by the depositional isopachs of the Central Rand Group. The asymmetry of the basin is shown by the distances between the zero isopachs and the depositional axis. Six fluvial fans, hosting the major gold fields, are all located on the short, shrinking side of the depository. (Reprinted by permission of the publisher from Pretorius, *Economic Geology, 75th Ann. Vol.,* Economic Geology Publishing Co., 1981.)

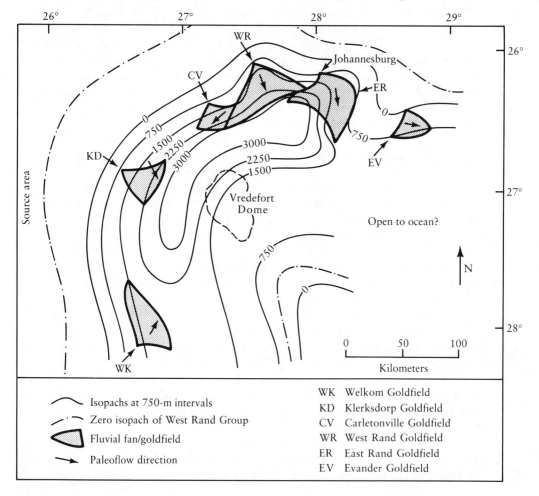

FIG. 12-5 • Large subangular clasts of the Ventersdorp Contact Reef in a proximal, wet alluvial fan, West Driefontein gold mine near Carletonville on the Far West Witwatersrand. (Adapted from Minter, *in* Armstrong, 1981.)

Bentonite

Bentonite is a rock composed almost entirely of colloidal montmorillonite and formed by the devitrification of volcanic glass, tuff, or ash. Minor amounts of euhedral brown biotite, zircon, feldspar, and quartz, commonly with beta-outlines, may also be present as residua from the original volcanogenic deposit. The most prominent use of bentonite is in drilling muds to increase the viscosity of the mud so that rock particles generated by the drill bit can be easily carried to the surface as the mud circulates. The high viscosity also aids in the prevention of blowouts that commonly result from penetration of a subsurface zone that has very high fluid pressure (gas or oil). Montmorillonite absorbs water readily and can increase in volume by as much as 20 times.

Bentonites may be either marine or nonmarine but are usually thin in either case; most are unfossiliferous. Nearly all economically valuable deposits are Cretaceous or younger because of the diagenetic change from smectite to illite and the resulting loss of swelling property that tends to occur with increasing time in clay minerals. Although bentonite beds up to 15 m thick are known, most beds are less than 0.3 m thick because of limitations on the amount of glass and mineral matter that can be ejected upward in a single eruptive episode. Such episodes are normally repeated numerous times within a geologically brief interval, however, so several dozen bentonitic units may occur in a formation, separated by either impure tuffs or other detrital materials. Compositionally, the bentonite beds may be either calcic, sodic, or potassic, depending on the composition of the original fragments. Individual beds that still contain recognizable detrital fragments typically are graded, both vertically and laterally, because the grains were deposited from atmospheric suspension. Bed thicknesses also decrease with increasing distance from the volcanic source. The beds may be laminated or massive. In general, the basal contact of bentonite with underlying beds is sharp, whereas the contact with overlying beds is gradational.

The best-studied bentonite sequence in North America occurs in a stratigraphic section of about 1300 m of mostly shallow-marine shales, marls, and argillaceous sandstones of Cretaceous age in northeastern Wyoming (Slaughter and Early, 1965). A single section may contain as many as 20 separate beds of bentonite. In the area as a whole, the bentonite beds vary in thickness from thin films to more than 3 m, the thicker beds commonly appearing to be composites of several individual ash falls rather than a single one. Many of the beds are lenticular, but some can be traced easily for more than 5 km. The base of the bentonite horizons is usually sharply distinct from the underlying sediment, usually a shale, although the upper part of the shale may be locally silicified to a chertlike layer up to 0.3 m thick. The thickness of the silicified zone is directly proportional to the thickness of the bentonite bed overlying it, indicating that the silica for silicification was obtained by downward percolation of aqueous solutions during diagenetic alteration of the original volcanic ash. The top of these bentonite beds is less distinct than their base and grades, often imperceptibly, into the overlying nonbentonitic bed.

As seen in outcrop, the bentonite beds are generally light yellow or green in color, the green being the natural color of the montmorillonite aggregates and the yellow resulting from oxidation and hydration of ferrous iron (limonite). Bed surfaces have a waxy character and weather to a flakey or granular form, repeated wetting and drying causing expansions and contractions that result in a popcornlike appearance on outcrop surfaces. Limestone concretions up to 5 m in diameter are associated with some of the bentonite horizons, and veins of gypsum frequently cut across the beds. Siderite concretions may also be present.

Kaolinite

In sediments as a whole, **kaolinite** is the least abundant of the three major clay mineral groups, and relatively pure deposits of kaolin are uncommon and volumetrically small. Although there are a few commercial deposits in the southeastern United States in Alabama, Georgia, and the Carolinas, the largest deposits occur in England, Czechoslovakia, Germany, France, and China. The purest, processed kaolins are used for fine-coated printing papers (50%) and in rubber, refractories, pottery, whiteware, and porcelain.

The origins of the relatively pure deposits of kaolin are varied and frequently controversial, but at least three possibilities are generally accepted (Jensen and Bateman, 1979). The most commercially important of the world's deposits are those in Cornwall and Devon, England, which contain 20–25% clay and have formed from porphyritic granite. These exceptionally pure kaolin deposits are believed by most investigators to have formed by hydrothermal alteration of the granite; but others maintain that it was the normal weathering of hydrothermally altered granite.

The kaolin deposits at Zettlitz, Czechoslovakia, lie in a graben (Fig. 12-6) and form a belt about 12 km long and up to 4 km wide. The deposit lies on the granite sides and bottom of the graben at depths of up to 60 m beneath Tertiary sediments that contain lignites and basalt flows. The deposit is about 14 m thick, and the upper part is the best; the lower part grades into granite. Clay forms 25–40% of the unit, the remainder being mostly quartz. It has resulted from normal weathering, possibly aided by thermal spring waters. The best areas of kaolinization are overlain by lig-

FIG. 12-6 • General section through the Zettlitz kaolin district, Germany. (Reprinted by permission of John Wiley & Sons, Inc., from Jensen and Bateman, *Economic Mineral Deposits*, copyright © 1979.)

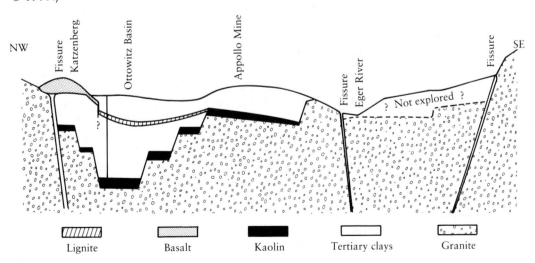

nite; therefore organic acids derived from the lignite are thought to be the important agents of kaolinization.

In the southeastern United States, according to Kesler (1956), the kaolin resulted from chemical weathering of feldspathic sands. Rapid erosion of crystalline rocks in the Piedmont led to formation of feldspathic sands that accumulated in a series of deltas. The kaolin formed by subsequent chemical decomposition of the feldspar during weathering when parts of the delta became exposed above sea level. Kaolin enrichment resulted from winnowing and deposition in pools and lakes formed from cut-off river meanders that were sometimes flooded by seawater. Water salinity affected the density of the deposit because of the effect of salinity on flocculation of clay minerals.

Aluminous (Bauxitic) and Ferruginous Laterite

Laterites are soils formed under humid tropical climatic conditions (Fig. 12-7) and composed almost entirely of only the most chemically resistant materials, generally hydrated ferric oxides and aluminum hydroxides (Butty and Chapallaz, *in* Jacob, 1984). These substances may be either amorphous or crystalline. Laterites are found in rocks from Proterozoic onward, but those of pre-Devonian age are rare because the rate of soil degradation is markedly enhanced by organic acids released from decaying vegetation. Plants did not become common on the Earth's land surface until Devonian time. Nearly all of the world's aluminum is obtained from bauxite deposits; ferruginous laterite is used as an ore of iron in underdeveloped areas and also as a building material, much as adobe is used in the southwestern United States. Because of the extreme climatic conditions required for the formation of areally extensive deposits, most modern laterites are located in low latitudes, but fossil laterites can occur in currently temperate regions because of continental drift and possible variations through time in the intensity of solar radiation.

Ferric iron and aluminum ions differ in ionic radius by 25%, so a complete solid solution between Fe_2O_3 and Al_2O_3 is not possible under Earth surface conditions. However, areally separable ferruginous and aluminous laterites may form at the same time, the Fe/Al ratio in the bedrock commonly being the decisive factor in determining which soil type is formed. The bauxite/ferruginous laterite separation may occur in the vertical dimension as well. For example, when bauxites are found overlying iron-rich rocks, the iron and aluminum tend to separate as an upper and a lower layer, respectively, probably as a result of upward leaching of iron under the influence of dense forest cover that provides the reducing agents and organic acids necessary for greatly enhanced mobility of the iron. Most lateritic bauxite deposits are found lying on erosion surfaces of relatively young age on cratons in South America, West and Central Africa, India, and Australia. In every case it can be shown that the erosion surfaces developed over a long period of time during the Tertiary, on the order of 5–10 m.y. in duration. The thickest blanket deposits are

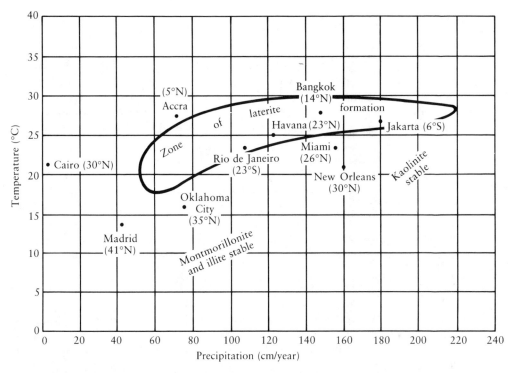

FIG. 12-7 ● Rainfall-temperature field in which laterite forms (latitude) in lowland areas. (Adapted from Persons, *Laterite*, Plenum Publishing Corp., 1970.)

25 m. A few examples of ancient bauxites are known in which the deposit is covered by later sediments. Obviously, such buried soils indicate unconformities.

Diatomite

Diatomite is a bioclastic rock composed largely or entirely of the microscopic shells of diatoms. Beds are up to tens of meters thick and range from yellowish tan to brown in outcrop, depending on impurities, but commonly they are so light in color that quarry faces appear brilliant white. The rock has a dull, earthy luster, is massive to finely laminated, is friable, and has an extremely high porosity, sometimes exceeding 75%. Commercial grades of diatomite range in chemical composition between 85% and 92% SiO_2, most of the remainder being Al_2O_3 derived from a few percent of clay mineral impurity. Gradations from diatomite through diatomaceous shale to normal shale are common along strike.

Diatoms evolved in the Jurassic but did not become numerous and/or widespread enough to form economically useful accumulations until Cretaceous time. Most commercial deposits are of Tertiary age. In general, the younger the accumulation of the opaline shells, the better the possibility that the diatomite will be of commercial importance because the diagenetic crystallization of the shells to chert destroys its most important physical property, its porosity. Although the specific gravity of opaline silica is 2.0–2.3, the shells are so loosely packed in diatomite that the specific gravity of the rock may be as low as 0.2 because more than 90% of the volume can consist of microscopic interconnected voids; 1 g of diatomite has an internal surface area of 20 m², similar to the specific surface area of clay shales. More than half of the diatomite quarried is used in filtering processes, and for this use a high surface area is required. The suitability of diatomite for filtration depends largely on the shapes and sizes of the constituent diatoms. Diatomite that contains a substantial percentage of long, hairlike diatoms (family Pennatae) together with fragments of discoid diatoms (family Centricae) generally is best for retention of fine impurities while allowing a high rate of flow. Rock of this type is mainly of marine origin. Other uses of diatomite include as fillers, insulating materials, mild abrasives, and absorbants.

Economic diatomite deposits are of three main types (Durham, *in* Brobst and Pratt, 1973): (1) marine rocks that accumulated near continental margins, (2) nonmarine rocks that formed in lakes or marshes, and (3) sediments in modern lakes, marshes, or bogs. The most extensive deposits of high-quality diatomite in the United States, and possibly the world, are located near Lompoc, California and exemplify the first type. The Lompoc accumulation (Fig. 12-8) formed near a convergent margin in a shallow-marine environment in Late Tertiary time and consists of more than 300 m of diatomite, both laminated and coarsely stratified, in the lower part of the Miocene Sisquoc Formation; the upper part of the formation is noncommercial diatomaceous claystone. In the commercial portion the diatomite consists of alternating layers of a pure, laminated variety, which is the material of value, and an impure, coarsely stratified variety, which is discarded as waste. The former is composed of laminae averaging about 0.8 mm in thickness and is light in color and highly porous. It consists almost entirely of well-preserved diatom tests, both discoid and needle-shaped. These types occur in varying proportions in different beds but in markedly uniform proportions in any one bed, so by selective quarrying it is possible to obtain pure accumulations with different characteristics useful for different commercial purposes. The commercially useless, coarsely stratified part of the diatomaceous unit contains up to 10% of clay, is of higher specific gravity, and is less porous than the laminated section, and the diatom tests in it are more likely to be broken.

Deposits of types 2 and 3 are of less economic importance than marine diatomites. Ancient nonmarine deposits include those in Nevada, Oregon, and Washington; modern lacustrine deposits are found in Florida, New Hampshire, and New York. Modern marine diatom accumulations are not of economic importance because of high mining costs.

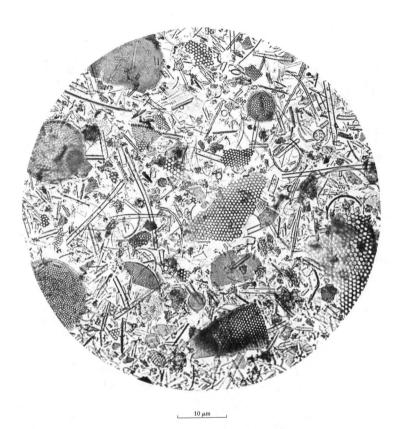

10 μm

FIG. 12-8 • Photomicrograph of marine diatom assemblage from Lompoc, California. Unbroken tests of both discoid and needle-shaped forms are abundant. (Reprinted by permission of the publisher from Kadey, *in* Lefond (ed.), *Industrial Minerals and Rocks*, 4th ed. American Institute of Mining, Metallurgical, and Petroleum Engineers, 1975.)

Evaporites

Evaporite beds are composed of the most soluble of naturally occurring substances at the Earth's surface and can be either marine or nonmarine. The thickest and most extensive deposits are marine. Most of the 70 or so minerals that constitute evaporites can be grouped for convenience as halite, gypsum/anhydrite, and potash salts.

Among the marine evaporites, salt (**halite**) is by far the most valuable mineral resource, ranking with coal, limestone, iron, and sulfur as a basic industrial raw material. It is mined in most countries of the world and is used as a source of sodium, chlorine, soda ash, hydrochloric acid, caustic soda, and other compounds that are indispensable in the manufacture of many other products and chemical reagents.

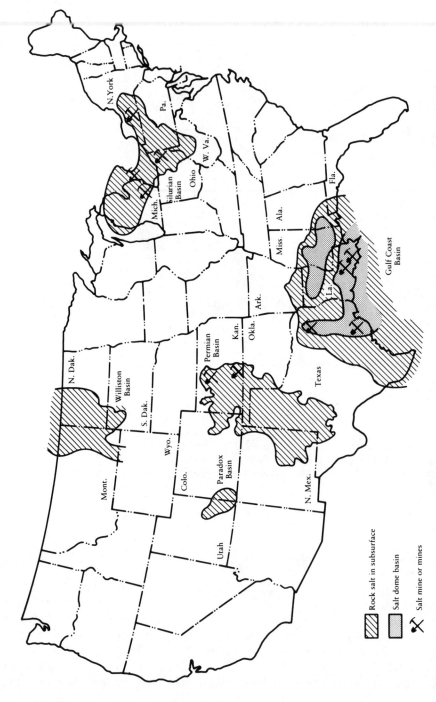

FIG. 12-9 ● Major salt basins of the United States. The names Permian and Silurian indicate the geologic age of the salt-bearing strata. Salt deposits in the Williston Basin are Devonian, those of the Paradox Basin are Pennsylvanian, and the deep bed from which the Gulf Coast salt domes arose is Jurassic. (Reprinted by permission of the publisher from Lefond, *Handbook of World Salt Resources*, Plenum Publishing Corp., 1969.)

Salt itself is important in the production and preservation of foodstuffs as well as for animal feed, water softening, snow and ice removal, and roadbed stabilization. Salt beds as thick as 400 m are known, but those pure enough to be of economic importance are less than 100 m thick. Naturally occurring salt contains between 1% and 4% impurities, which consist mainly of gypsum/anhydrite, shale, dolomite, or quartz, and some of these must be removed by chemical treatment or filtration to meet the specifications for some industrial uses.

Halite is a basic raw material used by many industries as well as an essential foodstuff. It is obtained commercially from several sources including bedded salt deposits, salt domes, evaporated seawater, salt lakes, and subsurface brines. As noted by Halbouty (1979), the potential supply of salt from domes in the Gulf coastal region of the United States alone is almost unlimited. Currently, the amount of minable salt from 156 known salt domes in the area is estimated to be about 1500 km^3; current annual production is less than 0.01 km^3.

Commercial gypsum deposits, like those of halite, may occur in any Phanerozoic system, but most gypsum deposits of economic significance are of Permian age. Commercial **gypsum** is a compact, massive, finely crystalline to granular rock; selenite and satin spar are common and of wide occurrence but are not of economic importance. Ore beds range in thickness from 1 m up to about 30 m; the best contain no anhydrite, which has little value at present and is thus considered gangue. More than 95% of mined gypsum is calcined (heated to drive off three-fourths of its water of crystallization) for use as plaster of Paris, for construction and industrial purposes, and for manufacturing wallboard and other prefabrication products widely used in the building industries. Gypsum is also used to retard setting time in portland cement and as an agricultural soil conditioner and fertilizer.

About 95% of the **potash** minerals mined in the United States goes into the manufacture of fertilizer for potassium-deficient soils. Sylvite (KCl) is the chief naturally occurring economic mineral. The mines that supply most of the potash minerals in the United States are latest Permian in age and are located east of Carlsbad, in the southeastern corner of New Mexico. In this deposit the most common and widespread potassium-bearing mineral is polyhalite, but the most extensively mined potash unit is a bed composed of about 40% sylvite and 60% halite, 2–4 m thick. Many masses of halite interrupt the lateral continuity of the potash because of differential solution of the potash bed during diagenesis and squeezing of halite both upward and downward into the space created.

Nonmarine, playa lake deposits are sometimes rich in economically valuable evaporite minerals, frequently minerals that are never found as precipitates from marine waters. These include trona (used to make sodium carbonate, bicarbonate, hydrazide, and nitrate), several borate minerals (used in the manufacture of glass, soaps and detergents, agricultural products, and other items), and nitrates (fertilizer).

Four large depositional basins totaling nearly 1.3×10^6 km^2 comprise the majority of U.S. resources of halite (Fig. 12-9). These basins are (1) the Gulf Coast (post-Permian salt diapirs based in a rifted-margin marine halite unit), (2) the

Permian (cratonic marginal marine setting), (3) the Salina (intracratonic Silurian basin) and (4) the Williston (intracratonic Devonian basin). As was noted previously, the bulk of the salt mined in the United States comes from the diapirs in the Gulf Coast region, but the parent bed of these pinnacles is many thousands of meters below the surface. Its stratigraphic relationships are unknown. Much better known are the stratigraphic relationships in the Salina Basin, in which salt is also mined and where lithologic relationships and sedimentology have been deciphered on the basis of several hundred well cuttings and cores and on outcrop observations.

Evaporites in the Salina Basin are of Silurian age (Cayugan) and, although most prominent in Michigan, they extend eastward into New York and southward into West Virginia over an area of several hundred thousand square kilometers (Fig. 12-10). The maximum thickness of the interval containing the evaporites is about 800 m in Michigan but usually less than half of this is evaporite beds; dolomite and shale are interbedded with dominant halite, subordinate sylvite, and gypsum/anhydrite. A typical stratigraphic column of the Salina Group in Michigan, traceable with some modification into the Appalachian Basin, illustrates the common lithologies (Alling and Briggs, 1961).

Unit H (Bass Island dolomite). 60–190 m thick, largely buff dolomite, some gray dolomite near base, anhydrite and salt beds near center of basin.

Unit G. Uppermost Salina Formation, 1–30 m thick, characteristically gray shaly dolomite, green and red shales near Mackinac Straits.

Unit F. Uppermost salt in the Salina section, 0–410 m thick. Thick beds of salt separated by shale, shaly dolomite, dolomite, and anhydrite.

Unit E. 10–41 m of gray or red shale with some dolomite, shaly dolomite, and anhydrite.

Unit D. 8–22 m, nearly pure salt, with thin partings of buff dolomite.

Unit C. 20–53 m, largely shale or shaly dolomite with anhydrite and buff dolomite in places.

Unit B. 8–133 m thick, almost pure salt with minor dolomite.

Unit A. 10–368 m thick, limestone, dolomite, salt, and anhydrite. This unit was subdivided into:

A_2: Buff to brown dolomite or limestone
    Dark gray dolomite
    Salt or anhydrite
A_1: Fine brown to grayish brown dolomite or limestone
    Fine gray dolomite or limestone
    Salt or anhydrite

Around the rim of the Michigan Basin the basal salt or anhydrite rests on light brown dolomite, whereas in the center of the basin the evaporites rest on reddish, argillaceous bioclastic limestone. Dolomite and shale overlie the evaporitic units.

The stratigraphic sequence of the Salina Group as shown above is characteristic of a **sabkha** environment, an arid climate tidal flat on which the supratidal area is

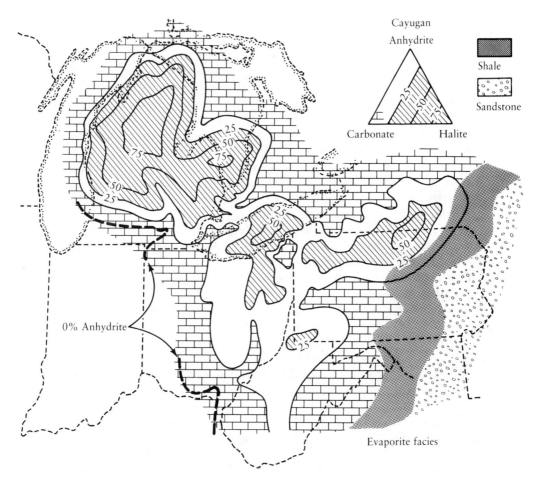

FIG. 12-10 ● Evaporite facies map showing distribution of three principal evaporite basin rock types: carbonate rock, anhydrite, and halite. Analyses are based on percentage thickness of each type in total Cayugan section at each well. (Reprinted by permission of the American Association of Petroleum Geologists from Alling and Briggs, *American Association of Petroleum Geologists Bulletin*, v. 45, 1961.)

host to isolated pools of seawater that, on evaporation, become evaporite-precipitating brines. Evaporation causes saturation with respect to calcium sulfate, which precipitates as gypsum. The remaining water has a much higher density and Mg/Ca ratio than the original seawater, and it sinks downward through underlying limestone, dolomitizing it. The resulting gypsum/anhydrite-dolomite sequence is easily recognizable in outcrop by lithology, textures, and sedimentary structures. Features such as nodular gypsum/anhydrite, chicken-wire texture, and intraformational dolomite flat-pebble rip-up clasts are quite diagnostic of this tidal flat setting. Detailed descriptions can be found in Blatt (1982).

Phosphate Rock

More than 80% of the minable deposits of phosphate occur in sedimentary rocks, rocks that contain between 5% and 35% P_2O_5, two orders of magnitude more than average crustal rock. The economic mineral in phosphate rock is carbonate fluorapatite, $Ca_5(PO_4,CO_3)_3F$, with carbonate ion substituting for phosphate ion in amounts up to 10%. Impurities are dominantly calcite and dolomite, quartz and chert, clay minerals, glauconite, and carbonaceous matter. The most productive deposits are located in North Africa, in Morocco, Algeria, Tunisia, and Egypt (Late Cretaceous–Eocene); the most widespread are in the western United States in the Phosphoria Formation (Permian). In the United States, about two-thirds of the phosphate consumed is used in agriculture, the remainder in cleaning compounds, foodstuffs, insecticides, and many other products.

Phosphate rock in outcrop varies considerably in hardness and color. Those of higher grade are typically massive to thickly bedded (up to 6 m) in contrast to the thinly bedded or laminated character of phosphatic shales. Secondary processes such as diagenetic phosphatization of limestone and interstitial precipitation, reworking of a primary deposit on the sea floor by waves and currents, and later weathering in outcrop have commonly played a part in forming deposits of minable quality. Leaching of the carbonate cementing material in surface outcrops results in a soft, porous rock that readily disintegrates with handling. Enrichment of the deposit is a characteristic feature of many deposits; in the western United States, mining of the Phosphoria Formation is restricted to the weathered zones, which may extend to depths of 400 m and contain up to 8% more P_2O_5 than the deeper, relatively unweathered rock.

Phosphate ore is known in rocks ranging from Proterozoic to Tertiary in age, and most deposits are of shallow marine origin, as revealed by stratigraphic relationships, fossils (commonly phosphatized), and sedimentary structures. The rocks consist of phosphatic chemical or biochemical mud and clastic grains, mostly intraclasts, pellets, oolites, and fossil skeletal material; the allochemical material dominates in Phanerozoic deposits, mud in Proterozoic ones. In many ways, phosphate rocks are phosphatic copies of limestones, and many may indeed be replacements of limestones, although no evidence has yet been found to support a replacement origin for extensive deposits such as the Phosphoria Formation. Most phosphorites are associated with black shales, cherts, and carbonate rocks; glauconitic units are not uncommon.

The Phosphoria Formation in the western United States is the most thoroughly studied phosphate deposit in the world and, because of this, has become an unofficial type section for such deposits (McKelvey et al., 1959). The Phosphoria phosphorites were deposited over an area of about 350,000 km² in both the platform and deeper shelf or geosynclinal portions of the Mesozoic Cordilleran structural belt (Fig. 12-11). Because the environments of deposition are bisected by the "hinge line" between shallow and "deep" waters, the rocks show pronounced changes in thickness and character from west to east (Fig. 12-12). In south-central Idaho they are more than 400 m thick and consist of bedded chert, dark carbonaceous mudstone, and phos-

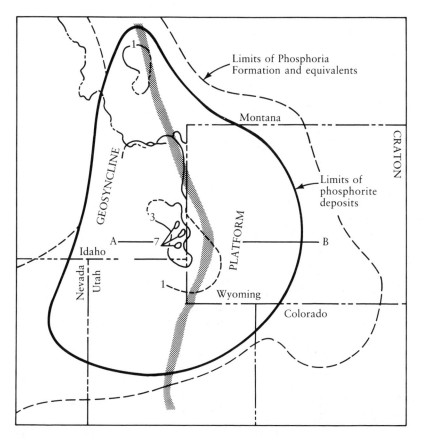

FIG. 12-11 • Geographic extent of the Phosphoria Formation and its phosphate rock deposits. Numbered lines are isopachs (in meters) of rocks containing at least 31% P_2O_5. Most mines are within the 3-m contour in Idaho and the 1-m contour in Montana. A-B is approximate line of cross-section of Fig. 12-12. (Reprinted by permission of the author from Bates, *Geology of the Industrial Rocks and Minerals,* Harper and Row, 1960.)

phorite. To the east these beds thin to about 100 m and merge into carbonate rocks on the southern part of the platform and into sandstone on the northern part. There is widespread intertonguing. Highly phosphatic beds occur in both the platform and geosynclinal facies and reach a maximum near the "hinge line," which also serves as a convenient dividing line between contrasting petrofacies of the Phosphoria. On the shelf to the east, noncarbonaceous, cross-bedded, allochemical facies dominate (pelletal, oolitic, bioclastic); to the west, carbonaceous, phosphatic mudstones and shales dominate. Notwithstanding the fact of regional facies changes and the naming of eleven members, strata within the Phosphoria are characterized by lateral continuity. The three members present in the type area in southeastern Idaho can be readily recognized over a wide area in Idaho and adjacent parts of Wyoming and

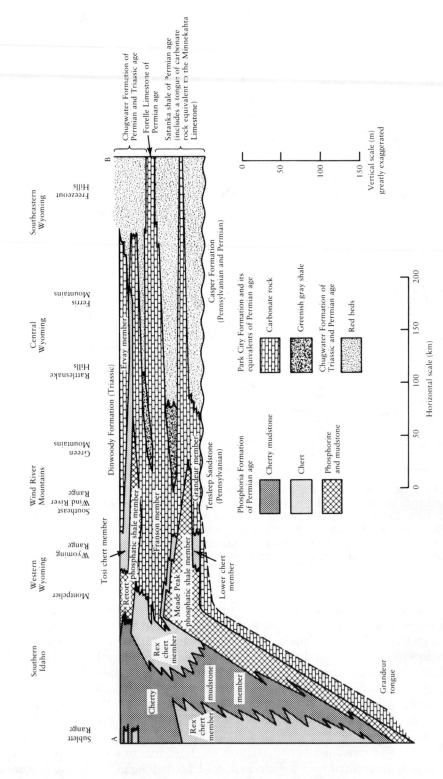

FIG. 12-12 ● Stratigraphic relations of the Phosphoria, Park City, and Chugwater Formations in Idaho and Wyoming. (Adapted from V. E. McKelvey et al., 1959.)

Utah. In parts of the region, beds less than 1 m thick can be traced for many tens of kilometers.

The origin of highly phosphatic units such as the Phosphoria Formation is still uncertain, although there is general agreement that they form mostly in oceanic areas characterized by upwelling of deep, cold, nutrient-rich waters, commonly along the west side of continents where offshore winds dominate. However, evidence that a major marine transgression preceded deposition of many phosphorite deposits (Mitchell and Garson, 1981) suggests a possible indirect plate tectonic control of some phosphorites, since many eustatic rises in sea level are probably related to development of ocean rises. The consequent flooding of shallow shelf areas may have resulted in higher organic productivity, with phosphate accumulation in mildly reducing environments produced by bacterial oxidation.

Bedded Iron Ore

The largest concentrations of iron ore are found in banded iron formations of Precambrian age (Gole and Klein, 1981), mostly 2.6–1.8 billion years (Fig. 12-13). They are mined extensively in the United States, Canada, South America, Africa,

FIG. 12-13 • Outcrop view, upper part of the Negaunee Iron Formation, Marquette District, Michigan. Dark bands are hematitic chert (jasper); light bands are specular hematite. (Reprinted by permission of the publisher from *Economic Geology*, v. 68, Economic Geology Publishing Co., 1973. Photograph courtesy of H. L. James.)

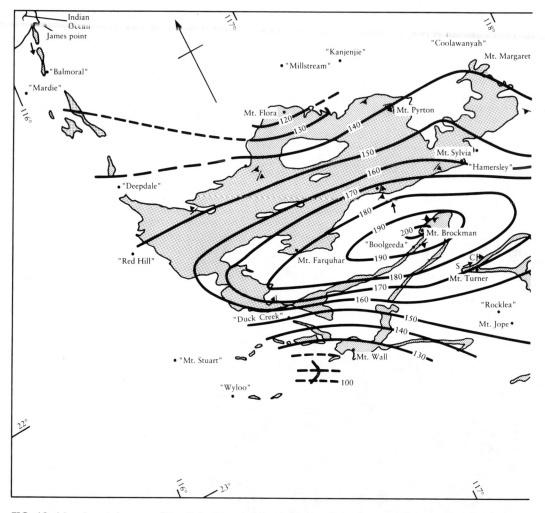

FIG. 12-14 • Isopach map of the Dales Gorge Member showing minor structures within the Hamersley Group. (Adapted from Trendell and Blockley, *Geological Survey of Western Australia Bulletin 119*, 1970.)

India, Australia, and the Far East. These unusual sedimentary rocks range in thickness from about 30 m to 700 m, can be traced along strike for distances up to 1500 km, and may be exposed over areas of many thousands of square kilometers. They usually consist of laminae of gray or red chert interbanded with cherty magnetite- and hematite-rich layers with accessory iron silicates and iron carbonate and contain 20–40% iron. In the Lake Superior region along the U.S.-Canada border these rocks are termed **taconite**. The formations have been locally enriched to high-grade deposits of hematite by oxidation of ferrous minerals and the removal of silica. These

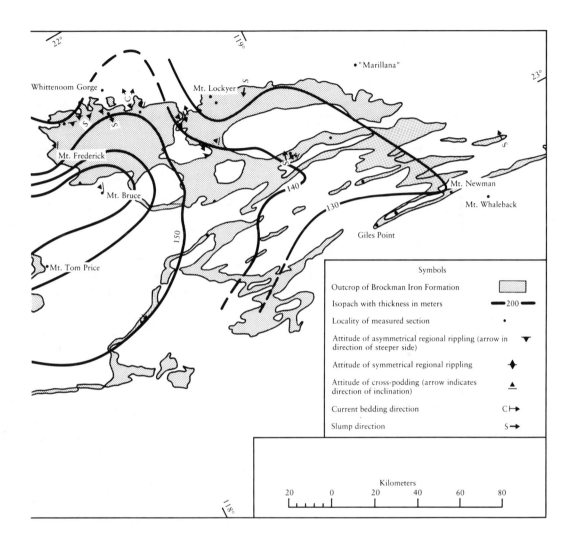

residual ores are usually earthy and porous and contain 50–60% iron in their natural state; examples occur in the Lake Superior district and the Krivoi Rog district of the Soviet Union.

Typical of the major, economically important banded iron formations is the Dales Gorge Member (200 m) of the Brockman Iron Formation (650 m) of the Hamersley Group (50,000 km², 1500 m) in Western Australia (Trendall and Blockley, 1970). The isopach map of the Dales Gorge Member (Fig. 12-14) is taken to represent the general form of the bottom of the Hamersley Basin, and more detailed isopachs of the upper, middle, and lower parts of the member indicate that the basin varied little in depth over wide areas at the onset of deposition. The major changes in basin contour during deposition of the member were deepenings around two centers, one at Mount Bruce and the other at Mount Brockman. The final stage in the deposition of the member was accompanied by deepening of the basin along a line joining these two centers, resulting in the shape that characterizes the isopach map of the whole member.

Trendall and Blockley (1970) recognized three scales of banding in the Dales Gorge Member, which they termed micro- (less than 4 mm), meso- (less than 80 mm), and macro- (80 mm−25 m), and they were able to "trace" mesobands of chert and magnetite over distances of hundreds of kilometers. Continuity of such thin laminae over such great distances is unknown in other types of rocks of any geologic age and suggests remarkable uniformity in environmental conditions over many millions of years. The mechanism for such uniformity is unclear, as is the depositional environment for the unit. Palinspastic reconstructions indicate that the sediments were deposited on a submarine, essentially volcanogenic platform or bank of Archean age. The bank either protruded into, or was marginal to, an ocean (Morris and Horwitz, 1983).

Banded iron formations are restricted to Precambrian rocks, the youngest being 600−800 m.y. old, and are exceedingly pure. They contain almost no detrital material such as clay minerals or quartz, a fact that has been interpreted by some workers to indicate that the chert and iron are chemical precipitates. If this is true, there can have been no oxygen in the atmosphere because only ferrous iron has a significant solubility in water; precipitation after transport of the ferrous iron would have been accomplished by oxygen-releasing plants in the depositional basin. All oxygen released by the plants would be used in oxidation of the ferrous iron ions, so none accumulated in the atmosphere. On the other hand, it is noteworthy that the age of the youngest banded iron formations is the same as the age of the Ediacara fauna that contains coelenterates and brachiopods. Modern representatives of these groups require a level of atmospheric oxygen at least 10% of that now present. Ferrous iron ion cannot exist under such atmospheric conditions. The dilemma is unresolved.

Another type of sedimentary iron deposit, of regional importance in the United States (Fig. 12-15) and western Europe, is the oolitic hematitic ironstones of Paleozoic to Cretaceous age. These formations are areally extensive and contain 20−40% iron, but they are very unlike the Precambrian banded iron formations. Ironstones are usually less than 20 m thick, and minable layers are often less than 7 m thick; interbeds are quartz-rich sandstones and shales. Locally, the beds are calcareous and grade into impure limestones, and normal marine shallow-water fossils are abundant. Oolitic nuclei are quartz grains and fossil fragments. Except for the enrichment in iron, the stratigraphic sequences in which the ores occur are identical to oolitic limestones that occur along some low-lying passive continental margins. The iron in the oolites probably is primary (Van Houten and Bhattacharyya, 1982), although some workers favor an origin by replacement during shallow burial.

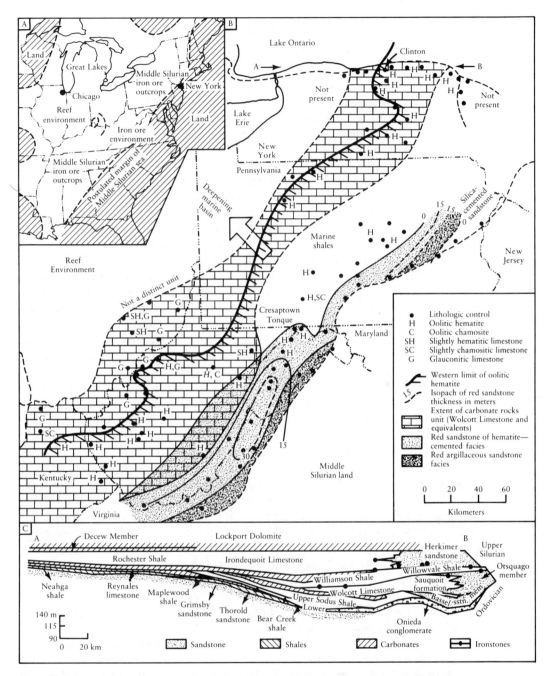

FIG. 12-15 ● (A) Distribution of the Silurian Clinton Group sedimentary ironstone ores of the United States. Relationship of the shallow-sea environment to the land masses. (B) Lithofacies map of the Upper, Lower and Middle Clinton Group. (C) East-west cross-section from southeast Lake Ontario through the town of Clinton. (Reprinted by permission of John Wiley & Sons, Inc., from Hutchison, *Economic Deposits and their Tectonic Setting*, copyright © 1983.)

Manganese

Most commercial manganese deposits are of sedimentary origin, the richest deposits containing about 50% manganese, a result of secondary enrichment of a primary deposit, usually a manganiferous sandstone. Few beds containing less than 25% manganese are being exploited at present. The major commercial manganiferous minerals are the oxides pyrolusite (MnO_2), manganite ($MnO(OH)$), and hausmannite (Mn_3O_4), the latter two minerals being the manganese analogs of goethite and magnetite. Sedimentary $MnCO_3$ is widely distributed but is seldom commercial. Manganese is used almost entirely in the steel industry to decrease brittleness and to increase resistance to abrasion or shock. It is not an overstatement to say that without manganese there would be very few useful steel products.

The manganese in the Earth's ore deposits originated from manganiferous mafic minerals in igneous rocks and was released into the sedimentary environment by normal weathering processes. The manganese was carried to the site of deposition in very dilute solutions, but a significant difference in characteristics is present between deposits of manganese laid down as carbonates and those deposited as oxides. Primary oxide deposits are more commonly associated with coarser clastic sediments without organic carbon, indicating near-shore and strongly oxidizing conditions with relatively free circulation of water. Oxide-facies manganese deposits in Precambrian and lower Paleozoic rocks may be closely associated in space and time with oxide-facies iron formation. Carbonate-facies manganese ores, however, are usually separated both stratigraphically and spatially from carbonate-facies iron-formation rocks.

Most large sedimentary deposits of manganese are oxides and are marine, as indicated by fossils and associated rocks. It is thought that the manganese in these deposits was precipitated from solution by oxidation from the relatively soluble divalent form in which the manganese was transported in streams to the less soluble quadrivalent form; the most common mineral is MnO_2. The locations of such deposits typically are strongly controlled by chemical facies, and highly oxidizing near-shore facies are common depositional sites (Force and Cannon, 1988). In many places these manganiferous lenses are much longer than they are wide. Some extend for many kilometers, but the lenses more commonly are a few hundred meters to a few kilometers long and a few tens of meters to a kilometer or more wide. In many deposits the manganese is in a single bed; in others it is in several separated beds. The thickness of the beds may be as much as 30 m, but few are more than 3 m thick.

Manganese ore in carbonate rocks occurs as impure rhodochrosite and is associated with carbonaceous rocks, often rather clayey, indicating a reducing environment with only weak current activity. The carbonate ore deposits are smaller and contain less manganese than the oxide beds. Pure $MnCO_3$ contains only 47.8% manganese, and 35% manganese is considered to be high-grade ore in carbonate deposits. Most beds contain 15–30% manganese and are less than 1 m thick.

Manganese ore is produced from the tropical to warmer temperate climatic zones of about 30 countries. The Soviet Union produces about one-third of the world's supply, followed by South Africa, Brazil, Australia, and Gabon, but there are few large deposits, and most are much smaller in size than sedimentary iron deposits. The United States is completely dependent on foreign sources of manganese; the last U.S. manganese mine closed in 1970.

Between 75% and 80% of the world's present reserves of manganese are contained in the deposits at Nikopol and Chiatura, on the northern and eastern fringes of the Black Sea in the Soviet Union. The Nikopol district includes an arc-shaped belt about 150 km long and 25 km wide and contains a single ore horizon of Oligocene age, 2–3 m thick. This bed contains marine fossils and rests on up to 20 m of fine sand and mud, which rests on granite. The ore thickens where it fills depressions in the underlying basement and is cut out in places by later erosion. To the north the ore bed wedges out; to the south it disappears because of an increase in clay interbeds and a gradual dilution of the ore by clastics. The ore grades from oxides in the north to carbonates in the south, corresponding to increasing depth of water. The top of the ore horizon is marked by iron hydroxides in the oxide ores but by glauconite in the carbonate ores; green montmorillonitic clays up to 25 m thick overlie the ore zone.

The ore horizon is divided into three facies: oxide, mixed oxide-carbonate, and carbonate. The carbonate facies occupies the most distal position and occurs in two varieties: (1) concretionary-nodular ores with nodules 1–25 cm in diameter and a mud matrix and (2) coarse-lumpy ores with a highly porous texture. Very fine-grained rhodochrosite and manganoan calcite are the ore minerals. Besides the carbonate minerals, the nodular variety contains sponge spicules, diatoms, and fishbones, while the lumpy variety is relatively free of such inclusions. Shoreward, the mixed facies consists of irregular spherical masses, 0.1–1.0 cm in diameter, of MnO_2 in a matrix of manganese carbonates. The oxides are corroded by the carbonates. Closest to the paleoshoreline is the oxide facies, which may be primary but may also be the result of near-surface alteration of carbonate ore. Texturally, the oxides in this facies consist of earthy masses or concretions up to 25 cm in diameter, sometimes with relics of carbonate or oxide-carbonate structures. There is well-developed concentric layering, and smaller concretions are often coalesced into larger, compound structures.

Coal

Coal is a combustible, stratified, organic rock composed largely of altered and/or metamorphosed plant remains mixed with a variable but subordinate amount of inorganic material. Such rocks underlie 13–14% of the land area of the United States and are present in 39 states, and reserves are estimated to be adequate for hundreds of years. Coal currently supplies about 25% of the energy consumed in the

United States (oil and natural gas supply 65%). In terms of domestic reserves, however, coal makes up about 70%, and oil and natural gas make up less than 10%. Approximately 20% of mined coal is used not for heat and power, but in the manufacture of iron and steel. If the environmental problems involving sulfurous emissions during coal combustion can be solved economically (coal averages 1–2% sulfur, mostly as pyrite and marcasite crystals), the commercial importance of coal as an energy fuel should rise dramatically during the next few decades. The United States is estimated to contain 20% of the world's reserves, and the Soviet Union 56% (Jensen and Bateman, 1979).

Coals extensive enough to be of commercial importance contain land plants as the essential constituents; therefore few minable beds are older than Devonian in age. By Carboniferous times the land flora included representatives of several of the major plant groups capable of producing both forest trees and ground cover vegetation. More than 3000 plant species have been identified from Carboniferous coal beds. Luxuriant swamp-forests developed, and thick peat deposits were laid down, which subsequently were transformed into the widespread Carboniferous coals in many parts of the world. Of greatest temporal importance are the Cretaceous-Tertiary coals, a result of the establishment during the Cretaceous of the flowering plants (Angiospermae) and their rapid ascendance to dominance in the plant world. Coal derived from the Tertiary flowering plants forms more than half of the total known world reserves.

The classification of coals for commercial purposes is based on three factors, termed rank, grade, and type (Blatt, 1982).

1. The **rank** of the coal is its position in the series peat, lignite, subbituminous coal, bituminous coal, anthracite coal, and meta-anthracite coal. Rank is established by determining the reflectance of a polished coal surface; the higher the reflectivity, the higher the rank.
2. The **grade** of a coal is determined by chemical analyses and refers to its content of impurities such as ash; higher and more valuable grades of coal have fewer impurities.
3. The **type** of coal refers to the kinds of plant particles (called *macerals*) of which the coal is composed. The three groups of particle types are termed *vitrinite* (usually 50–90%), formed by woody and cortical tissues; *exinite* or *liptinite* (5–15%), formed by the waxy and resinous parts of plants such as spores, needle and leaf cuticles, and wound resins; and *inertinite*, formed by fungal remains, oxidized wood or bark (perhaps fire-burned), and plant fragments so altered that their origin is uncertain. Each of these three groups of particles has been extensively subdivided by coal petrographers.

Luxuriant growth of land plants and subsequent accumulation of peat requires a humid climate, a high water table, and a delicate balance between basin settling and plant accumulation so that the dead vegetation is buried, not oxidized to carbon dioxide and water and not eroded after burial. Such conditions are most easily met along subsiding low-relief coastlines in temperate to subtropical climates, but smaller plant accumulations can occur in more arctic conditions, as in Canada (muskeg), Alaska, and ancient Gondwana during Permian time. At present in the United States, prime peat-forming settings are located in the Louisiana Gulf Coastal region and in the swampy areas along the Atlantic coast from Florida northward to Virginia (Dapples and Hopkins, 1969). Although these modern swamps seem large to a person lost in one, ancient swamps were much more extensive and persisted for longer periods of time because the sea level was not subject to the wide fluctuations associated with today's rapid and repeated continental glaciations. The extensive swamps of the past, in which through-flowing streams were rare, permitted development of laterally extensive peat beds with relatively constant thicknesses over great distances. Many of our most valuable coal beds have resulted from the persistence of such stable coal-forming settings. For example, the Pittsburgh seam in Pennsylvania underlies 38,000 km^2.

Most of the large peat deposits of Pennsylvanian age that were the precursors of Appalachian coal were formed near sea level, in estuaries or coastal lagoons, on large deltas or coalescing deltas, or on low-lying, broad coastal plains. Slowly transgressing seas ultimately covered the peat-forming swamp and terminated plant growth. In the very delicate balance between sedimentation, subsidence, and uplift of the land, the sea also regressed from time to time. Peat swamps also formed during the regressive phase of the cycle, but these were subject to oxidation and are less commonly preserved. These cyclic repetitions of coal-forming conditions are documented in many of the world's coalfields, as in West Virginia, where 117 coal beds of sufficient geologic and economic interest have been described and named.

The Pennsylvanian coals in the Appalachian basin contain nearly all of the anthracite coal and about 25% of the bituminous coal in the United States. The facies in the central part of the region in western Pennsylvania, West Virginia, and Kentucky are exceedingly complex in detail and range from non-coal-bearing alluvial fans to alluvial plains both with and without coal beds, to upper and lower delta plain facies that contain the most economically valuable coal seams, to the coal-free tidal flat and marine barrier island facies (Ferm and Horne, 1979). Peat deposits in the Appalachian basin formed in any standing body of water in which vegetation was able to accumulate, environments such as swamps, oxbow lakes in abandoned meanders, and interdistributary bays that exist in the transition zone between the upper and lower deltaic plains. All these types of environments, however, are characterized by geologically very rapid facies changes because fluvial channels change position frequently on the low-lying near-coastal surface. Figure 12-16 is a generalized map and cross-section of this type of paralic setting; Figure 12-17 is a photograph showing the thin and discontinuous coal seams that commonly are formed because of the instability of depositional environments. An extensive discussion of coal facies is given by Galloway and Hobday (1983).

The most extensive coal deposits are formed in paralic settings along stable rifted continental margins. The present-day Niger, Amazon, and Mississippi River deltas are excellent examples. Ancient examples include the Carboniferous coals of northwestern Europe and the North American Pennsylvanian coals. The rivers responsible for paralic coal basins flowed off stable low-lying cratons toward the sea, and the transition from continental to oceanic crust is not marked by tectonic activity. Climates were at least subtropical, as indicated by paleomagnetic data that indicate paleolatitudes of major coal basins to lie within 20° of the paleoequator.

FIG. 12-16 • Depositional model for peat-forming (coal) environments in coastal regions. The upper part of the figure is a plan view showing sites of peat formation in modern environments: the lower part is a cross-section (AA′) showing, in relative terms, thickness and extent of coal beds and their relations to sandstones and shales in different environments. (Adapted from Horne et al., *in* Ferm and Horne, 1978.)

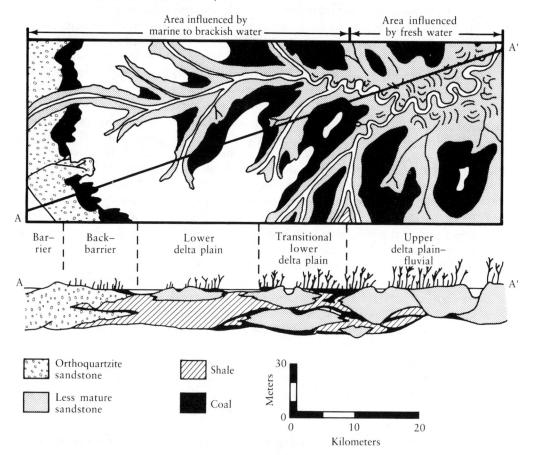

FIG. 12-17 • Discontinuous layers of coal about 5–10 cm thick interbedded with sandstone, Petersburgh Formation (Pennsylvanian), Indiana. The continuity of coal seams is frequently interrupted by lenticular sandstones, e.g., upper right. Environment of deposition of coal was apparently a floodplain adjacent to a meandering, low-gradient stream. (Reprinted by permission of the publisher from Pettijohn and Potter, *Atlas and Glossary of Primary Sedimentary Structures*, Springer-Verlag New York, Publishers, 1964.)

Oil Shale

Oil shale is a brown to black fine-grained sedimentary rock containing abundant organic matter that is insoluble in ordinary petroleum solvents but yields oil when heated in a closed container (destructive distillation). Such organic matter is termed **kerogen**. The fraction of organic matter that *is* soluble in such solvents is termed **bitumen**. The amount of oil that can be extracted ranges from about 4% to more than 50% of the weight of the rock (50–700 liters/metric ton), and in the United States, which has the largest proven reserves in the world (Fig. 12-18), oil shales contain a potential two trillion barrels of oil. The figure of two trillion barrels of oil in these shales may be compared with the total world production of petroleum to date, which is less than 0.3 trillion barrels. Nevertheless, there is no commercial

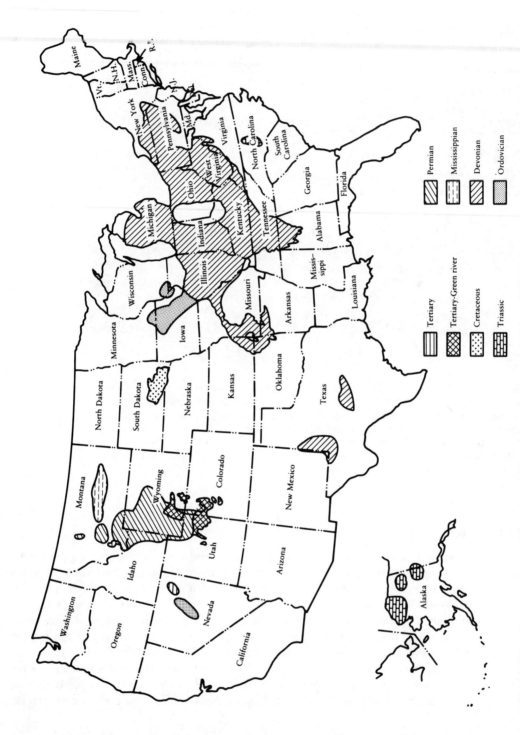

FIG. 12-18 ● Oil shale deposits of the United States. (Reprinted by permission of the publisher from Smith, *Mineral and Energy Resources*, v. 23, Colorado School of Mines Press, 1980.)

Legend:
- Permian
- Mississippian
- Devonian
- Ordovician
- Tertiary
- Tertiary-Green river
- Cretaceous
- Triassic

production of oil from oil shales in the United States at present because of the relatively abundant supplies of petroleum still available at lower prices from porous and permeable sandstones and carbonate rocks. For oil shale to be economically competitive with petroleum and natural gas, the price of a barrel of crude oil needs to rise from the present cost of $15–20 to about $35–40. The startup cost for oil shale generation is very high.

Rocks called oil shales grade into coals as the amount of mineral matter in them decreases; normally, 40% minerals is taken as the dividing line between the two rock types. Much of the organic matter in oil shales is finely disseminated and so altered by diagenetic degradation that the organisms from which it formed cannot be identified, but in many rocks the remains of algae and algal spores are common, so much of the organic matter is assumed to be of algal origin. Fine-grained debris of higher plants as well as megaspores may also be important constituents. The common sedimentary feature of many oil shales is a distinct lamination on a millimeter scale of alternating clastic and organic laminae, generally attributed to annual algal blooms in lacustrine environments. The kerogen in the laminae is typically yellow to amber-colored, is structureless, and occurs in bands and stringers parallel to stratification.

The principal oil shale deposit of lacustrine origin in the United States is the Green River Formation (Eocene), which underlies about 44,000 km² in western United States (Brobst and Tucker, 1973). The formation was deposited principally in two large lakes but, because of subsequent tectonic activity, is now preserved in seven basins (Fig. 12-18). It consists of oil shale interbedded with varying amounts of tuff, siltstone, sandstone, claystone and locally with halite, trona, or nahcolite (sodium bicarbonate); the maximum thickness is close to 1000 m, and the formation intertongues laterally with the predominantly fluvial beds of the Wasatch Formation. The chief oil shale unit in the Green River Formation is the Parachute Creek Member, which contains up to 600 m of oil shale in alternating rich and lean beds. The richest oil shale sequence in the member is called the Mahogany ledge in surface section (Fig. 12-19) and the Mahogany zone in the subsurface, after the red-brown color of the rich oil shales. This zone underlies more than 5200 km² in the Piceance and Uinta Basins and locally exceeds 60 m in thickness; the richest bed in the zone is 1–3 m thick and has an average oil content of 190 liters/metric ton of rock. Mineralogically, the Mahogany ledge consists of about 46% dolomite, 22% calcite, 14% analcime, 11% quartz, 3% each of potassium feldspar and albite, and 1% illite and is thus an impure calcitic dolostone. The kerogen-rich carbonate unit is usually calcite; the kerogen-poor carbonate units are dolomitic. Kerogen forms about 80% of the organic fraction of the Green River oil shales.

The sites of deposition of the oil shale were very flat. The lakes were large in area, and the relatively shallow water ranged from fresh to highly alkaline. The surrounding highlands contributed clastic sediment to the margins of the lakes, but during most of their history, little or none of the clastic material was carried into the central part of the lakes. Conditions were such that algae grew abundantly on the surface of the lakes, and their remains were preserved in the bottom sediments.

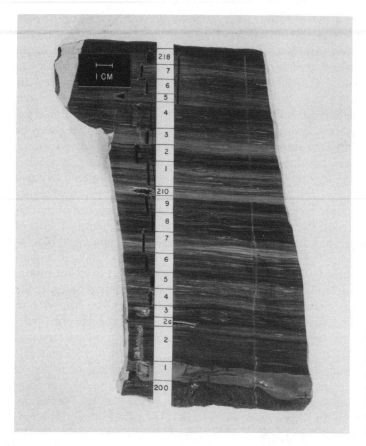

FIG. 12-19 • Laminated oil shale from the Mahogany ledge in the lower Piceance Creek section. Dark spots show where samples were taken. Numbers are sample numbers used in the measured section. (Adapted from Brobst and Tucker, 1973.)

The basins sank slowly and at irregular rates, causing the centers of the deposition of oil shale to shift during Eocene time. In general, the areas were tectonically quiescent during deposition and for million of years after the sediments were buried. Subsequently, uplift caused local tilting and erosion of these beds.

Summary

Most of the Earth's mineral wealth occurs in sedimentary rocks and includes materials as varied as building stone, gold, aluminum ore, lead, zinc, and sulfur, as well as the more commonly discussed coal, petroleum, and natural gas. The occurrences of these materials involve most of the same sedimentologic and stratigraphic variables that form non-economic sedimentary rocks and should be understood by stratigraphers. Many of the economic mineral deposits found in sedimentary rocks are being rapidly depleted by the world's rapidly growing industrial economy and the next few decades will witness a growing need for stratigraphers familiar with the origin and occurrence of these materials.

The locations of primary sedimentary ores are controlled mostly by the physical and chemical conditions of sedimentation at the depositional site. Climate and paleogeography are particularly important influences on ore formation. Sand and gravel deposits and heavy mineral placers reflect precipitation and hydraulic conditions. Economic deposits of kaolinite, bauxite, ferruginous laterite, evaporites, banded iron ores, manganese ore, and coal reflect the composition of the atmosphere and climate at the site of deposition. Diatomite, phosphate rock, and oil shale reflect specific oceanographic conditions and locations.

There are temporal restrictions for the occurrence of some of the primary ores. Banded iron ores are restricted to Precambrian rocks. Evaporite deposits are absent from Precambrian sections because of the great solubility of evaporite minerals. Economic coal deposits do not occur until the spread of land plants in the late Paleozoic. Diatomite cannot occur in pre-Cretaceous rocks, and bentonite is found largely in Cretaceous and Tertiary sections because montmorillonite alters to non-swelling illite during diagenesis.

References

General (for Chapters 12 and 13)

These books and articles in some of the leading geological journals deal with most types of commercially valuable rocks, minerals, and ores, irrespective of whether they occur in igneous, metamorphic, or sedimentary rocks. All of these publications, however, have extensive sections that are concerned with accumulations in sedimentary rocks.

Bates, R. L., 1960. Geology of the Industrial Rocks and Minerals. New York, Harper & Bros., 441 pp.

Bates, R. L. and Jackson, J. A., 1982. Our Modern Stone Age. Los Altos, Calif., William Kaufmann, Inc., 132 pp.

Brobst, D. A. and Pratt, W. P. (eds.), 1973. United States Mineral Resources. U.S. Geological Survey, Prof. Paper 820, 722 pp.

Bureau of Mines Staff, 1980. Mineral Facts and Problems. Washington, D.C., U.S. Bureau of Mines, Bull. 671, 1060 pp.

Cronin, T. M., Cannon, W. F., and Poore, R. Z., 1983. Paleoclimate and Mineral Deposits. Washington, D.C., U.S. Geological Survey, Circ. 822, 59 pp.

Dixon, C. J., 1979. Atlas of Economic Mineral Deposits. Ithaca, New York, Cornell University Press, 143 pp.

Edwards, R. and Atkinson, K., 1986. Ore Deposit Geology. London, Chapman and Hall, 466 pp.

Guilbert, J. M. and Park, C. F., Jr., 1986. The Geology of Ore Deposits. New York, W. H. Freeman & Co., 985 pp.

Harben, P. W. and Bates, R. L., 1984. Geology of the Nonmetallics. New York, Metal Bulletin, Inc., 392 pp.

Hutchison, C. S., 1983. Economic Deposits and their Tectonic Setting. New York, John Wiley & Sons, 365 pp.

Jensen, M. L. and Bateman, A. M., 1979. Economic Mineral Deposits, 3rd ed. New York, John Wiley & Sons, 593 pp.

Lefond, S. J. (ed.), 1975. Industrial Minerals and Rocks, 4th ed. New York, American Institute of Mining, Metal, and Petroleum Engineering, 1360 pp.

Maynard, J. B., 1983. Geochemistry of Sedimentary Ore Deposits. New York, Springer-Verlag, 305 pp.

Mitchell, A. H. G. and Garson, M. S., 1981. Mineral Deposits and Global Tectonic Settings. New York, Academic Press, 405 pp.

Parrish, J. T. and Barron, E. J., 1986. Paleoclimates and Economic Geology. Tulsa, Society of Economic Paleontologists and Mineralogists, Short Course Notes No. 18, 162 pp.

Tarling, D. H., 1981. Economic Geology and Geotectonics. New York, John Wiley & Sons, 213 pp.

Watson, J. V., 1984. Continental crustal regimes as factors in the formation of sedimentary ore deposits. J. Geol. Soc., v. 141, pp. 215–220.

Wauschkuhn, A. and Zimmermann, R. A. (eds.), 1984. Syngenesis and Epigenesis in the Formation of Mineral Deposits. New York, Springer-Verlag, 653 pp.

Wolf, K. H. (ed.), 1976–1986. Handbook of Stratabound and Stratiform Ore Deposits. New York, Elsevier Science Publishers, 14 vols.

Primary Deposits

Alling, H. L. and Briggs, L. I., 1961. Stratigraphy of Upper Silurian Cayugan evaporites. Amer. Assoc. Petroleum Geol. Bull., v. 45, pp. 515–547.

Altschuler, Z. S., Jaffe, E. B., Cuttitta, F., 1956. The Aluminum Phosphate Zone of the Bone Valley Formation, Florida, and Its Uranium Deposits. Washington, D.C., U.S. Geological Survey, Prof. Paper 300, pp. 495–504.

Armstrong, F. C. (ed.), 1981. Genesis of Uranium- and Gold-Bearing Precambrian Quartz-Pebble Conglomerates. Washington, D.C., U.S. Geological Survey, Prof. Paper 1161A-BB. (A group of 28 papers by different authors; papers D-S deal with the Witwatersrand gold placers.)

Barron, J. A., 1987. Diatomite: Environmental and geologic factors affecting its distribution. In J. R. Hein (ed.), Siliceous Sedimentary Rock-Hosted Ores and Petroleum. New York, Van Nostrand Reinhold, pp. 164–178.

Blatt, H., 1982. Sedimentary Petrology. New York, W. H. Freeman & Co., 564 pp., esp. Chapter 13 on coal.

Boyle, R. W., 1979. The Geochemistry of Gold and Its Deposits. Toronto, Geological Survey of Canada, Bull. 280, particularly pp. 333–386 on placer gold.

Brady, L. L. and Jobson, H. E., 1973. An Experimental Study of Heavy-Mineral Segregation Under Alluvial-Flow Conditions. Washington, D.C., U.S. Geological Survey, Prof. Paper 562-K, 39 pp.

Brobst, D. A. and Tucker, J. D., 1973. X-ray Mineralogy of the Parachute Creek Member, Green River Formation, in the Northern Piceance Creek Basin, Colorado. Washington, D.C., U.S. Geological Survey, Prof. Paper 803, 53 pp.

Cook, P. J. and McElhinny, M. W., 1979. Reevaluation of the spatial and temporal distribution of sedimentary phosphate deposits in the light of plate tectonics. Econ. Geol., v. 74, pp. 315–330.

Dapples, E. C. and Hopkins, M. E. (eds.), 1969. Environments of Coal Deposition. Boulder, Geological Society of America, Spec. Paper 114, 204 pp.

Ferm, J. C., Staub, J. R., Baganz, B. P., et al., 1979. The shape of coal bodies. *In* J. C. Ferm and J. C. Horne (eds.), Carboniferous Depositional Environments in the Appalachian Region. Columbia, S.C., University of South Carolina, pp. 605–619.

Flores, R. M. and Ethridge, F. G., 1981. Nonmarine deposits and the search for energy resources and minerals. *In* Recent and Ancient Nonmarine Depositional Environments: Models for Exploration. Tulsa, Society of Economic Paleontologists and Mineralogists, Spec. Pub. No. 31, pp. 1–17.

Force, E. R. and Cannon, W. F., 1988. Depositional model for shallow-marine manganese deposits around black shale basins. Econ. Geol., v. 83, pp. 93–117.

Frakes, L. A. and Bolton, B. R., 1984. Origin of manganese giants: Sea-level change and anoxic-oxic history. Geology, v. 12, pp. 83–86.

Galloway, W. E. and Hobday, D. K., 1983. Terrigenous Clastic Depositional Systems. New York, Springer-Verlag, 423 pp. Chapter 11 deals with coal facies.

Gole, M. J. and Klein, C., 1981. Banded iron-formations through much of Precambrian time. J. Geol., v. 89, pp. 169–183.

Gordon, M., Jr., Tracey, J. I., Jr., and Ellis, M. S., 1958. Geology of the Arkansas Bauxite Region. Washington, D.C., U.S. Geological Survey, Prof. Paper 299, 268 pp.

Grim, R. E., 1962. Applied Clay Mineralogy. New York, McGraw-Hill, 422 pp.

Grim, R. E. and Güven, N., 1978. Bentonites. New York, Elsevier, 256 pp.

Halbouty, M. T., 1979. Salt Domes, Gulf Region, United States and Mexico. 2nd ed. Houston, Gulf Publishing Co., 561 pp.

Horne, J. C., Ferm, J. C., Carrucio, F. T., and Baganz, B. P., 1979. Depositional models in coal exploration and mine planning in the Appalachian region. *In* J. C. Ferm and J. C. Horne (eds.), Carboniferous Depositional Environments in the Appalachian Region, Columbia, S.C., University of South Carolina, pp. 544–575.

Hutchinson, R. W., 1987. Metallogeny of Precambrian gold deposits: Space and time relationships. Econ. Geol., v. 82, pp. 1993–2007.

Jacob, L., Jr. (ed.), 1984. Bauxite. New York, American Institute of Mining, Metal, and Petroleum Engineering, 918 pp.

Journal of the Geological Society (London), 1985. Thematic group of ten papers on placer deposits, including depositional mechanics, diamonds, and tin (cassiterite). J. Geol. Soc. (London), v. 142, pp. 725–848.

Kadey, F. L., Jr., 1975. Diatomite. *In* S. J. Lefond (ed.), Industrial Minerals and Rocks, 4th ed. New York, American Institute of Mining, Metallurgical, and Petroleum Engineers, pp. 605–635.

Kartashov, I. P., 1971. Geological features of alluvial placers. Econ. Geol., v. 66, pp. 879–885.

Keller, W. D. and Stevens, R. P., 1983. Physical arrangement of high-alumina clay types in a Missouri clay deposit and implications for their genesis. Clays and Clay Mins., v. 31, pp. 422–434.

Kesler, T. L., 1956. Environment and origin of the Cretaceous kaolin deposits of Georgia and South Carolina. Econ. Geol., v. 51, pp. 541–554.

Lefond, S. J., 1969. Handbook of World Salt Resources. New York, Plenum Press, 384 pp.

Macdonald, E. H., 1983. Alluvial Mining. New York, Chapman and Hall, 508 pp.

Mark, H., 1963. High-Alumina Kaolinitic Clay in the United States. Washington, D.C., U.S. Geological Survey, Mineral Inv. Resource Map MR-37.

McKelvey, V. E., 1967. Phosphate Deposits. Washington, D.C., U.S. Geological Survey, Bull. 1252-D, 21 pp.

McKelvey, V. E. and others, 1959. The Phosphoria, Park City and Shedhorn Formations in the Western Phosphate Field. Washington, D.C., U.S. Geological Survey, Prof. Paper 313-A, 47 pp.

Morris, R. C. and Horwitz, R. C., 1983. The origin of the iron-formation-rich Hamersley Group of Western Australia—Deposition on a platform. Precambrian Res., v. 21, pp. 273–297.

Mossman, D. J. and Harron, G. A., 1984. Witwatersrand paleoplacer gold in the Huronian Supergroup of Ontario, Canada. Geosci. Canada, v. 11, pp. 33–40.

Notholt, A. J. G., 1980. Economic phosphatic sediments: Mode of occurrence and stratigraphical distribution. J. Geol. Soc., v. 137, pp. 793–805.

Nurmi, R. D. and Friedman, G. M., 1977. Sedimentology and depositional environments of basin-center evaporites, lower Salina Group (Upper Silurian), Michigan Basin. *In* AAPG Studies in Geology, No. 5, Reefs and Evaporites—Concepts and Depositional Models. Tulsa, American Association of Petroleum Geologists, pp. 23–52.

Patterson, S. H., Kurtz, H. F., Olson, J. C., and Neeley, C. L., 1986. World Bauxite Resources. Washington, D.C., U.S. Geological Survey, Prof. Paper 1076-B, 151 pp.

Patterson, S. H. and Murray, H. H., 1984. Kaolin, Refractory Clay, Ball Clay, and Halloysite in North America, Hawaii, and the Caribbean Region. Washington, D.C., U.S. Geological Survey, Prof. Paper 1306, 56 pp.

Patton, J. B. and Carr, D. D., 1982. The Salem Limestone in the Indiana Building-Stone District. Bloomington, Ind., Indiana Geological Survey, Occasional Paper 38, 31 pp.

Pedlow, G. W., III, 1979. A depositional analysis of the anthracite coal basins of Pennsylvania. *In* J. C. Ferm and J. C. Horne (eds.), Carboniferous Depositional Environments in the Appalachian Region. Columbia, S.C., University of South Carolina, pp. 530–542.

Persons, B. S., 1970. Laterite. New York, Plenum Press, 103 pp.

Pettijohn, F. J. and Potter, P. E., 1964. Atlas and Glossary of Primary Sedimentary Structures. New York, Springer-Verlag.

Pirkle, E. C. and Yoho, W. H., 1970. The heavy mineral ore body of Trail Ridge, Florida. Econ. Geol., v. 65, pp. 17–30.

Pretorius, D. A., 1981. Gold and uranium in quartz-pebble conglomerates. Econ. Geol., 75th Anniversary vol., pp. 117–138.

Puffer, J. H. and Cousminer, H. L., 1982. Factors controlling the accumulation of titanium-iron oxide-rich sands in the Cohansey Formation, Lakehurst area, New Jersey. Econ. Geol., v. 77, pp. 379–391.

Reimer, T. O., 1984. Alternative model for the derivation of gold in the Witwatersrand Supergroup. J. Geol. Soc., v. 141, pp. 263–272.

Roy, S., 1981. Manganese Deposits. New York, Academic Press, 458 pp.

Ruhl, W., 1982. Tar (Extra Heavy Oil) Sands and Oil Shales. Stuttgart, West Germany, Ferdinand Enke Publishers, 149 pp.

Scott, A. C. (ed.), 1987. Coal and Coal-Bearing Strata: Recent Advances. Oxford, Blackwell Scientific Publications, 332 pp.

Slaughter, M. and Early, J. W., 1965. Mineralogy and Geological Significance of the Mowry Bentonites, Wyoming. Boulder, Geological Society of America, Spec. Paper 83, 95 pp.

Slingerland, R. and Smith, N. D., 1986. Occurrence and formation of water-laid placers. Ann. Rev. Earth Planet. Sci., v. 14, pp. 113–147.

Smith, J. W., 1980. Oil shale resources of the United States. Mineral and Energy Resources, v. 23, no. 6, 20 pp.

Smith, N. D. and Minter, W. E. L., 1980. Sedimentological controls of gold and uranium in two Witwatersrand paleoplacers. Econ. Geol., v. 75, pp. 1–14.

Sutherland, D. G., 1982. The transport and sorting of diamonds by fluvial and marine processes. Econ. Geol., v. 77, pp. 1613–1620.

Trendall, A. F. and Blockley, J. G., 1970. The iron formations of the Precambrian Hamersley Group, Western Australia. Perth, Geological Survey of Western Australia, Bull. 119, 366 pp.

United Nations, 1976. The Development Potential of Dimension Stone. New York, United Nations, Department of Economic and Social Affairs, 95 pp.

Valeton, I., 1972. Bauxites. New York, Elsevier Science Publishers, 226 pp.

Van Houten, F. B. and Bhattacharyya, D. P., 1982. Phanerozoic oolitic ironstones—Geologic record and facies model. Ann. Rev. Earth Planet. Sci., v. 10, pp. 411–457.

Winkler, E. M., 1973. Stone: Properties, Durability in Man's Environment. New York, Springer-Verlag, 230 pp.

Winkler, E. M. (ed.), 1978. Decay and Preservation of Stone. Boulder, Geological Society of America, Engin. Geol. Case Hist. No. 11, 104 pp.

> *If facts conflict with a theory, either the theory must be changed or the facts.*
>
> Benedict Spinoza

13
Economic Stratigraphy II: Deposits Formed by Secondary Mineralization

Mineral deposits formed during diagenesis result from the interaction of several factors during the normal evolution of a sedimentary basin (Baskov, 1987). Fluids move both during basin formation and long afterward in response to hydrostatic gradients established by regional thermal gradients, local magmatic heat, topographic relief, deformation, and rock/water factors such as compaction, water composition, O_2/CO_2 ratio, and, perhaps most important, the porosity and permeability of the rocks. If the rock has no permeability, there can be no movement of fluids through it and no secondary mineralization. Because of this fact, it is useful to consider some basic principles of fluid flow in sedimentary rocks before proceeding with considerations of specific types of secondary economic mineral deposits.

Porosity

Sandstones, mudrocks, and nearly all carbonate rocks are formed initially of clastic particles. In sandstones, about two-thirds of the particles are quartz on the average; in mudrocks, 60% are clay minerals; in carbonate rocks, 90–95% are bioclastic grains composed of aragonite or calcite. Because of the obvious differences in physical and chemical properties among the three dominant grain types, different processes dominate in the early diagenetic environment. In sandstones the major event is a loss of perhaps half of the original porosity of 40–50%; in mudrocks, compaction of the clayey sediment reduces porosity from an initial 60–80% to 20–

30%; in carbonate rocks, recrystallization of relatively unstable aragonite and high-magnesium calcite particles may destroy nearly all the initial porosity.

Later diagenesis at depths between 500 m and 5000 m normally causes further reductions in porosity as a result of chemical processes. Pores in sandstones typically become partially or completely clogged with precipitates of calcite or quartz; mudrocks undergo clay mineral recrystallization and porosity reduction to perhaps 10–20%; limestones suffer pressure solution and probably numerous episodes of recrystallization, which may result in porosity loss. On the positive side, however, the cements in sandstones are commonly leached to a greater or lesser extent during later diagenesis to generate appreciable secondary porosity. Limestones can be similarly leached to produce secondary porosity; and dolomitization may cause an increase in porosity of carbonate rocks. Mudrocks, on the other hand, do not recover their initial porosity, and the amount of pore space they contain continues to decrease. However, many mudrocks may contain fracture porosity, as also can sandstones and carbonates.

Permeability

Variations in the permeability of sedimentary rocks are much more subtle and complex than those of porosity because permeability variations depend not only on rock properties but also on properties of the fluid. Permeability is a measure of the ability of a fluid to flow through a rock. It is defined mathematically by an empirical relationship first recognized by the French hydrologist Henri Darcy in 1856 and may be written

$$V = \frac{Q}{A} = k \frac{\Delta p}{\mu l},$$

where

V = apparent fluid velocity (cm/s)
Q = fluid discharge (cm^3/s)
A = cross-sectional area of rock (cm^2)
k = permeability (darcies = cm^2 × 10^8)
μ = fluid viscosity (centipoises, gm/cm·s × 10^{-2})
l = distance of flow (cm)
p = fluid pressure and acceleration (dynes/cm^2)

Permeability is generally much greater parallel to bedding surfaces than in other directions, the major exception occurring in fractured rocks. Fractures in sedimentary rocks are nearly always perpendicular to bedding surfaces, and in rocks with little permeability because of cementation or other factors these fractures can be the major conduit for fluid migration. The permeability along fractures can be orders of magnitude greater than the permeability along bedding planes.

High porosity does not necessarily imply high permeability. In some rocks, particularly carbonates, a high proportion of the pores may be of the dead-end variety,

vugs, so fluid flow is much less than would otherwise occur. Even in sandstones, some pores may not be interconnected, but the proportion of such pores normally is small. Another important factor is pore diameter. Although sandstones and mudrocks often have similar porosities, fluid flow is not possible through unfractured mudrocks (except well-sorted siltstones) because of the small size of the pores, their high surface to volume ratio, and resultant capillarity. The empirical relationship between permeability and interconnected porosity is known as the Kozeny equation, one form of which is

$$k = \frac{10^8 \ \phi^a}{2t^2 S^2},$$

where

k = permeability (darcies = cm² × 10⁸)
ϕ = porosity (proportion)
t = tortuosity (flow length/straight line length)
S = specific surface area (cm²/cm³ of rock)
a = exponent greater than unity

The effect of grain size on permeability of unfractured sandstones is well shown in Figure 13-1. For example, a coarse-grained sand with 14% porosity will have a permeability of about 1300 md; fine-grained sand, 100 md; clayey sand, 2 md; a difference of almost three orders of magnitude. Typical shales have permeabilities of perhaps 10^{-5} md. The permeabilities of carbonate rocks vary from essentially zero to values much greater than those in sandstones because of cavernous leaching in some limestones.

To evaluate the amount of precipitate (calcite, quartz, uranium, copper, etc.) possible from fluid flow, four pieces of data are needed: (1) the molarity of the required ion in the water, (2) the amount of water passing through the rock per unit time, (3) the efficiency of the precipitation process (not all the ions present in excess of the saturation amount will be precipitated), and (4) the length of time involved. In most cases, none of these variables is known very accurately.

Groundwater

Although not a mineral accumulation in the usual sense, groundwater is perhaps the world's most important subsurface resource. It is an important source of water in nearly all inhabited places on the Earth and the only dependable source of water in most arid and semiarid regions, regions that total 34% of the land surface. Even in the United States, 40% of the fresh water used for all purposes except hydropower generation and electric powerplant cooling is groundwater (Heath, 1984). The widespread use of groundwater results not only from its general availability, but also from economic and public health considerations. It is usually present where needed, eliminating transportation costs. It is usually of good quality, normally free of sus-

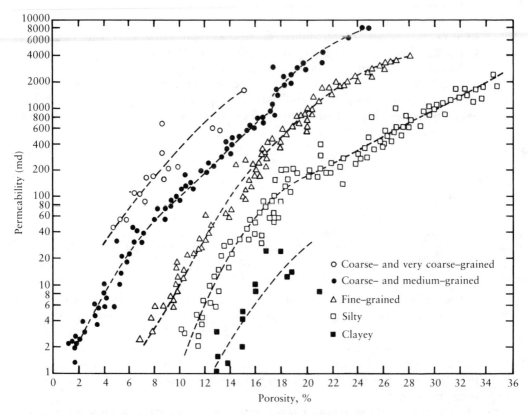

FIG. 13-1 • Relationship between porosity and permeability of cores of very coarse, coarse, and medium grained sands, silty sands, and clayey sands. (Reprinted by permission of the publisher from Chilingar, *in: Deltaic and Shallow Marine Deposits*, Elsevier Science Publishers B.V., 1964.)

pended sediment and, except in areas where it has been polluted (a growing problem), free of pathogenic organisms and harmful chemicals.

The conterminous United States has been divided into 11 groundwater regions (Fig. 13-2) in which the factors that affect water accumulation are generally similar. These factors include (1) the balance between precipitation and evapotranspiration; (2) similarities in the composition, arrangement, and structure of rocks at and near the surface; and (3) the nature and extent of the dominant aquifers and their relations to other units of the groundwater system.

As an example of a groundwater province, we will consider the High Plains Province of the midcontinent (U.S. Geological Survey, 1984–1988), an area of about 450,000 km² extending from Texas to South Dakota (Fig. 13-3). The plains are a remnant of a larger alluvial surface built in Neogene time by streams flowing east from the Rocky Mountains, but subsequent erosion has removed almost all of the plain adjacent to the mountains. The High Plains region is underlain by one of

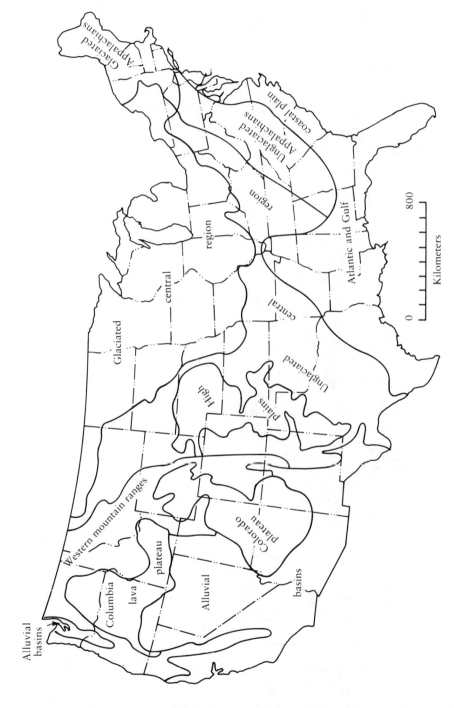

FIG. 13-2 ● Groundwater regions of the conterminous United States. (Adapted from Heath, 1984, p. 13.)

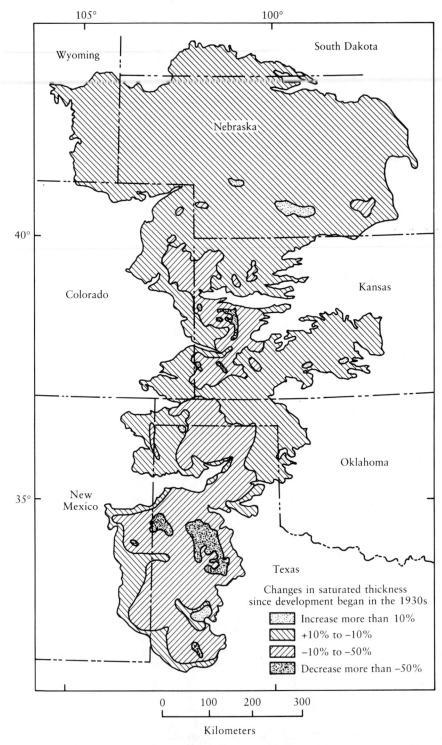

FIG. 13-3 • Change in saturated thickness of the High Plains aquifer as of 1980. (Adapted from Heath, 1984.)

the most productive and most intensively developed aquifers in the United States, the Ogallala Formation (Pliocene). The formation ranges in thickness from a few meters to more than 200 m and consists of poorly sorted and generally unconsolidated mud, feldspathic sand, and gravel. In most parts of the High Plains, Quaternary alluvial materials derived from the Ogallala overlie the formation, and where saturated with water, they are included as part of the Ogallala hydrostratigraphic unit. In the northern half of the High Plains area the Ogallala is underlain by the Arikaree Group (Miocene), a massive fine-grained sandstone up to 300 m thick that is hydrologically connected to the Ogallala and is therefore also part of the hydrostratigraphic unit. In the southern half of the High Plains groundwater region the underlying rocks range from Cretaceous to Permian in age. The Permian rocks contain gypsum, anhydrite, and halite and, where connected hydrologically to the Ogallala Formation, make the Ogallala water too highly mineralized for most human uses.

Before the erosion that removed most of the western part of the Ogallala near the Rocky Mountains, the High Plains aquifer was recharged by the streams flowing eastward as well as by local precipitation. The only source of recharge now is local precipitation, which ranges from about 400 mm along the western boundary of the region to about 600 mm along the eastern boundary. Recharge rates range from about 5 mm/year in Texas and New Mexico to about 100 mm/year in western Nebraska. The large difference results from differences in evapotranspiration and differences in permeability of the surficial materials. In western Nebraska the lower evapotranspiration and permeable sandy soil result in about 20% of the precipitation reaching the water table as recharge.

The water table of the High Plains aquifer slopes toward the east at about 2–3 m/km, and it has been estimated that water moves through the aquifer at a rate of about 0.3 m/day (Heath, 1984).

Natural discharge from the aquifer occurs along the eastern boundary of the plains by evapotranspiration and to streams, springs, and seeps, but the largest discharge is probably through wells. The extensive agricultural economy that has developed because of the Ogallala aquifer is rapidly depleting it, resulting in a long-term continuing decline in groundwater levels of as much as 1 mm/year. The lowering of the water table has resulted in a 10–50% reduction in the saturated thickness of the High Plains aquifer over an area of 130,000 km², the greatest reductions occurring in drier climatic areas such as panhandle Texas (Fig. 13-3). It is estimated that the volume of water in the aquifer will be decreased by 19% by the year 2000 and 40% by 2030, compared to the volume in 1980.

Simple depletion of fresh water by irrigation and direct human consumption is not the only problem caused by groundwater withdrawal. As fresh water is withdrawn, saline water tends to move upward from deeper sources or laterally from an adjacent ocean to replace it in the rock pores (Johnston and Bush, 1988). In about half of the United States the depth to saline water is less than 150 m (Fig. 13-4), and in many areas this critical interface is rising rapidly, as in southern Florida, where seawater incursion has become a significant problem in recent years (Atkinson et al., 1986).

A related problem caused by the withdrawal of large amounts of groundwater

FIG. 13-4 • Depth to saline groundwater in the conterminous United States. (Adapted from Heath, 1984.)

Depth to saline
groundwater

Less than 150 m

150 – 300 m

More than 300 m

Not present

Approximate

0   200 400 600
Kilometers

is land subsidence. In the Santa Clara Valley of California, for example, the land surface subsided about 4 m between 1915 and 1970 as the depth to the water table increased by 30 m.

Petroleum and Natural Gas

Petroleum and **natural gas** are a mixture of solutions with dissolved gas that are composed predominantly of organic compounds of carbon and hydrogen. Most commercial accumulations of hydrocarbons are recovered from reservoir rocks less than 100 m.y. old; on a worldwide basis the reservoir rocks are 29% Cenozoic, 57% Mesozoic, 8% Permian, and 6% Precambrian. The dominance of Mesozoic rocks reflects oil from the Persian Gulf area, which contains about two-thirds of the world's proven reserves; less than 10% of the world's oil is in the United States, and this percentage is declining. About two-thirds of the world's oil is produced from depths of less than 1500 m (Jensen and Bateman, 1979), which reflects a variety of economic and scientific factors such as the logarithmic increase in drilling costs with increased depth, the difficulty of locating potential traps at greater depths, and the destruction (volatilization) of hydrocarbons at high subsurface temperatures.

The occurrence of fluid hydrocarbons in rocks depends on four critical properties of a sedimentary basin.

1. There must be organic-rich source beds within the sedimentary sequence.
2. There must have been appropriate heat for sufficient time for thermal maturation of petroleum or gas.
3. There must have been permeable migration paths for the fluid hydrocarbons to move from source beds to reservoirs.
4. There must be porous reservoir beds confined within some trapping configuration by impermeable capping beds, which prevent the hydrocarbons from escaping.

Source Beds

Fine-grained rocks such as shales and mudstones contain about 95% of the organic matter buried in rocks, so the nature of the source beds is clear—organic-rich mudrocks. A lower limit of 0.5% organic matter is taken as the minimum required for a rock to serve as a source, but most recognized source beds contain more than this amount; some contain as much as 10% (Hutchison, 1983).

Thermal History

The most important factor in the origin of petroleum is the thermal history of the source rock. Temperatures in the Earth increase at a rate of 1°–5°C/100 m of depth, but the near-linear increase in temperature causes a logarithmic increase in reaction rate for most reactions involving petroleum formation. Increasing temperatures also

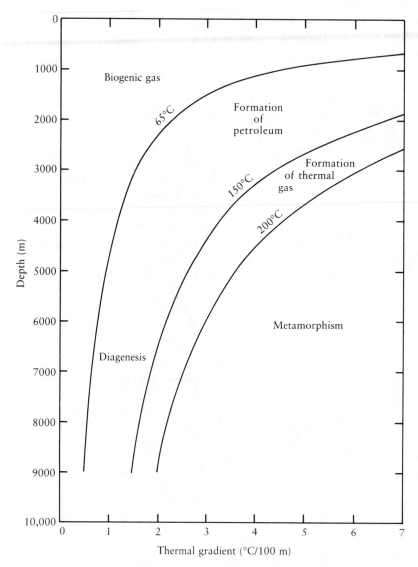

FIG. 13-5 • Liquid window concept of hydrocarbon formation from organic matter in sediments. (Reprinted by permission of the publisher from Pusey, *World Oil,* v. 176, 1973.)

increase the solubility of some organic compounds in the sediment fluids and convert solids to liquids and liquids to gases, thereby increasing their ability to migrate. Because temperature gradients vary in both space and time, the depth at which petroleum and natural gas are generated also varies (Fig. 13-5), but most oil is formed at depths between 1.5 km and 3.5 km; most natural gas occurs below 2.5 km. The thermal gradient–depth zone in which petroleum is formed has become known as the "liquid window," a concept that has been of great importance in the search for increasingly hard-to-find petroleum reserves in the United States. Oil reservoirs found at temperatures less than 65°C are invariably associated with either significant uplift after heating or, less commonly, long-distance migration updip.

Fluid Migration

The method by which liquid hydrocarbons migrate from their fine-grained, apparently rather impermeable source beds (primary migration) is obscure (Roberts and Cordell, 1980). By the time the hydrocarbons are generated, most connate pore water has been compacted out of the muds and so cannot serve as a forcing mechanism. Further, liquid hydrocarbons have notably low solubilities in water, even at the temperatures existing in the subsurface where the oil is formed; and solubility decreases even more with the increased water salinity that typically accompanies increased burial depth. On the other hand, it might be significant that the temperatures of the liquid window are substantially also those at which smectitic clays are illitized, because a very large amount of relatively fresh water is released during this process. Because of the difficulties associated with solubilizing large amounts of liquid hydrocarbons, many petroleum geologists and geochemists believe that the oil migrated as a separate phase, perhaps moving through fractures in the parent shale. The problem has received intense study for two decades but is still unresolved.

Reservoir Rocks

Once the oil leaves the shale and enters coarser-grained and more easily permeable rocks, it continues to move because of its buoyancy relative to water contained in the pores (secondary migration). Subsurface oil densities range from 0.5 g/ml to 1.0 g/ml and water from 1.0 g/ml to 1.2 g/ml, so appreciable buoyant force is normally present. Because pores in the new host rock are larger, oil globules will connect and enlarge, enabling oil to migrate easily to regions of lower pressure.

The final trapping mechanism for the oil and gas may be structural (for example, a dome capped by impermeable rocks), stratigraphic (facies change from sandstone to shale), or a combination of the two (sandstone faulted against a shale). Unconformities can also generate hydrocarbon traps, as occurs when tilted, porous rocks are overlain by flat-lying impermeable rocks. Other possibilities for trapping mechanisms include buried paleotopographic irregularities and updip seals by asphalt.

During the past 20 years a method has been developed for determining the age of the oil in a reservoir rock, based on changes in percentage of napthenes, paraffins, and aromatics in the gasoline-range hydrocarbons (Hunt, 1979). A linear relationship has been developed between the hydrocarbon composition of crude oils and the product of the oil's reservoir temperatures and ages. Results using this technique have shown, for example, that most of the oil in some Miocene reservoir rocks in the Middle East originated in Cretaceous or Jurassic source beds and that the ages of the offshore Gulf Coast oils are, on average, 8.7 m.y. older than the reservoirs in which they occur, and this age difference indicates an average vertical migration of 3350 m. In the Denver and Williston basins, correlations between source rock and oil composition indicate lateral migrations of as much as 160 km from the thermally mature basin-center sourcerocks.

Petroleum and natural gas can form in any type of tectonic setting. Convergent plate margin oil is found, for example, in Tertiary sandstones of California; rift margin oil occurs in Cretaceous and Tertiary sandstones and carbonate rocks of the Sirte Basin in Libya and the Mesozoic rocks of the North Sea; oil in intracontinental basins is represented by carbonate rock reservoirs in the Devonian of the Williston Basin in the United States and Canada and in Silurian reefs of the Illinois Basin. Klemme (1980) presents a detailed classification of major petroleum fields in relation to their tectonic setting.

Oil in the Niger Delta

The occurrence of petroleum in relation to sedimentary facies and depositional systems is discussed by Galloway and Hobday (1983), and an excellent illustration is given by Weber (1971) for the Afiesere and Eriemu oil fields of the Tertiary Niger delta system. In the delta rocks, almost all the oil is contained in folds associated with and caused by abundant growth faults. The sediments form a typical deltaic offlap sequence consisting of a wedge of continental sands grading downward into marine clayey sediments that, at greater depth, are undercompacted. The depositional setting is similar to that of the central Gulf coastal region of the United States. Most of the oil accumulations in the Niger section occur in the transition zone between continental and fully marine sediments. This paralic sequence consists of a large number of sedimentary offlap sequences, each cycle generally less than 50 m thick and starting with a marine clay and changing upward into proximal fluvio-marine interlaminated silt, sand, and clay, which are usually succeeded by various types of more sandy barrier bar and coastal plain deposits. The cycles are terminated by transgressions, which erode away part of the offlap sequence and which are generally represented by a thin, very fossiliferous gravelly sand.

The Afiesere and Eriemu fields are located in a series of four sedimentary cycles (Fig. 13-6) and have several types of reservoirs and a complex oil distribution. The variable trend and continuity of individual depositional units has resulted in multiple oil-water contacts and areal configurations for the productive horizons. The sands of cycle I are typically poorly consolidated, submature, medium-grained sands (mean diameter = 1.8ϕ, standard deviation = 1.4ϕ); many have cut almost to the basal marine clay of this cycle and are interpreted to be point bars of meandering distributary channels. Discontinuous, thin muddy layers may partially separate vertically adjacent point bars, so two or three bars typically can be distinguished on electric logs.

Cycle II is more complex, consisting of two submature, fine-grained barrier bar sands (mean diameter = 2.6ϕ, standard deviation = 1.5ϕ), in places eroded by a

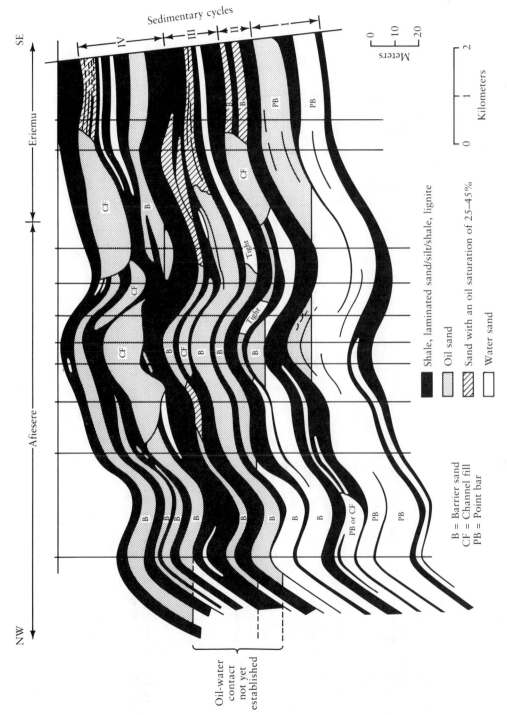

FIG. 13-6 • Complex genetic facies assemblage displayed by stacked reservoirs within the Afiesere and Eriemu oil fields of the Tertiary Niger delta system. (Reprinted by permission of the publisher from Weber, *Geologie en Mijnbouw*, v. 50, Martinus Nijhoff/ Dr. W. Junk Publishers, 1971.)

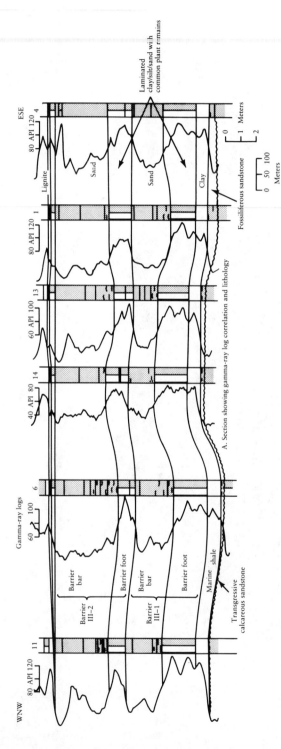

FIG. 13-7 ● Example of a depositional cycle including two barrier bar sands (cycle III of Fig. 13-6). (Reprinted by permission of the publisher from Weber, *Geologie en Mijnbouw*, v. 50, Martinus Nijhoff/Dr. W. Junk Publishers, 1971.)

distributary channel fill in the Eriemu field. The upper barrier sand is cemented by carbonate derived from the disconformably overlying fossil-rich transgressive deposits and contains little oil. The channel fill is quite coarse-grained, has good oil saturation, and produces easily.

Cycle III contains important oil reserves in a series of three barrier sands. Permeability and oil saturation deteriorate updip as the barrier bar pinches out into more clayey deposits to the side of the channel center during sediment deposition. A detailed correlation of the lower two barriers in cycle III is shown in Figure 13-7, illustrating the use of gamma-ray logs in correlation.

The uppermost cycle, IV, is the most complex one but, like the other cycles, is composed of an association of barrier and channel fill sands. The channel fills generally produce ten times as much oil per day as the barrier bars because of their coarser grain size and consequent much higher permeabilities.

Tar Sands

A **tar sand** is defined as a sedimentary rock that contains hydrocarbons so viscous that they cannot be recovered by conventional petroleum recovery methods. The terms "heavy oil" and "tar" are defined as petroleum that has a viscosity greater than 10,000 cp (cp = mPa·s) at room temperature; normal subsurface petroleum accumulations have viscosities of less than 1 cp. Either heat or hydrocarbon solvents are required to reduce the viscosity of the bitumen or oil so that it will flow and become a commercial-grade deposit, and at present the costs of such recovery are too high for general feasibility. As examples of the viscosities of tar sands at 25°C, the heavy oils of the Lloydminster area (Lower Cretaceous, Saskatchewan, Canada) have viscosities of about 1000 cp; the Peace River and Cold Lake tars of Alberta, about 25,000 cp; Athabasca tars in Alberta, about 5,000,000 cp (Fig. 13-8).

Total world in-place heavy oils or tars total at least 2800×10^9 barrels (Meyer, 1987), of which Canada contains 1390×10^9 barrels and Venezuela contains $1200-2000 \times 10^9$ barrels. In terms of *recoverable* reserves, at least two-thirds are in Venezuela, reflecting the fact that Venezuela's deposits are in the subsurface and exist at higher temperatures than the Canadian deposits. Higher temperature equates with lower viscosity and easier recovery of the oil.

Tar sands have porosities in the range 15–35% and permeabilities between 100 md and 15,000 md depending on mean grain size and sorting. They contain little water, and the water they do have is of low salinity compared to conventional subsurface oil reservoirs. In general, the tar sands occur near the surface but may reach to depths of at least 3000 m.

The source beds of the very viscous oils or tars are the same as those of normal petroleum, but in the case of the tars, all the volatile hydrocarbons have escaped, and at shallow depths, oxygen contained in the pore water or in the atmosphere has oxidized the hydrocarbons, generating oxygen-rich molecules of aromatic-asphaltic composition with high molecular weights and high viscosities. It also is possible that anaerobic bacteria have attacked preferentially the light paraffinic hydrocarbons and

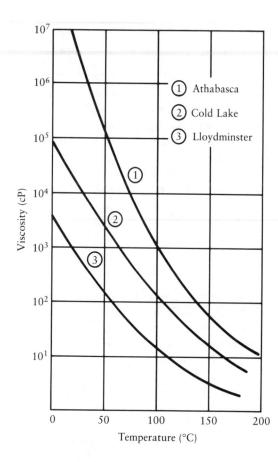

① Athabasca

② Cold Lake

③ Lloydminster

FIG. 13-8 • Viscosity-temperature relationship of very heavy oils and tar sands. (Reprinted by permission of the publisher from Rühl, *Tar Sands and Oil Shales*, Ferdinand Enke Verlag, 1982.)

degraded them, leaving the heavier components and enriching the residue in resins and asphaltenes, which are high-molecular-weight, viscous substances. The effect of enrichment in these two types of organic compounds is seen by comparing the content of asphaltenes and resins in a normal petroleum such as occurs in the Ellenberger Formation in Texas, 4.4%, with the amount in the Athabasca tars, 43%.

As an example of the stratigraphy of an oil sand, we can consider the Athabasca tar sand, the largest of the Cretaceous deposits in Alberta, Canada, with in-place reserves of 870×10^9 barrels, probably the largest oil deposit in the world. Approximately 10% of the reserves crop out and are easily mined at the surface; the remaining 90% will have to be recovered by *in situ* methods. In 1989, production was 205,000 barrels/day (G. D. Mossop, personal communication). Nearly all the reserves are in the McMurray Formation (Albian), which averages 40–60 m in thickness but reaches 70 m in isolated locations; in most areas the McMurray is tar-bearing from top to base. Porosity in the cleaner sands varies between 25% and 35%, and oil saturations of 10–18 weight percent are very common.

The McMurray lies in a broad depression cut into westward-dipping Devonian rocks (Fig. 13-9) with significant topographic relief, and the McMurray thins over paleohighs. The source of the sediment was to the south and southeast in the Precambrian Shield. Initially, the topographic lows were filled with nonmarine, conglomeratic and coarse sandy, lenticular, cross-bedded fluvial deposits (Fig. 13-10) composed of 90–95% quartz, 5% feldspar, and minor amounts of heavy minerals, largely tourmaline and zircon; a typical mineral assemblage for sediments derived from a cratonic shield area. These fluvial sediments (Lower McMurray) are locally richly impregnated with tarry oil. The overlying middle zone of the McMurray is composed of medium-grained, uniformly oil-impregnated, uncemented quartz sands and lenticular beds of barren siltstones and shales. Much of this sediment was deposited in a fresh-water deltaic to lagoonal environment. The upper zone of the formation is largely marine and consists of fine-grained sandstone locally impregnated with oil; siderite is the main cementing material. The Clearwater Formation conformably or paraconformably overlies the McMurray and is fully a shallow-marine shale.

Uranium

Approximately two-thirds of the world's economically valuable deposits of uranium minerals occur in sedimentary rocks, about 45% in Phanerozoic sandstone, 20% in Precambrian quartz-pebble conglomerates, and 2% in Quaternary calcretes (Nash et al., 1981). The highest percentage of well-documented resources is located in the United States (31.4% of the total), almost entirely in sandstone-type deposits. Currently, nuclear reactors fueled by uranium produce about 10% of electric power in the United States, and nuclear power accounts for substantial portions of the generating capacity of several European countries as well as Japan. However, the future of uranium as a fuel source is problematic because of concerns about operational safety and waste-disposal problems.

The ultimate source of the uranium atoms that relocate into sedimentary rocks is silicic igneous rocks. Granites average 4 ppm (0.0004%) of elemental uranium; granodiorites, 2.6; gabbros, 0.9; and ultramafic rocks contain still less. Most of the uranium occurs as a trace element in minerals such as zircon, monazite, and titanite but can also occur in primary uranium minerals such as uraninite and coffinite. The volcanic equivalents of plutonic rocks contain 1.5–2 times more uranium than their plutonic counterparts because of additional fractionation during the volcanic process. Many uraniferous sandstones are believed on regional stratigraphic criteria to have obtained their uranium from leaching and alteration of silicic tuffs topographically higher in the paleodrainage basin. Commercial uraniferous sandstones contain up to 10,000 ppm U_3O_8 and average 1000–2000 ppm, an enrichment of about 400 times over the amount in granites (1000 ppm U_3O_8 = 850 ppm U). Black shales, which are not now commercial sources of uranium, can contain several hundred parts per million U_3O_8 and may serve as proximate sources for the uranium atoms

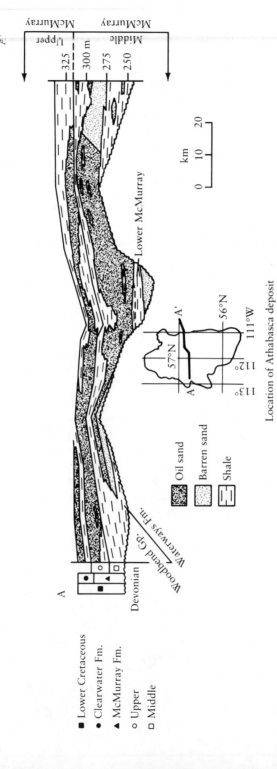

FIG. 13-9 ● Lithologic cross-section of the Athabasca tar sands, showing generalized distribution of reservoir sands in the McMurray Formation, Alberta, Canada. (Reprinted by permission of the publishers from Meyer and Steele, *The Future of Heavy Crude Oils and Tar Sands*, McGraw-Hill, 1981.)

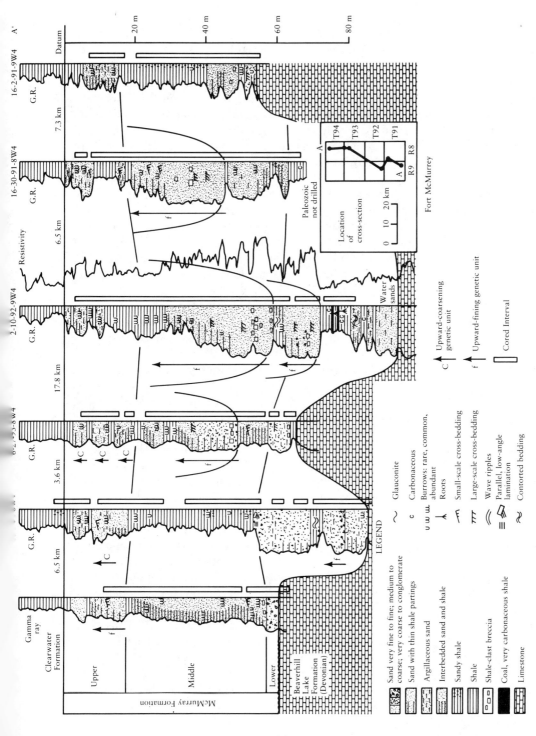

FIG. 13-10 ● Stratigraphic cross-section of McMurray Formation, southeast part of study area, with lithology shown on gamma-ray log. Fluvial deposits of lower member fill in lows on erosional unconformity. Thick (up to 30 m) upward-fining channel deposits are laterally juxtaposed to thick shales in the middle member. Small upward-coarsening sequences are common in the upper member. (Reprinted by permission of the American Association of Petroleum Geologists from Flach and Mossop, *American Association of Petroleum Geologists Bulletin*, v. 69, 1985.)

461

in uraniferous sandstones. For example, the Kulm Shale (Cambrian) in Sweden contains 350 ppm U_3O_8; the Chattanooga Shale (Devonian) in Tennessee, 50–60 ppm. Marine phosphorites commonly contain 50–150 ppm U_3O_8.

The oldest commercially uraniferous sedimentary rocks are the Proterozoic quartz-pebble conglomerates, which lie exclusively on Archean cratons rather than on post-Archean mobile belts. The two world-class deposits are in the Elliot Lake district of southeastern Ontario, Canada and in the Witwatersrand gold-uranium district in South Africa, and these deposits supply about 20% of known world reserves. Other uraniferous conglomerates occur in Australia, North and South America, and Eurasia. The depositional setting of at least the major deposits is a series of basins containing coarse alluvial fan–fluvial gravel and sandstone deposited by braided streams and composed largely of quartz but with smaller amounts of feldspar, sericite, and chlorite. The latter two minerals are a product of post-depositional low-grade metamorphism. The deposits are up to 7500 m thick and contain many unconformities and nonconformities of varying temporal and areal extent (Galloway and Hobday, 1983). The highest percentages of uranium tend to follow paleostream channels and paleotopographic lows in the depositional basin (Fig. 13-11).

The primary uranium in the quartz-pebble conglomerate ores occurs as discrete, rounded grains of thorian uraninite and apparently detrital pyrite of coarse silt to fine sand size. The matrix of the conglomerate is the ore portion. Associated heavy minerals show hydraulic equivalency, supporting a detrital origin for the uraninite. Because of the relative abundances of detrital uraninite and pyrite, many geologists have inferred that the conglomerates formed in an atmosphere lacking oxygen because the U^{4+} in uraninite would rapidly oxidize to U^{6+} in the presence of oxygen, destroying the mineral in the primary weathering process.

Almost all of the U.S. production of uranium has come from small orebodies in Mesozoic-Cenozoic nonmarine sandstones of Wyoming, Colorado, New Mexico, and Texas (Maynard, 1983). Commonly, they are of the roll-front type; ore occurs at the interface between oxidized and unoxidized sandstone in a crescentic or "roll" shape, as seen in cross-section (Fig. 13-12). It is generally agreed that U^{6+}, carried by oxidizing groundwater, is precipitated at this interface by reduction to U^{4+}. In plan view the front is elongate and sinuous. Remnants or islands of mineralization may lie behind the mineralization front within unaltered portions of the sand aquifer. In some geochemical settings, continuous mineralization fronts break up into isolated pods within beds characterized by poorly defined iron-alteration zonation.

Most sandstone-hosted ore occurs within facies of terrigenous systems deposited in closed continental basins. Marginal marine coastal plains are a secondary but important setting for sandstone deposits (Galloway and Hobday, 1983). Thus sandstone uranium host systems provide good examples of fluvial, alluvial fan, and strandline depositional facies, as well as of ore deposit morphologies and mineralization histories typical of diagenetic deposits.

The geology of the south Texas Gulf coastal plain deposits has been described by Galloway and Kaiser (1980). Predominantly fluvial sands and shales of Eocene

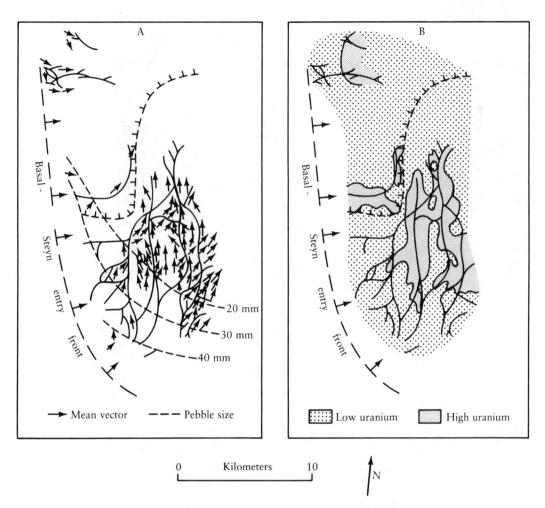

FIG. 13-11 ● Maps of the Basal-Steyn placer deposit, Central Rand Group, Witwatersrand Basin, showing: (A) paleocurrent vector means and downflow decrease in maximum pebble size and (B) distribution of uranium within the placer. In both maps, combined gold-uranium mineralization trends are shown by the anastomosing solid lines, and the "T" pattern outlines the junction of the Basal and Steyn placers. (Reprinted by permission of the publisher from Galloway and Hobday, *Terrigenous Clastic Depositional Systems,* Springer-Verlag New York, Publishers, 1983.)

to Pliocene age, interbedded with silicic tuffs, have an outcrop thickness of up to several hundred meters and dip seaward at less than 1°. All units show a strong volcanic influence; sandstones are rich in plagioclase and volcanic glass and silicic felsite fragments; shales contain abundant smectite and occasionally zeolites. Uranium mines are concentrated in the southern part of the area, probably because of higher proportions of volcanic debris present there. Commonly, deposits are located

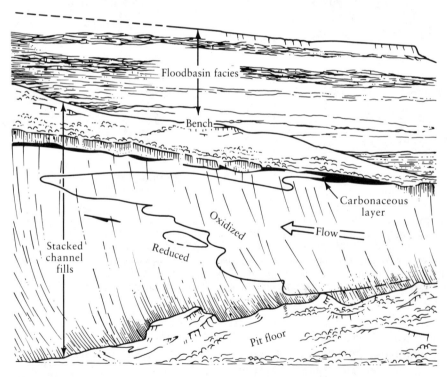

FIG. 13-12 • Mineralization front (roll front) of uranium as exposed by mining, Shirley Basin, Wyoming. (Reprinted by permission of John Wiley & Sons, Inc., from Galloway and Hobday, *Economic Deposits and their Tectonic Setting*, copyright © 1983.)

at facies boundaries that juxtapose rocks of different permeability, for example, the edges of crevasse splays, where the permeable channel sands finger out into impermeable overbank muds. The glassy volcanic material contains 7–9 ppm U_3O_8, paleosol profiles 2–3 ppm. Apparently, leaching occurred in the vadose zone initially, but there is evidence for subsequent alteration of the volcanic materials in phreatic conditions.

The mobilized uranium enters shallow groundwater circulation and is then precipitated by a reducing agent as coffinite or an amorphous uranium-rich phase. The reductant is either organic matter in the sediment or H_2S gas or aqueous sulfide species moving upward along fault planes from hydrocarbon accumulations in the strata below. Abundant petroleum is present in many of the Tertiary units in the Texas Gulf coastal region.

Copper

Copper has had a tremendous increase in usage in recent years, as indicated by the fact that of the total world production during the last 100 years, about 80% was mined since 1960, more than half of that since 1975. Sedimentary copper deposits, which range in age from Middle Proterozoic to Holocene, contribute 25–30% of world production, and about two-thirds of this amount comes from one district, the Central African Copperbelt in Zaire and Zambia. These exceptionally rich deposits average 4% copper, and some units contain as much as 13%. Cobalt may form a primary or secondary ore. Dark-colored shales and siltstones are the most common host rocks for sedimentary copper ores; copper ore in carbonate rocks is less common, although dolomite is the host rock for some of the ore in Zaire and in northwestern Canada (Chartrand and Brown, 1984).

The ultimate source of the copper that relocates into sedimentary rocks is intermediate and mafic igneous rocks; basalt averages 90 ppm Cu; andesite, 53; and granite only 13 (Maynard, 1983). Copper is a strongly sulfur-seeking (chalcophile) element, and it tends to be concentrated in sulfide deposits; in igneous rocks it occurs mainly as sulfides, principally chalcopyrite ($CuFeS_2$), bornite (Cu_5FeS_4), and chalcocite (Cu_2S).

The Central African copperbelt extends over 10,000–12,000 km² on the southern edge of the Congo craton and belongs to a vast Late Proterozoic metallogenic province. The copper is confined to the younger orogens that have been repeatedly deformed during the last 1000 m.y. The major geochemical problem in the copperbelt is to account for 170 million tons of copper metal in an area of only about 10,000 km²; an amount of 15,000 tons of copper/km² is a unique geochemical anomaly on the Earth.

The copperbelt ore deposits are the most outstanding examples of stratiform base-metal sulfide deposition in detrital sedimentary rocks. They are characterized by their enormous lateral continuity parallel to the depositional strike, their fine dissemination in sediments, their relation to certain restricted stratigraphic horizons,

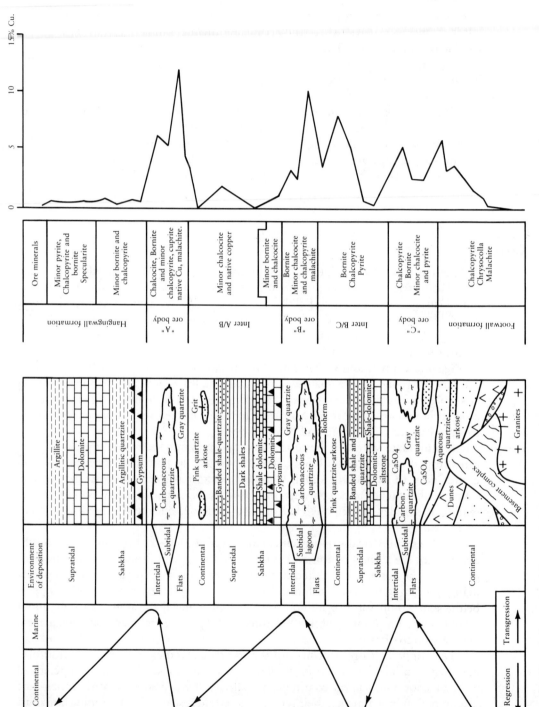

FIG. 13-13 • Relationship between lithology, environment, and tenor of mineralization in the eastern basin as represented by the classic section at Muflira, Zambia. (Reprinted by permission of the publisher from Bowen and Gunatilaka, *Copper: Its Geology and Economics,* Elsevier Science Publishers, 1977.)

and association with particular sedimentary facies (Bowen and Gunatilaka, 1977). All these features are consistent with a syngenetic or very early diagenetic origin for the ore deposits. The deposits in Zambia can be used as an example of the relationship between sedimentology and the location of the copper ores.

The ores occur in the Late Precambrian Katangan sequence of sediments, which were deposited on an eroded crystalline basement complex with up to 300 m of topographic relief. The thickness of the sediments ranges up to about 1500 m of mostly shallow-marine sandstones, shales, and dolomites, with occasional alluvial fan–fluvial arkosic continental facies sandstones interspersed.

In the eastern ore-containing basin at Mufulira (Fig. 13-13), as in other basins in the copperbelt, each ore horizon is

1. contained within strata that accumulated in a marginal marine basin, shallow semirestricted lagoon, or tidal flat;
2. hosted by markedly carbonaceous rocks of various lithologies;
3. underlain by red beds above an unconformable surface;
4. overlain by evaporites (gypsum, anhydrite) or sabkha dolomite; and
5. laterally and vertically zoned with respect to metal content, the typical sequence being chalcocite-bornite-chalcopyrite-pyrite in the direction of transport of the copper (seaward).

These characteristics are also possessed by most of the well-known stratiform sulfide occurrences elsewhere in the world.

The period of ore formation (Fig. 13-13, ore body C) commenced with a marine transgression over the alluvial deposits that covered the basement complex, and the deposition of a well-sorted cross-bedded marine sandstone. The sandstone most likely is a shoal or offshore bar. Overlying it is a carbonaceous, clayey sandstone perhaps representing a back-bar, quieter water facies. Three basin depressions (eastern at Mufulira, central, western) developed on the shallow sea floor and were separated from each other and from the open ocean by shoals. The shoal sediments were coarse-grained and cross-bedded, while laminated fine-grained sands accumulated in the deeper parts of the depressions. This environment of lower kinetic energy sheltered from the open sea was the site of deposition of the copper sulfides.

The two main lithologies hosting the copper are a carbonaceous quartzite and a laterally equivalent fine-grained mottled sandstone. The quartzite is texturally immature and carbonaceous, indicating restricted, anoxic bottom waters. The mottled sandstone consists of dark (ore-rich) and light (carbonate) colored patches and is highly porous. The mottled sandstones are overlain by algal bioherms, which may be the source of the carbonate.

At Mufulira, three superimposed ore bodies occur, and each one has the same rock sequence from top to bottom:

1. dolomite with gypsum (evaporitic and supratidal),
2. clayey sandstone and carbonaceous quartzite (subtidal-intertidal),
3. conglomeratic quartzite (continental alluvial fan–fluvial).

The three ore bodies thus correspond to three transgressive-regressive cycles (Fig. 13-13). Modern coastal sabkhas routinely generate the typical lithologic assemblage, oxidized bed/reduced bed/evaporitic bed, that characterizes the evaporite/base metal ore association in sedimentary sequences.

In all stratiform copper deposits, the source of the copper is believed to be from the landward direction, from stream waters unusually enriched in copper by leaching of rocks exposed higher in the drainage basin. Continental red sandstones would be a good source of copper (adsorbed onto hematite), and the typical association of copper deposits with redbeds supports hematite as a copper source. Concentrations of copper in stream or connate waters can be at least as high as several hundred parts per billion. An oxidizing environment is required to transport appreciable amounts of copper, since the element is rather insoluble as cuprous ion. To transport large amounts of copper ion, chloride-bearing solutions are important, perhaps critical, because the solubility of copper is greatly increased by the formation of such complexes as $CuCl_2^-$ and $CuCl_3^{-2}$. The need for chloride ion may explain the near-ubiquitous association of stratiform copper deposits and evaporites. Copper is precipitated when the environment becomes reducing. (Black shales contain three times as much copper as other shales.)

Lead and Zinc

Lead and zinc normally occur together in sedimentary economic deposits, almost entirely as galena (PbS) and sphalerite (ZnS). The richest accumulations contain up to 20% lead and 12% zinc. About half the world's lead and zinc comes from North America; in the United States the main uses of lead are in batteries, gasoline, paint, and ammunition. Zinc is used mostly in galvanizing metals and in alloys.

Lead is derived mostly from granitic rocks as a result of the decay of uranium and thorium and shows a regular decrease in abundance from 24 ppm in granite to 3.2 ppm in gabbro; zinc abundance is highest in gabbro (100 ppm) and lowest in granite (48 ppm).

Lead and zinc are found in two distinct associations in sedimentary rocks: carbonate-hosted and shale-hosted. The ores in carbonate rocks are generally agreed to be diagenetic in origin; those in shales are more controversial, but most appear to be at least partly syngenetic (Tarling, 1981, pp. 177–184). Both lead and zinc in natural waters are abundant only in highly saline brines such as are commonly associated with evaporites in the subsurface.

Sediment-hosted stratiform lead-zinc deposits range in age from about 2000 m.y. to Cenozoic and occur in tectonically active intracratonic settings, commonly in fault-controlled sedimentary basins. The controlling characteristics are stratigraphic facies and fluid migration (Sverjensky, 1984).

Most of the lead that has been produced in the world or is known as reserves occurs in tabular stratigraphic units of Phanerozoic, shallow-water marine limestone or dolomite. These ores, commonly called Mississippi Valley type because of their abundance in central United States (Fig. 13-14), characteristically consist of sphal-

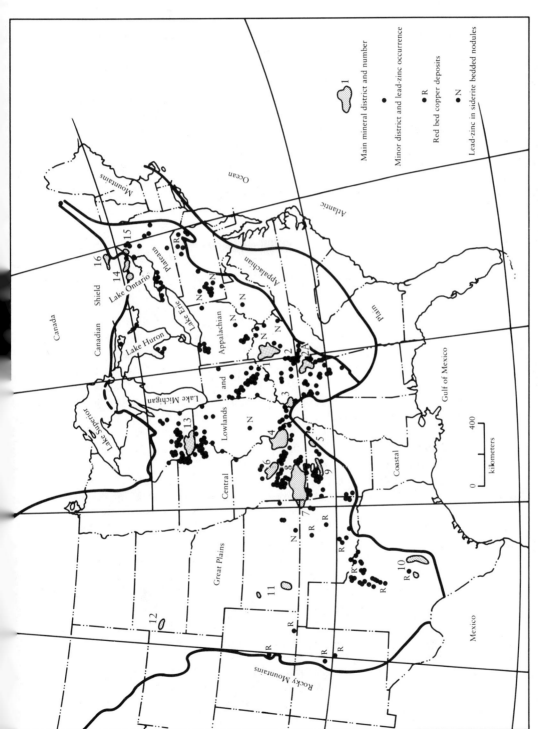

FIG. 13-14 • Map showing the central and eastern United States and the Mississippi Valley-type districts. 1, Central Kentucky; 2, Cumberland River; 2A, Central Tennessee; 3, Illino s; Kentucky; 4, Southwest Missouri; 5, Northwest Arkansas; 6, Central Missouri; 7, Tri-State; 8, Seymour; 9, Northern Arkansas; 10, Central Texas; 11, Western Kansas; 12, North Black Hills; 13, Upper Mississippi Valley; 14, Mades and Kingston; 15, Rosse; and 16, Ottawa. (Reprinted by permission of John Wiley & Sons, Inc., from Jensen and Bateman, *Economic Mineral Deposits*, copyright © 1979.)

erite and galena with marcasite/pyrite, barite, and fluorite in dolomitic carbonates. The mineralization occurs as vug fillings in either the dolomitic portions or in solution-brecciated limestone and occurs close to facies fronts with basinal argillaceous rocks. There is also a strong spatial connection with evaporites. The ore bodies are characteristically large and commonly are localized by stratigraphic pinchout zones, by masses of breccia of diverse origin, by zones of minor faults and fractures, or by bioherms. The prevalent occurrence of the ore bodies in carbonate rocks is probably related to the tendency of such rocks to brecciate readily, to develop zones of secondary porosity and karst during periods of preore groundwater movement, and to be highly soluble and reactive. Many individual ore bodies are 100 m wide and 50 m high and have been followed for several kilometers.

Fluid inclusion data suggest that the lead and zinc sulfides were precipitated from saline chloride brines at temperatures between 50°C and 100°C. Like copper, solubilities of zinc and lead increase with increasing salinity as measured by the activity of chloride ion. Base-metal chloride complexes, like those of copper, solubilize the metals, and evaporites are commonly present in stratigraphic sections that contain carbonate-hosted lead-zinc ores. Basinal brines associated with evaporites have been described that contain 15% NaCl, have temperatures of 90–160°C, and contain about 100 ppm lead and 350 ppm zinc. The sulfur may be derived from reduction of seawater sulfate or evaporite sulfate or from H_2S associated with nearby petroleum accumulations. Many stratiform lead-zinc deposits contain degraded hydrocarbons, and fluid inclusions in minerals often contain hydrocarbons as well. Various metal zonation parameters and the close association of oil with ore have led to the model of essentially lateral migration of waters and oil out of shale basins at different times during lithification and compaction and into stratigraphic and/or structural traps. Oilfield brines are frequently metalliferous and of the same composition that precipitated the ores. The main difference between the behavior of oil and ore in these systems is that the ore requires a precipitant (reduced sulfur) and is then fixed, whereas the oil can continue maturing and migrating.

One well-described example of a Mississippi Valley type of deposit is that in southeastern Missouri (Fig. 13-15). The district is composed of two parts, one trending northwest-southeast, the other (Viburnum Trend) trending north-south, but both areas of lead-zinc deposits are hosted by the Bonneterre Dolomite (Cambrian). The Bonneterre is composed of a horseshoe-shaped oolite bar and shoreward algal reef complex located on the flanks of the exposed Precambrian rocks of the St. Francis Mountains. Mineralization is concentrated near the facies change from oolite bar to stromatolite reef, primary porosity and brecciation controlling the movement of ore-bearing fluids. The ore is mostly lead with minor amounts of zinc; copper, nickel, cobalt, cadmium, and silver occur in small amounts. Lead-zinc mineralization is present in every formation from the Upper Cambrian Lamotte Sandstone to Lower Ordovician Jefferson City Dolomite.

Most aspects of the formation of Mississippi Valley base-metal deposits are uncertain and controversial. A Late Pennsylvanian paleomagnetic date (pole position) has been obtained for a deposit hosted by Cambrian carbonate rock in the Vibur-

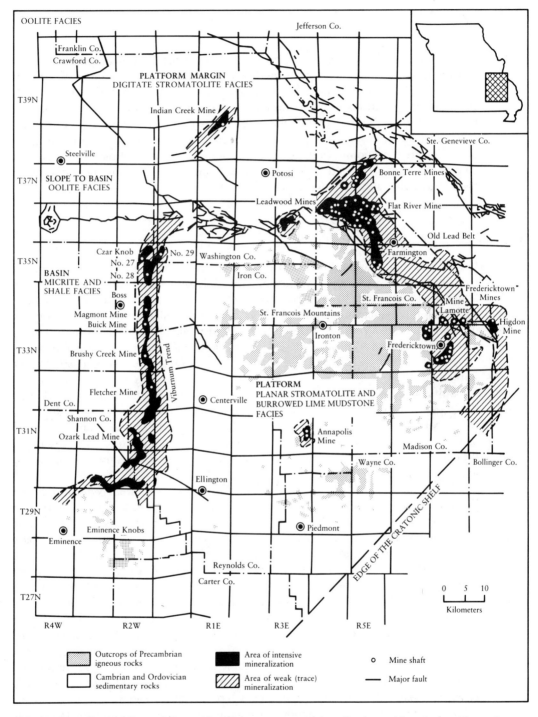

FIG. 13-15 • The southeast Missouri lead-zinc-copper mining district and its relationship to the Bonneterre oolite shoal-reef complex. (Adapted from Kisvarsanyi, 1977.)

num trend, possibly suggesting a genetic relationship between metal generation and orogenic activity associated with formation of the Ouachita foldbelt located 500 km to the southwest. What might be the nature of such a relationship? Do the high chloride concentrations in the inclusions in ore minerals require that evaporites be part of the stratigraphic section below the mineralized zone? Does the sulfur travel with the metal atoms, or is it added from elsewhere at the site of mineral precipitation? The great economic value of the Mississippi Valley type of base-metal deposit continues to generate a great deal of research aimed at answering these questions, but no satisfying answers have appeared so far.

Sulfur

On a worldwide basis, sulfides (mostly pyrite), sedimentary rocks, and oil and gas contribute 40%, 30%, and 25%, respectively, of commercial sulfur. In the United States at present, sulfur is obtained commercially almost entirely from sedimentary rocks, but there is reason to anticipate that this source will become minor by the end of this century as a result of federally mandated removal of sulfurous solid, liquid, and gaseous wastes to protect the environment. The long-range expectation is that 75% of our sulfur will be obtained from such wastes regardless of cost (Shelton, *in* Bureau of Mines, 1980).

Sulfur is unusual in comparison with most mineral commodities in that by far the largest portion of it is used as a chemical reagent (sulfuric acid) rather than as a component of a finished product. About 84% of the sulfur consumed in the United States in 1979 was either converted to sulfuric acid or produced directly in this form. Principal uses of the acid are for soluble fertilizers, in nonferrous metal production, and in petroleum refining.

Sedimentary sulfur deposits are essentially restricted to areas where extensive gypsum or anhydrite deposits are present. These may occur as bedded evaporites in areas such as west Texas (Permian) or as cap rock on diapiric salt domes (Gulf coastal region of the United States). The diapiric salt and anhydrite in the Gulf coast has been derived from beds of Late Triassic–Early Jurassic age, now thousands of meters below the surface, the beds having been formed during the early stages of the rifting process that formed the Gulf of Mexico. The bedded salt is 90–95% halite and 5–10% anhydrite with only traces of other materials such as dolomite, calcite, and quartz. Posthalite salts are rare. As the salt rose into zones of active groundwater circulation, halite preferentially dissolved to leave the included anhydrite as cap rock. Locally, the anhydrite hydrated to form gypsum.

Actual recovery of organisms and isotopic data reveal that the sulfur in the salt domes was formed by three species of anaerobic bacteria: *Desulfovibrio desulfuricans*, the most widely distributed and most active at surface temperatures; *Desulfovibrio orientis*, a form that is happier at medium temperatures; and *Clostridium nigrificans*, a heat-loving form. All these anaerobes can consume hydrocarbons or carbon dioxide as a source of energy but use sulfur instead of oxygen as a hydrogen

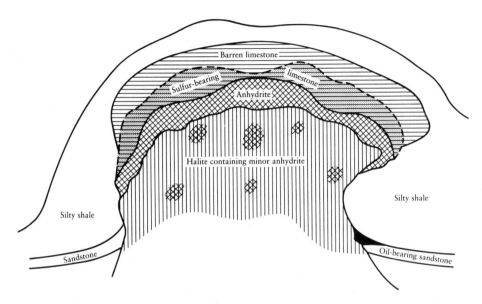

FIG. 13-16 • Typical cross-section of Gulf Coast salt dome showing cap rock with development of sulfur and limestone. (Blatt, Middleton and Murray, *Origin of Sedimentary Rocks*, 2e, © 1980, p. 564. Reprinted by permission of Prentice Hall, Inc., Englewood Cliffs, NJ.)

acceptor to produce calcite, hydrogen sulfide, sulfur, and oxygen in a series of reactions that are summarized by

$$CaSO_4 + H_2O + CO_2 \rightarrow CaCO_3 + H_2S + 2\,O_2$$

and

$$2\,H_2S + O_2 \rightarrow 2\,S + 2\,H_2O$$

The stratigraphic result is shown in Figure 13-16. The sulfur occurs in the lower part of the cavernous limestone and also in the upper part of the gypsum, but only that in the limestone is of commercial importance. It occurs in seams, crystals, and groups of crystals and forms about 30% of the volume. The sulfur zone ranges from 8 m to 100 m in thickness with a mode at perhaps 30 m (Jensen and Bateman, 1979). The mineralized cap rock is overlain by 2–70 m of barren cap rock, and this is overlain by unconsolidated sediments up to the surface, sometimes several thousand meters above the cap rock.

Although the structures produced by intrusion of the salt diapirs (domes, upturned flank beds) favor migration of groundwater and petroleum toward the salt masses, the critical factors needed to produce sulfur deposits seem to be stratigraphic. Permeable strata are needed up to the apexes of the diapirs to permit the flow of·fluids, whereas impermeable strata are needed above cap rocks to retain

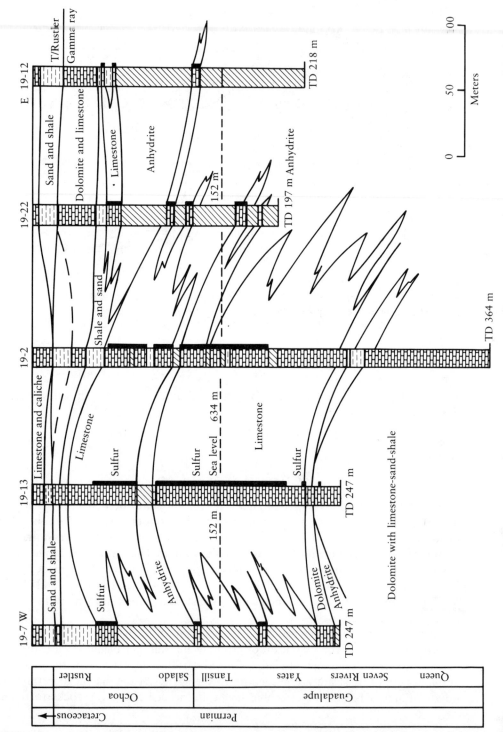

FIG. 13-17 • Stratigraphic distribution of sulfur, Sinclair Oil Corporation's Fort Stockton area, Pecos County, Texas. (Reprinted by permission of the publisher from Zimmerman and Thomas, *University of Texas Bureau of Economic Geology Circular*, 1969.)

hydrogen sulfide. The relatively uncommon occurrence of sulfur deposits in the Gulf coast illustrates how infrequently all conditions leading to their generation are met. Only 10–15% of the salt domes contain commercial deposits of sulfur.

The sulfur deposits in west Texas (Fig. 13-17) occur in thick and extensive units of impermeable, bedded anhydrite that underlies the entire region where the sulfur occurs. Permeability to permit water circulation was generated by the formation of anhydrite collapse breccias made possible by upward movement of groundwater along joints or faults from underlying aquifers. The upward-moving water dissolved anhydrite at the base of the unit to produce an underground, cavernous karst system into which the overlying anhydrite collapsed. Hydrocarbons ascended along these same fissures, and the broken anhydrite was altered by bacterial activity in the same manner as in the Gulf Coast sulfur deposits. In the west Texas deposits, the seals needed to retain hydrogen sulfide in breccia masses until the gas was oxidized to elemental sulfur differ from deposit to deposit, consisting of either overlying anhydrite, overlying impermeable beds, or alluvial clay and silt within the breccia masses. However, many breccia masses contain no sulfur, indicating that they were not sealed against loss of hydrogen sulfide gas (Brobst and Pratt, 1973).

Summary

Economic deposits formed in sedimentary rocks by secondary mineralization include petroleum, natural gas, tar sands, uranium, copper, lead-zinc, and sulfur. Each of these materials is the product of diagenesis and, therefore, requires that the host rocks be porous and permeable. Petroleum, natural gas, and tar sands all are formed from accumulations of microscopic marine organisms on the sea floor which, during diagenesis, are transformed into hydrocarbons and migrate into porous host rocks. The tar sands represent petroleum from which the more volatile hydrocarbons have escaped.

Uranium, copper, and lead-zinc deposits precipitate from diagenetic waters with specific chemical characteristics. Uranium precipitates under reducing conditions, as does copper. Lead and zinc occur together in solution and are abundant only in highly saline brines typically associated with buried evaporite deposits. Most economic sulfur deposits in the United States occur at the crests of salt diapirs and result from bacterial reduction of gypsum and anhydrite associated with the diapir.

References

Adams, S. S., Curtis, H. S., Hafen, P. L., and Salek-Nejad, H., 1978. Interpretation of post-depositional processes related to the formation and destruction of the Jackpile-Paguate uranium deposit, northwest New Mexico. Econ. Geol., v. 73, pp. 1635–1654.

Anderson, G. M. and Macqueen, R. W., 1982. Ore deposit models. 6: Mississippi Valley-type lead-zinc deposits. Geosci. Canada, v. 9, pp. 108–117.

Atkinson, S. F., Miller, G. D., Currey, D. S., and Lee, S. B., 1986. Salt Water Intrusion. Chelsea, Mich., Lewis Publications, 390 pp.

Baskov, E. A., 1987. The Fundamentals of Paleohydrogeology of Ore Deposits. New York, Springer-Verlag, 253 pp.

Bjørlykke, A. and Sangster, D. F., 1981. An overview of sandstone lead deposits and their relation to red-bed copper and carbonate-hosted lead-zinc deposits. Econ. Geol., 75th Anniversary vol., pp. 179–213.

Blatt, H., Middleton, G. V., and Murray, R. C., 1980. Origin of Sedimentary Rocks, 2nd ed. Englewood Cliffs, N.J., Prentice-Hall, 782 pp. Chapter 12 deals with porosity and permeability of detrital rocks.

Bowen, R. and Gunatilaka, A., 1977. Copper: Its Geology and Economics. New York, John Wiley & Sons, 366 pp.

Brobst, D. A. and Pratt, W. P. (eds.), 1973. United States Mineral Resources. U.S. Geological Survey, Prof. Paper 820, 722 pp.

Bureau of Mines Staff, 1980. Mineral Facts and Problems. Washington, D.C., U.S. Bureau of Mines, Bull. 671, 1060 pp.

Carrigy, M. A., 1971. Deltaic sedimentation in Athabaska tar sands. Amer. Assoc. Petroleum Geol. Bull., v. 55, pp. 1155–1169.

Chartrand, F. M. and Brown, A. C., 1984. Preliminary comparison of diagenetic stratiform copper mineralization from Redstone, NW Territories, Canada, and Kamoto, Shaban copperbelt, Zaire. J. Geol. Soc., v. 141, pp. 291–297.

Chilingar, G. V., 1964. Relationship between porosity, permeability, and grain-size distribution. In: Deltaic and Shallow Marine Deposits. New York, Elsevier Science Publishers.

Chilingarian, G. V. and Yen, T. F. (eds.), 1978. Bitumens, Asphalts and Tar Sands. New York, Elsevier Science Publishers, 331 pp.

Eugster, H. P., 1985. Oil shales, evaporites and ore deposits. Geochim. Cosmochim. Acta, v. 49, pp. 619–635.

Flach, P. D. and Mossop, G. D., 1985. Depositional environments of Lower Cretaceous McMurray Formation, Athabaska oil sands, Alberta. Amer. Assoc. Petroleum Geol. Bull., v. 69, pp. 1195–1207.

Friedrich, G. H., Genkin, A. D., Naldrett, A. J., Ridge, J. D., Sillitoe, R. H., and Vokes, F. M., 1986. Geology and Metallogeny of Copper Deposits. New York, Springer-Verlag, 592 pp.

Galloway, W. E. and Hobday, D. K., 1983. Terrigenous Clastic Depositional Systems. New York, Springer-Verlag, 423 pp.

Galloway, W. E. and Kaiser, W. R., 1980. Catahoula Formation of the Texas Gulf Coastal Plain: Origin, Geochemical Evolution, and Characteristics of Uranium Deposits. Austin, Texas Bureau of Economic Geology, Rept. Inv. No. 100, 81 pp.

Garlick, W. G., 1981. Sabkhas, slumping, and compaction at Mufulira, Zambia. Econ. Geol., v. 76, pp. 1817–1847.

Gustafson, L. B. and Williams, N., 1981. Sediment-hosted stratiform deposits of copper, lead, and zinc. Econ. Geol., 75th Anniversary vol., pp. 139–178.

Hanor, J. S., 1979. The sedimentary genesis of hydrothermal ore fluids. In H. L. Barnes (ed.), Geochemistry of Hydrothermal Ore Deposits, 2nd ed. New York, John Wiley & Sons, pp. 137–172.

Heath, R. C., 1984. Ground-Water Regions of the United States. Washington, D.C., U.S. Geological Survey, Water-Supply Paper 2242, 78 pp.

Hunt, J. M., 1979. Petroleum Geochemistry and Geology. New York, W. H. Freeman & Co., 617 pp.

Interstate Oil Compact Commission, 1984. Major Tar Sand and Heavy Oil Deposits of the United States. Oklahoma City, 272 pp.

Jensen, M. L. and Bateman, A. M., 1979. Economic Mineral Deposits, 3rd ed. New York, John Wiley & Sons, 593 pp.

Johnson, K. S. and Croy, R. L. (eds.), 1976. Stratiform Copper Deposits of the Midcontinent Region: A Symposium. Norman, Okla., Oklahoma Geological Survey, Circ. 77, 99 pp.

Johnston, R. H. and Bush, P. W., 1988. Summary of the Hydrology of the Floridan Aquifer System in Florida and in Parts of Georgia, South Carolina, and Alabama. Washington, D.C., U.S. Geological Survey, Prof. Paper 1403-A, 24 pp.

Kimberley, M. M. (ed.), 1978. Short Course in Uranium Deposits: Their Mineralogy and Origin. Toronto, Mineralogical Association of Canada, 523 pp.

Kisvarsanyi, G., 1977. The role of the Precambrian igneous basement in the formation of the stratabound lead-zinc-copper deposits in southeast Missouri. Econ. Geol., v. 72, pp. 435–442. (See also other papers in this issue, which is devoted to the Mississippi Valley Viburnum trend in southeast Missouri.)

Klemme, H. D., 1980. Petroleum basins—classifications and characteristics. J. Petroleum Geol., v. 3, pp. 187–207.

Larsen, K. G., 1977. Sedimentology of the Bonneterre Formation, southeast Missouri. Econ. Geol., v. 72, pp. 408–419.

Maynard, J. B., 1983. Geochemistry of Sedimentary Ore Deposits. New York, Springer-Verlag, 305 pp.

Meyer, R. F. (ed.), 1987. Exploration for Heavy Crude Oil and Natural Bitumen. Tulsa, American Association of Petroleum Geologists, Studies in Geology No. 25, 731 pp.

Meyer, R. F. and Steele, C. T. (eds.), 1981. The Future of Heavy Crude Oils and Tar Sands. New York, McGraw-Hill, 915 pp.

Miller, J. A., 1986. Hydrogeologic Framework of the Floridan Aquifer System in Florida and in Parts of Georgia, Alabama, and South Carolina. Washington, D.C., U.S. Geological Survey, Prof. Paper 1403-B, 91 pp.

Morganti, J. M., 1981. Ore deposit models. 4: Sedimentary-type stratiform ore deposits: Some models and a new classification. Geosci. Canada, v. 8, pp. 65–75. (See also discussion in v. 8, pp. 170–172.)

Nash, J. T., Granger, H. C., and Adams, S. S., 1981. Geology and concepts of genesis of important types of uranium deposits. Econ. Geol., 75th Anniversary vol., pp. 63–116.

Pusey, W. C. III, 1973. How to evaluate potential gas and oil source rocks. World Oil, April, pp. 71–75.

Rhodes, D., Lantos, E. A., Lantos, J. A., Webb, R. J., and Owens, D. C., 1984. Pine Point orebodies and their relationship to the stratigraphy, structure, dolomitization, and karstification of the Middle Devonian barrier complex. Econ. Geol., v. 79, pp. 991–1055.

Roberts, W. H. and Cordell, R. J. (eds.), 1980. Problems of Petroleum Migration. Tulsa, American Association of Petroleum Geologists Studies in Geology No. 10, 273 pp.

Rühl, W., 1982. Tar (Extra Heavy Oil) Sands and Oil Shales. Stuttgart, West Germany, Ferdinand Enke Publishers, 149 pp.

Sangster, D. F. (ed.), 1983. Short Course in Sediment-Hosted Stratiform Lead-Zinc Deposits. Toronto, Mineralogical Association of Canada, 309 pp.

Sawkins, F. J., 1984. Ore genesis by episodic dewatering of sedimentary basins: Application to giant Proterozoic lead-zinc deposits. Geology, v. 12, pp. 451–454.

Schowalter, T. T., 1979. Mechanics of secondary hydrocarbon migration and entrapment. Amer. Assoc. Petroleum Geol. Bull., v. 63, pp. 723–760.

Sverjensky, D. A., 1984. Oil field brines as ore-forming solutions. Econ. Geol., v. 79, pp. 23–37.

Sverjensky, D. A., 1986. Genesis of Mississippi Valley-type lead-zinc deposits. Ann. Rev. Earth Planet. Sci., v. 14, pp. 177–199.

Tarling, D. H., 1981. Economic Geology and Geotectonics. New York, John Wiley & Sons, 213 pp.

Tilsley, J. E., 1981. Ore deposit models. 3: Genetic considerations relating to some uranium deposits, Part II. Geosci. Canada, v. 8, pp. 3–7.

Tissot, B. P. and Welte, D. H., 1984. Petroleum Formation and Occurrence, 2nd ed. New York, Springer-Verlag, 699 pp.

U.S. Geological Survey, 1984–1988. A Series of Papers on the High Plains Aquifer. Washington, D.C., U.S. Geological Survey, Prof. Paper 1400 in several sections.

Weber, K. J., 1971. Sedimentological aspects of oil fields in the Niger Delta. Geol. Mijnbouw, v. 50, pp. 559–576.

Wilson, R. C. L. (ed.), 1983. Residual Deposits. Oxford, Blackwell Scientific Publications, 258 pp.

Zimmerman, J. B. and Thomas, E., 1969. Sulfur in West Texas: Its Geology and Economics. Austin, University of Texas Bureau of Economic Geology, Circ. 69-2, 35 pp.

• Subject Index

• **Author Index**